Memoirs of the American Academy in Rome

Supplementary Volume XII

Support for this publication was provided by the Lucy Shoe Meritt, FAAR '37, '50, Publication Fund of the American Academy in Rome, as well as by research grants from the American Council of Learned Societies, the University of Massachusetts, and the National Endowment for the Humanities, a Federal Agency.

The Hellenistic, Roman, and Medieval Glass from Cosa

David Frederick Grose

PUBLISHED FOR THE AMERICAN ACADEMY IN ROME
by
The University of Michigan Press
Ann Arbor, Michigan
2017

Published in the United States of America by the
University of Michigan Press
Manufactured in the United States of America
∞ Printed on acid-free paper

2020 2019 2018 2017 4 3 2 1

A CIP catalog record for this book is available from the British Library.

ISBN 978-0-172-13062-7 (hardcover)
978-0-472-12343-8 (e-book)

To the memory of

Mason Hammond,
Frank E. Brown, and
Donald B. Harden

Vitrum: Neque est alia nunc sequacior materia.
—Pliny, *Naturalis Historiae* 36.198

Contents

Plates

Preface

His publications forcefully show that David Grose was a remarkable scholar of Roman glass, but he was also remarkably loyal to the colleagues who worked in the same vineyard and the institutions he and they served. He took on the study of the glass from Cosa at the invitation of its chief excavator, the late Frank Edward Brown, in 1969, not long before financial difficulties interrupted the regular flow of publications by the American Academy in Rome that included the results of these excavations. Despite the climate of uncertainty, however, he persevered with the Cosa glass, as he did with other projects, even though it did not receive the *ultima manus* before his death in 2004.

That was to come from the Director of the Corning Museum of Glass, David B. Whitehouse, who had generously agreed to see it readied for publication thereafter. But a cruel stroke of fate suddenly deprived the world of late Roman and medieval studies of one of its champions, *cuius desiderio etiam nunc angimur*, in early 2013. In these circumstances, the Cosa glass is being published as written and so reflects the state of the discipline at that time. It also reflects the author's long and profitable association with Professor Jennifer Price (Durham University) from the early 1970s until his death. Her contributions to the work in gestation were many and various. Of the originality and abiding relevance of David's work she has written as follows:

> This work was commenced at a time when the study of glass from archaeological sites was in its infancy, and no comparable assemblages had yet been published from Roman towns in Italy or elsewhere in the western provinces. It is now difficult to comprehend that even the forms of glass vessels were then not widely recognised, and that there was no tradition of studying techniques of manufacture and decoration or of standard terms to describe glass fragments. Thus David's study of the Cosa glass was truly pioneering and a major advance in the subject. He had to devise a suitable means of sorting, examining, recording and identifying the large quantity of very broken glass vessels before he could begin to set the material in context. To a great extent he invented his own methods, with guidance from Donald Harden and Gladys Weinberg.
>
> Despite the near-absence of comparable published data, David's approach to the study of the Cosa glass was confident and mature. He wrote fluently and the finished manuscript demonstrates excellent scholarship and wide research, presenting a detailed study of a diverse body of material surviving from more than a millennium of occupation. His acute observations and his careful interpretation of problem pieces were also ahead of his time. From the outset he recognised the value of studying fragments in settlements as well as vessels deposited in burials, and was innovative in demonstrating that examination of groups of glass in closed deposits could lead to the better understanding of their economic and social significance as well as their chronology.

> David rapidly became a leading authority on Hellenistic and early Roman glass. His work always aimed to show what the study of glass could contribute to wider issues of social and economic activity, investigating topics then much debated but now commonplace, such as the western evidence for the introduction of blown glass, and the formation and innovative nature of the early Roman glass industry. His argument that glass production became an important industry in Italy during the reign of Augustus is now taken for granted, and much of our current understanding of the diversity and complexity of the traditions of glass production in the early empire has also developed from his writings in the 1970s and 1980s. *The Hellenistic, Roman, and Medieval Glass from Cosa* will now take its place as a foundation publication among his works.

Professor Price also made suggestions for the improvement of this final product, including the listing of the most important works on Italian glass that have appeared since Grose's own publication of the ancient glass in the Toledo Museum of Art in 1989, and Katherine Larson (Ph.D. University of Michigan, 2015) has been most important to realizing this end. Karol Wight and Tom Knotts of the Corning Museum of Glass, along with Elizabeth Whitehouse, have also rendered great service in tracking down materials pertinent to the work that came to the museum along with other papers of David Grose after his death.

These efforts led to the recovery of the graphic illustrations for the book originally prepared by John Huffstot, but it has proved impossible to recover the photographs by the late Barbara Bini of Rome that should also have illustrated it. Other careful eyes and hands have made significant editorial contributions to the final text: in addition to Paul Hays (M.A. Bryn Mawr) and Katherine Larson, E. Marianne Stern should be singled out for her welcome, careful assistance. Finally, it is a pleasure to note that, once again, Oliva Cardona of Bryn Mawr has generously employed her formidable talents to bring these several strands together elegantly for the University of Michigan Press.

R. T. Scott
Bryn Mawr College

Jennifer Price
Durham University, UK

Abbreviations

A.P.	Atrium Publicum I and its Annex
A Th	average uniform thickness
Bas.	Basilica
ca.	circa
cat.	catalogue
cat. no.	catalogue number
cm	centimeter
col.	column
Coll.	Collection
D	diameter (rim implied unless otherwise stated)
DAI	Deutsches Archäologisches Institut, Rome
ed.	edition, edited by
esp.	especially
fig.	figure
H	height
H.T.	House of the Treasure
ill.	illustration, illustrated
inv. no.	inventory number
L	length
Max	maximum
Min	minimum
n.	note, footnote
n.d.	no date
neg. no.	negative number
no.	number
No.	number in the present catalogue
n. s.	new series
p.	page
pl.	plate
S.H.	State University of New York (SUNY) House (also called House of the Skeleton)
suppl.	supplement
Th	thickness
trans.	translation, translated by
vol.	volume
W	width

1 ♦ Cosa: The Site and Its Glass

General Scope and Selection of Material

This study presents a conspectus of the glass discovered at Cosa (Etruria) from the beginning of the excavations there on 1 May 1948 until the end of July 1975. During these years approximately 15,000 fragments of glass, overwhelmingly Roman, were unearthed.[1] The vast majority of the Roman fragments belong to undecorated, free-blown vessels; the remainder come from decorated blown vessels or from mold-blown or cast vessels. Some core-formed and cast glass of Hellenistic date was also discovered, as were fragments of medieval glass dating from the thirteenth or early fourteenth century. The site also yielded a considerable amount of Roman window glass and a small number of glass objects, mostly disk-shaped gaming pieces, insets for jewelry, and mosaic tesserae.

The chronological span of the ancient glass ranges from the mid-second century B.C. through the first half of the fifth century A.D., although the bulk of the datable material can be assigned to the late first century B.C. through the second century A.D. Of particular importance is the large quantity of glassware attributed to Augustan or Julio-Claudian contexts. This glass measurably enhances our understanding of the development of the Roman glass industry in Italy and in the western empire during the first century.

The catalogue describes all significant types and groups of glassware found at Cosa, including most variants and unusual pieces. All recognizable fragments of core-made, cast, mold-blown, and decorated free-blown vessels are either catalogued or listed. Representative samplings of free-blown undecorated fragments are included, but only when individual pieces are sufficiently preserved to make their shapes or types intelligible. Window glass and glass objects are treated similarly. Because several diverse areas of the town were carefully excavated, the temples of the arx, the public buildings and shops fronting on the forum, and a group of private houses and their adjoining streets, the catalogue can be considered an accurate reflection of the fine and decorated glass in use in the town throughout its long history.

The conclusions drawn in the text concerning the historical development of the glass are provisional, *experimenti causa*. Although the material is discussed with frequent references to the existing literature, current knowledge of glass is both sparse and speculative. New excavations will undoubtedly improve and enlarge upon these results. Indeed, additional studies of Italian glass would be most welcome if only because this volume represents the first comprehensive study of ancient glass from a major excavation in Italy.[2] The absence of studies of other Italian assemblages

[1] A preliminary survey of the glass from Cosa appears in Grose 1973, with the glass from Deposit VII published separately in Grose 1975. Additional finds are discussed in Grose 1977, esp. 20–21, 23–24, 26, and in Grose 1983 with color plates on 44. Glassware from the harbor excavations is published in A. M. McCann 1987, ch. 12, p. 266.

is especially regrettable because Italy in the Hellenistic Age was a major importer of fine glass and under the early Empire became a foremost manufacturing center, one that exported quantities of tablewares to the provinces of North Africa and Europe and determined trends in taste and style everywhere in the Mediterranean world. As a result, studies of Italian glass should provide a useful basis for revisiting our understanding of the development of the Roman glass industry, especially in the western half of the empire. At present, our knowledge is overly dependent on information provided by finds from sites in the outlying provinces of central and northwestern Europe, principally those of Britain and along the Rhine–Danube frontier. This has led to an exhaustive literature on the styles of provincial imperial glassware but has done little to clarify the types of glass manufactured or used in Italy during the Republican or Imperial epochs or to explain how and when the Romano-Italian industry was created.

History of the Site

Cosa was planted in territory seized from neighboring Etruscan Vulci, defeated by the Romans in 280 B.C. Following its foundation early in the third century, its development was arrested by the First and Second Punic Wars (264–244 B.C. and 218–201 B.C.) In 199 the colony petitioned Rome for an additional draft of a thousand colonist families, and these arrived in 197. Thereafter, until the first decades of the first century the colony prospered, and most of its public buildings were either erected or remodeled in this period. Then, about 72–70 B.C. the town inexplicably was sacked, partially burned, and its population reduced. The precise source of trouble is unknown, but during the late Republic, Etruria was beset by disorders and its coasts harassed by pirates. Under Augustus the territory was resettled and the town repopulated. The exact date for this repopulation is not known, but the literary and archaeological evidence would suggest the years immediately following the Augustan victories of 30 or 18 B.C. and the subsequent resettlement of his veterans. In any case, Cosa was reoccupied before the final decade of the first century. Thereafter, the community, now with the status of a *municipium*, enjoyed a modest but declining occupation into the third and fourth centuries A.D. Recent work in the forum and on the arx has produced additional information for sixth-century occupation, and, while there is little to suggest continuity thereafter, from the eleventh to the fourteenth centuries A.D. the old town site passed from the hands of one monastic order to another, and finally to vassals of the Aldobrandeschi. In August 1329 a punitive expedition from Siena expelled them.[3] Thereafter Cosa was given over to thickets of myrtle, laurel, and wild olive, in George Dennis's words, "a chaos of crumbling walls, overturned masonry and subterranean vaults."[4] The excavators arriving at Cosa shortly after the Second World War found it precisely as Dennis had described it a century earlier.

[2] For references to glass already published from Italian sites and museums, consult "Cosa: A Select Bibliography," pp. 231–234, and "Works Cited," pp. 235–245 below.

[3] The published volumes of the excavations at Cosa are: F. E. Brown, *Cosa I: History and Topography* (*MAAR* 20, 1951), F. E. Brown, E. H. Richardson, L. Richardson, jr., *Cosa II: The Temples of the Arx* (*MAAR* 26, 1960), F. E. Brown, E. H. Richardson, L. Richardson, jr., *Cosa III: The Buildings of the Forum* (*MAAR* 37, 1993), V. J. Bruno and R. T. Scott, *Cosa IV: The Houses* (*MAAR* 38, 1993), E. Fentress et al., *Cosa V: An Intermittent Town* (*SMAAR* 2, 2003). Hereafter they will be referenced as *Cosa I*, *Cosa II*, *Cosa III*, *Cosa IV*, *Cosa V*.

[4] George Dennis, *The Cities and Cemeteries of Etruria* (London 1848) 2:276.

Areas and Levels Excavated

The initial survey of the site in 1948 divided the town into a grid of squares, fifty meters on a side. This grid-plan provides the coordinates (in Roman numerals and capital letters) employed in describing the locations of areas and buildings and of the objects found within them (fig. 1). The catalogue entries in this volume refer to these coordinates, unless the object was discovered in a building or area subsequently designated by name, such as the basilica or arx. Although fragments of glass were found everywhere on the site, the majority of the catalogued pieces come from a relatively small number of *loci*. The most common are the Atrium Publicum I (A.P.), located in VIII-D, E in the north corner of the forum; the basilica (Bas.) in VIII, IX-D, E, also in the north corner of the forum; and SUNY House (S.H.), a late Republican house in V, VI-D.

Cosa Inventory Numbers and Catalogue Numbers

During the first series of campaigns small finds were assigned inventory numbers that combined the letter "C" (signifying Cosa) and a capital letter (A through G) that consecutively designated the years between 1948 and 1954. Thus CA 100 refers to the hundredth object given an inventory number during the first season of excavation. In the second series of campaigns, which began in 1965, each season was designated by the last two digits of the year. Thus C66.100 represents the hundredth object to have been inventoried at Cosa in 1966. On occasion a single inventory number was assigned to several fragments. In such instances each piece is distinguished by the addition of a capital letter: thus C66.100A and C66.100B.

In addition to an inventory number, each entry in this catalogue is assigned a consecutive catalogue number. The numbers on the plates of drawings correspond to the catalogue numbers in the text. Fragments of glass that are not catalogued may be listed in an abbreviated fashion beneath a catalogued entry. In such cases the listed entries are differentiated by a lower case letter: thus No. 110a, 110b and so forth. Concordances of inventory numbers, catalogue numbers, and plate numbers appear on pages 215–226 below.

Find-Spots and Dating

In this catalogue the find-spot and dated context, if known, of each object is placed immediately after the catalogue and inventory numbers. As noted above, the find-spot represents the coordinates keyed to the site survey or, more commonly, to a named building or area in which the fragment was discovered. The stratum or level from which the object was uncovered is also cited. Levels are designated by Roman numerals: I, II, and so forth. Thus, S.H. Room 4, I signifies a find from Level I of Room 4 in SUNY House. Unstratified or surface finds are designated by level "0."

The date assigned to a particular fragment depends on the archaeological context from which it was recovered. Most fragments at Cosa come from unstratified or disturbed levels. Several hundred, however, are from sealed deposits with inclusive dates or at least with a *terminus post* or *ante quem*. Such deposits are discussed in chapter 2.

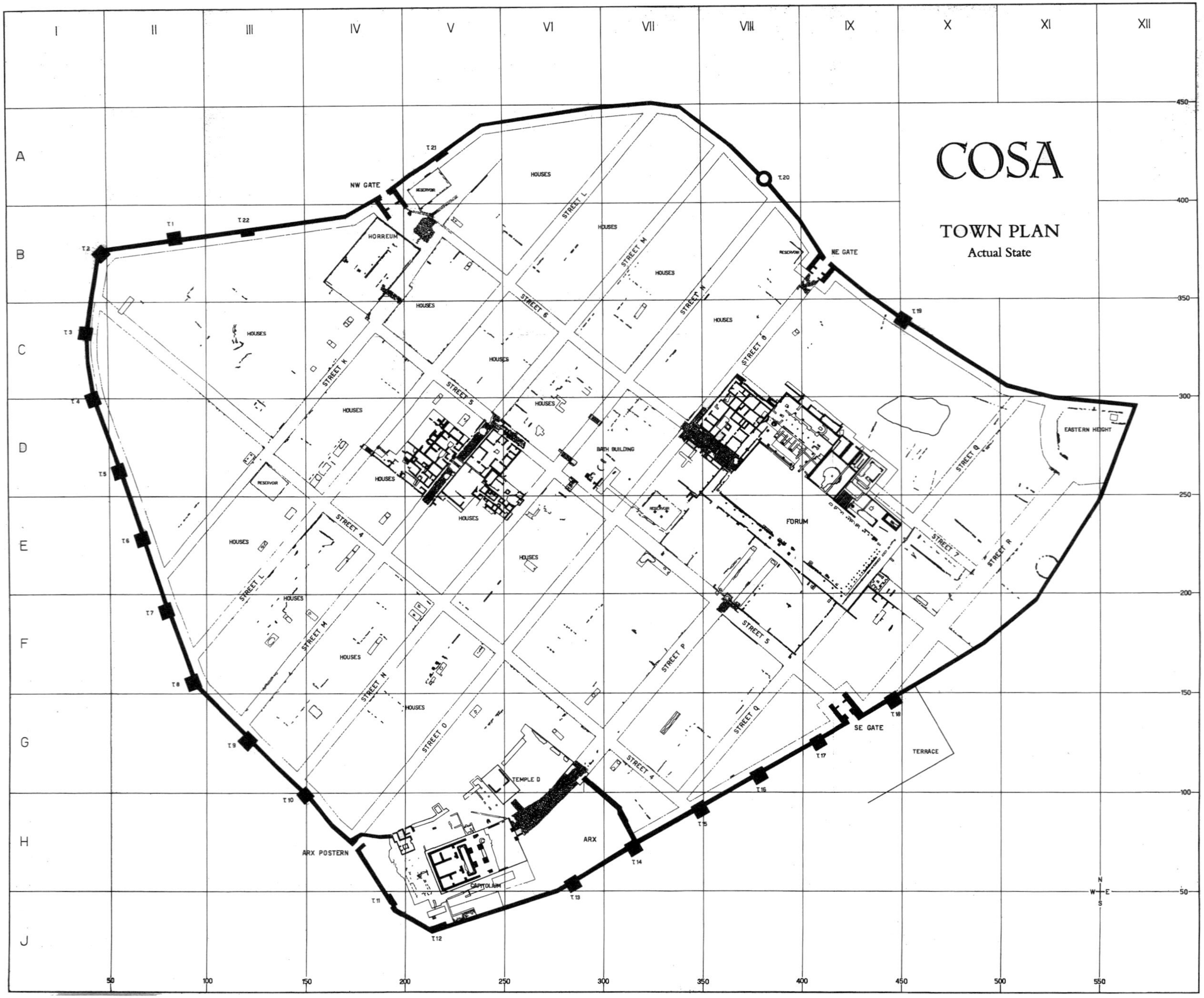

Fig. 1. Cosa, Town Plan. Actual State (adapted from Brown 1980, foldout plan).

Preservation and Disposition of the Glass

Almost without exception glass from Cosa was found in a highly fragmentary state. Scarcely any intact vessels were discovered, and few could be reconstructed in their entirety from excavated fragments. Of those that are largely intact, most come from cisterns, cesspools, or sewers or were recovered from various destruction levels. Otherwise, most vessels are represented by only a single fragment of their entire form, usually a portion of rim or base. This is scarcely surprising considering that Cosa was an active urban community inhabited for several centuries. Other classes of small finds, including the pottery, are similarly fragmentary. Uncertain quantities of tiny fragments were discarded by the excavators prior to study of the material. At present, the glass from Cosa is housed in the Antiquarium on the site.

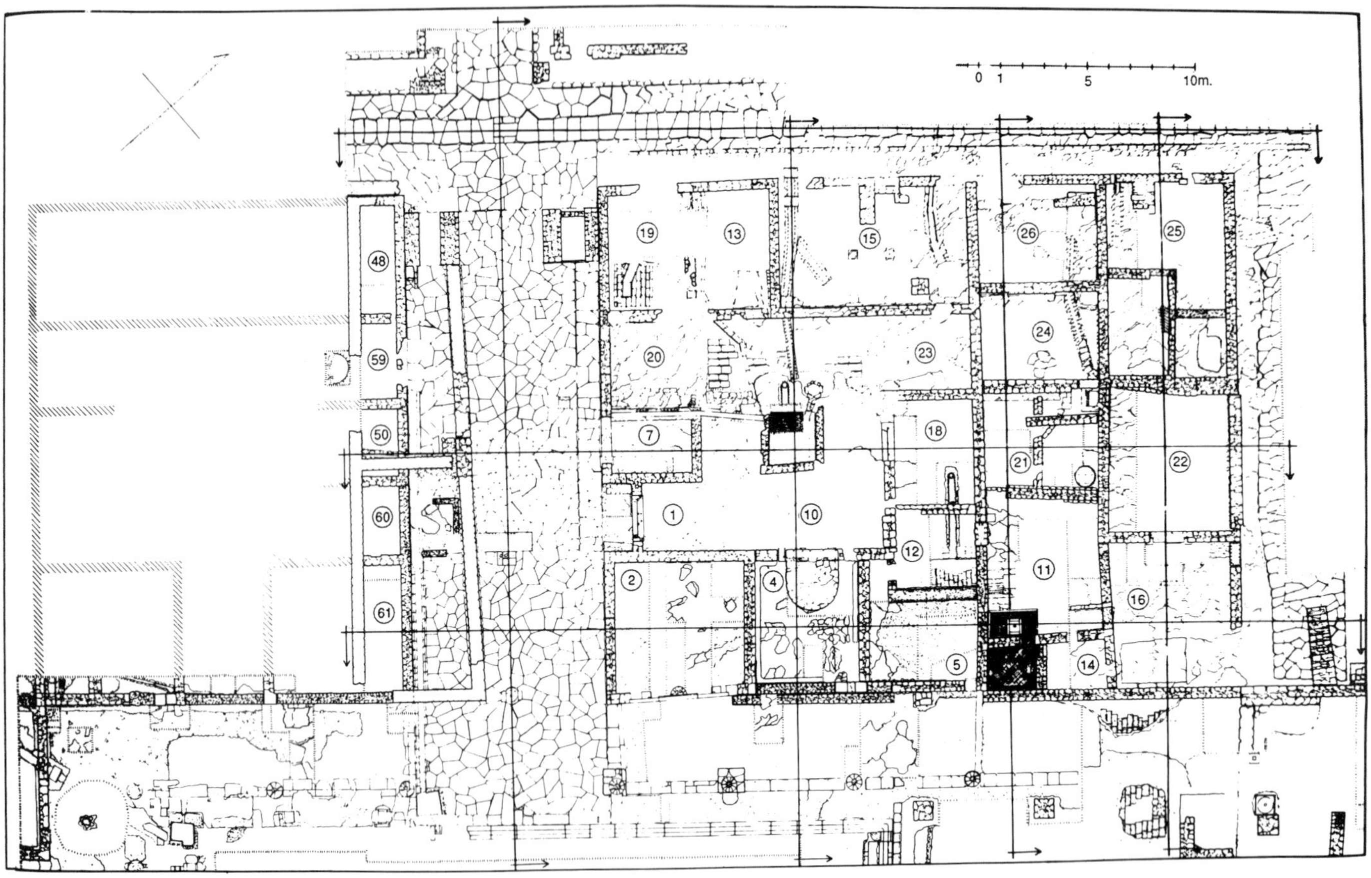

Fig. 2. Cosa, Forum. State Plan of Atrium Buildings I and II, with rooms numbered (source Brown 1993, pl. V) .

2 ♦ Dated Deposits of Glassware

It is unfortunate that Cosa, a site so prolific in glass, has produced only a few securely dated deposits of glassware. Over most of the site the living rock lies less than a meter below the surface, and centuries of inhabitation have greatly disturbed the soil and its contents. Nonetheless, some areas of the town do preserve groups of glasses in sealed and datable contexts.

The dating of these deposits relies on different types of evidence: namely, certain historical and archaeological considerations that provide fixed points in time; reliably dated objects found in the deposits, usually coins or inscriptions; and special classes of objects that provide more general but corroborative dating information, chiefly lamps, amphorae, Arretine, and other types of pottery.

Crucial events in the history of Cosa that provide a framework for the colony's development begin with the founding of the town in 273 B.C., a fact recorded in Livy and Velleius Paterculus,[1] but also include the partial destruction of the site ca. 72/70 B.C.; the reoccupation of the colony under Augustus, thought to have occurred between 25 and 15 B.C.; and finally, the collapse of the forum basilica ca. A.D. 40/45 and concomitant destruction of adjacent rooms in Atrium Publicum I and its Annex (fig. 2).

Whereas the founding date is fixed by the historical record, evidence for the other dates is based on excavated material. The destruction of the town in the first century, for example, is determined generally by datable debris found in destruction levels at many locations about the site, and specifically by a hoard of 2004 denarii found hidden in the House of the Treasure. This hoard had accumulated over several decades, and it contains almost every issue of denarius from the later second century B.C. down to 73/71, when the series abruptly comes to an end.[2] Since the latest coins in the hoard have seen no circulation, it appears that the house, together with the rest of the colony, was sacked around 72/70 B.C. or shortly thereafter. The fact that no one returned to recover the hoard suggests the severity of the disaster.

The date for the Augustan resettlement of the town is less precisely fixed, but the pottery (especially the early Arretine) associated with the rebuilding and refurbishing of the colony suggests a date around 20 B.C. If the new settlers were veterans of Augustus's legions, a date in the 20s or slightly later would be appropriate.[3] In lieu of more exact evidence, a date of 25/15 B.C. is employed here for the reoccupation of Cosa.

[1] Livy, *Per.* 11, Velleius 1.14.7; cf. Pliny *HN* 3.51. Also see chapter 1, n. 1.

[2] Concerning the hoard and its date, consult Buttrey 1980, 79–147, esp. 82, where burial of the hoard is assigned to 73–71 B.C. Further discussion in Brown 1980, 73–74, where the abandonment of the hoard is placed shortly after 71–70 B.C. Here, a date of 72/70 B.C. is used as a terminus for the hoard and the destruction level in the town.

[3] Information furnished by Professor Frank E. Brown, director of the excavations at Cosa. Also consult *Cosa IV*.

The date (ca. A.D. 40/45) for the collapse of the northwestern wall of the basilica and the accompanying destruction of several neighboring rooms in Atrium Publicum I and its Annex is set by the coins, Arretine, lamps, amphorae, and other material trapped beneath tons of fallen masonry and roofing tiles.[4] This date is further confirmed by two inscriptions attesting the reconstruction and conversion of the basilica into an odeon by Nero following his adoption by Claudius but before his accession to the throne—that is, between A.D. 50 and A.D. 55.[5]

Again, it should be recalled that the dates given in the catalogue represent either inclusive or terminal dates for deposits containing glass. It is understood that such dates represent the period when the deposit was formed or sealed and not necessarily the time when the vessels were made or the period of their use. Several deposits cover broad spans of time (Deposits IX and X, for example) and are included only because the site lacks more precisely dated contexts for the middle and late Empire. Others, however, are closely dated, especially Deposits IV and VII, where the overall homogeneity of the material and the possibility of reconstructing the vessels indicate undisturbed strata that can be safely assigned to a brief span of time. Since most of these deposits have been described in detail in other Cosan volumes, only previously uncited contexts are discussed at length here.

Deposit I: Basilica ENE 6, Level III and ENE 7, Level IV (ca. 150/145 B.C. to ca. 72/70 B.C.)

Level III marks a sealed layer of debris immediately outside the rear wall of the basilica but associated with the use of the basilica, ca. 150/140 to ca. 72/70 B.C. (see fig. 2) Level IV in the same location represents the fill put down immediately prior to the construction of the building. Based in part on a denarius now assigned to 157 B.C.,[6] found underneath the lowest step on the northeast side of the building, the basilica appears to have been erected ca. 150/140 B.C. The terminal date of the fill in Level III coincides with the sack of the colony ca. 72/70 B.C.

At the point where Levels III and IV converge, three fragments of a core-made alabastron (**No. 1**) were discovered. Since the fragments were unearthed at the bottom of Level III and in Level IV, it is probable that they were discarded during or shortly after the construction of the building.

Deposit II: SUNY House Room 3, Cesspool, Sediment Level (prior to ca. 25/15 B.C.)

Room 3 of SUNY House is actually a part of a contiguous Republican house (called the House of Salvius), constructed in the second century B.C. and occupied until it was destroyed ca. 72/70.[7] As stated above, the date for that calamity is set primarily by the accumulation hoard of denarii found in the House of the Treasure, yet another Republican building in an adjacent block.

[4] Moevs 1973, 27–28; Dyson 1976, 115–116; Grose 1975; *Cosa III*, 241.

[5] *Cosa III*, 243–244. Inscriptions: CC 875 and CC 876.

[6] Coin: CD 1050 = Buttrey 1980 41, no. 67. Owing to a revision in the dating of Roman denarii, the forum basilica at Cosa is now thought to have been erected ca. 150–145 B.C.; see Brown 1980, 58, contra Taylor 1957, 92; *Cosa III*, 207–227.

[7] Moevs 1980b, 243–245; also see *Cosa IV*, 31–42.

When the cesspool below Room 3 was excavated, it yielded an almost intact, cast "grooved bowl" (**No. 92**) similar to others found in Hellenistic contexts throughout the Mediterranean area. Although the cesspool had been reopened and used in the last quarter of the first century B.C., the bowl was discovered in the lowest part of a greenish-gray sedimentation level and probably dates to the period prior to 72/70 when the house was in use.

Deposit III: Atrium Publicum Room 16, Level III (prior to ca. 25/15 B.C.)

Room 16 is a part of a house that was built into Atrium Publicum I (see fig. 2) on the northeastern side of the forum when the town was resettled under Augustus, ca. 25/15 B.C. (see Deposit V). At that time an *opus signinum* pavement was put down in Room 16. The layer of fill beneath the pavement was labeled Level III. The fill contained Roman black-glaze and thin-walled pottery "of a prevailing late Republican type."[8] It also contained a fragment of a cast, blue linear-cut bowl (**No. 100**). This fragment, consequently, antedates 25/15 B.C., the time when it was sealed under the pavement.

Deposit IV. SUNY House Room 4, Levels 0-B and I (prior to ca. 25/15 B.C.)

Like Room 3, Room 4 of the SUNY House is actually part of an adjoining Republican house, termed Northeast House, which abuts the House of Salvius (see Deposit II).[9] Room 4 is a walled, sunken courtyard, probably a bathroom, originally reached by a flight of steps at the rear of the house. The Courtyard was abandoned when the house was destroyed ca. 72/70 B.C. When the town was resettled under Augustus, a new house (House of the Birds) was built into its Republican predecessor, and the courtyard was filled with debris and sealed with a *signinum* pavement. Thus the material in the fill antedates the construction of the Augustan house, ca. 25/15 B.C.

The date of the material in the fill varies from middle Republican to early Augustan, but the pottery, lamps, and other objects belong mainly either to the second quarter of the first century B.C. (that is, the decades just prior to the destruction of Northeast House) or to the final decades of that century but prior to the construction of the *signinum* pavement, ca. 25/15 B.C. On the basis of analogy with comparable pieces, the glass would appear to belong to the closing quarter of the century, just prior to the laying of the *signinum* floor. The fact that several of the vessels are reconstructible, rarely possible at Cosa, is also significant. It seems likely that they were broken and discarded just before the courtyard was filled in. In fact a large plate (**No. 145**) was found in more than sixty pieces lying on successive steps leading from the house to the foot of the courtyard.

This deposit contains the earliest and perhaps the most significant group of cast tablewares found at Cosa. It yielded a partially intact striped mosaic-glass dish (**No. 12**), another mosaic-glass dish in an opaque white ground (**No. 20**), two fragments of a dark blue lidded pyxis (**No. 135**), a large circular opaque white plate (**No. 145**), and a highly fragmentary opaque green rectangular dish (**No. 147**). It also contained the rim of a small blown bottle (**No. 364**), one of the earliest securely dated examples of blown glass from Roman Italy.

[8] Moevs 1973, 28–30; *Cosa III*, 238–240.

[9] Moevs 1980b, 245–246; *Cosa IV*, 43–49.

Deposit V: Atrium Publicum Rooms 14 and 16, Level I
(ca. 25/15 B.C. to ca. A.D. 40/45)

Rooms 6, 11, 14, 16, and 21 in Atrium Publicum I (see fig. 2) were unified to create the Augustan house discussed in Deposit III.[10] These rooms were wholly destroyed ca. A.D. 40/45 by the collapse of the abutting basilica wall, and Rooms 14 and 16 were left undisturbed thereafter. Level I designates the stratum below the fallen wall and roofing tiles but above the floor of the house. It contains debris trapped in the house when it was destroyed, all of which had accumulated after the conversion of Atrium Publicum I into a home, ca. 25/15 B.C. Room 16 may have been an open courtyard used by the occupants as a dumping ground. This would explain the tremendous amount of fragmentary pottery and glass found strewn on the paved floor. The bulk of the datable pottery, primarily Arretine, is attributed to the Augustan or Tiberian period.

The glass from Level I of Rooms 14 and 16 is both cast and blown and represents a great many types. Among cast vessels are six fragments of network mosaic-glass bowls (**Nos. 5–10**); three examples of striped mosaic-glass bowls (**Nos. 14–14a**); two fragments of marbled mosaic-glass, one a jar (**No. 30**), the other a ribbed bowl (**No. 32**); five fragments of purposely colored and naturally colored monochrome ribbed bowls (**Nos. 57**, **65**, **83–84**, **90f**); two examples of linear-cut bowls (**Nos. 101**, **104**), one naturally colored and the other colored; three colored rotary-polished vessels of varying shapes (**Nos. 121**, **136–137**, **204**); and a fragment of a large colorless plate with overhanging rim (**No. 161**), presumably intrusive in a context of this early date. Among blown vessels are examples of small colored bowls with pinched ribs (**No. 251**), a small dark blue amphoriskos (**No. 360**), three beakers with horizontal cut grooves (**Nos. 426**, **430**, **436**), and eight colored and naturally colored flasks and small bottles (**Nos. 371**, **372**, **382**, **512**, **514**, **549**, **555**, **565**). Glass objects from the deposit include a fragmentary rod (**No. 592**) and a dark blue tessera (**No. 595**), presumably part of a pavement or wall decoration.

Deposit VI: Atrium Publicum Room 11, Cesspool
(ca. 25/15 B.C. to ca. A.D. 40/45)

The cesspool was built for the Augustan house described in Deposits III and V (see fig. 2). Accordingly, it was not constructed prior to the building of the house ca. 25/15 B.C., and it was no longer in use after the destruction of the home ca. A.D. 40/45.[11] It contained two joining fragments of a cast, opaque red bowl (**No. 154**).

Deposit VII: Atrium Publicum Room 22, Level II
(decade prior to ca. A.D. 40/45, with some earlier material to ca. 25/15 B.C.)

Room 22, Level II, in the Annex of Atrium Publicum I marks the latest and most closely dated deposit of glassware at Cosa (see fig. 2).[12] The room served as the storeroom of a shop (Room 25) that faced Street "O." Like the room at the Augustan house that it abutted (Deposit V), Room 22

[10] Moevs 1973, 28–30; *Cosa III*, 238–240.

[11] Information furnished by Professor Frank E. Brown.

[12] The glass from the deposit has been published in Grose 1975; also see citations in n. 4 above. Deposit VII = Deposit 6 (Atrium Publicum 22, II) in Dyson 1976.

was wholly destroyed by the collapse of the northwestern wall of the basilica. Under tons of rubble and tiles, the excavators discovered more than two hundred pottery vessels in Arretine, thin-walled, and other fabrics, a dozen amphorae, two coins, forty or more lamps, marble statuary, a large assortment of metal fittings and implements, in addition to seventy-six glass vessels. Unlike most of the finds at Cosa, many of these vessels could either be reconstructed or restored on paper. Given the fact that they were housed in a shop, they may well have been of recent manufacture and have constituted merchandise awaiting sale at the time of the calamity. If new at that time, they would date to the years immediately before A.D. 40/45.

Deposit VII also produced some fragmentary glass, mostly cast, that was already broken and pressed into the earthen floor prior to the destruction of the shop. This glass is associated with Arretine and thin-walled pottery fragments that date primarily to the late Augustan or Tiberian periods.

The majority of pieces from Level II are blown and include fine colored tablewares, as well as naturally colored tableware and containers. Most common are dark blue or emerald green plates and dishes (**Nos. 308–311e**) or dark blue or purple bowls (**Nos. 315–320**, **323**, **333**, **337**), but there also are fragments of a yellow bowl with pinched ribs (**No. 257**), a tall purple beaker (**No. 342**), a small green beaker (**No. 351**), an unusual spherical blue flask (**No. 353**), a purple two-handled jug with applied blobs of opaque white decoration (**No. 354**), a small aquamarine bottle (**No. 358**), and a graceful blue ladle (**No. 373**). Naturally colored bluish-green and light green tablewares appear in many of the same shapes—plates and bowls (**Nos. 374–381**), beakers (**Nos. 445–446a**, **450**, **452**, **454**, **463**), flasks (**Nos. 509–510**), jugs (**No. 517**), and small unguentaria (**Nos. 550–551**, **558–559**, **562–564**). Mold-blown prismatic bottles make their appearance for the first time (**Nos. 534–535**, **537–538**, **545–547**), and there is a fragmentary example of a mold-blown "Romano-Syrian" beaker (**No. 207**) bearing a Greek inscription.

Cast glass is uncommon, but the few extant pieces are of special interest. They include a fragmentary mosaic-glass vessel (**No. 22**), two fragmentary bluish-green ribbed bowls (**Nos. 69**, **85**), a purple linear-cut bowl (**No. 107**), an elegant aquamarine dish (**No. 125**), an emerald green pyxis (**No. 134**), and a unique blue carinated bowl (**No. 139**).

Deposit VIII: Basilica ENE I, Level III; ENE 3, Level III; ENE 4, Level II (prior to ca. A.D. 50/55, but mainly Tiberian)

Following the collapse of the basilica, the building was reconstructed as an odeon (fig. 3). Two building inscriptions attribute the project to Nero after his adoption by Claudius and prior to his accession to the throne, as noted on page 8 above. These events place the construction program around A.D. 50/55. During the renovation a deep layer of fill was laid down on the northeast side of the building. It contained a considerable quantity of pottery, glass, and other objects and was designated Level II or III depending on the sector (ENE 1, ENE 3, ENE 4, etc.). The datable material, principally Arretine and Italian thin-wailed pottery, can be assigned to the Tiberian or early Claudian periods, though some of the material is Augustan or even as early as the resettlement of the site ca. 25/15 B.C.[13] The glass in the fill is principally cast and would be consistent with dates in the Augustan or Tiberian age. Since it was already fragmentary when the fill was laid down, much

[13] Information furnished by Professor Frank E. Brown; see *Cosa III*, 241–244.

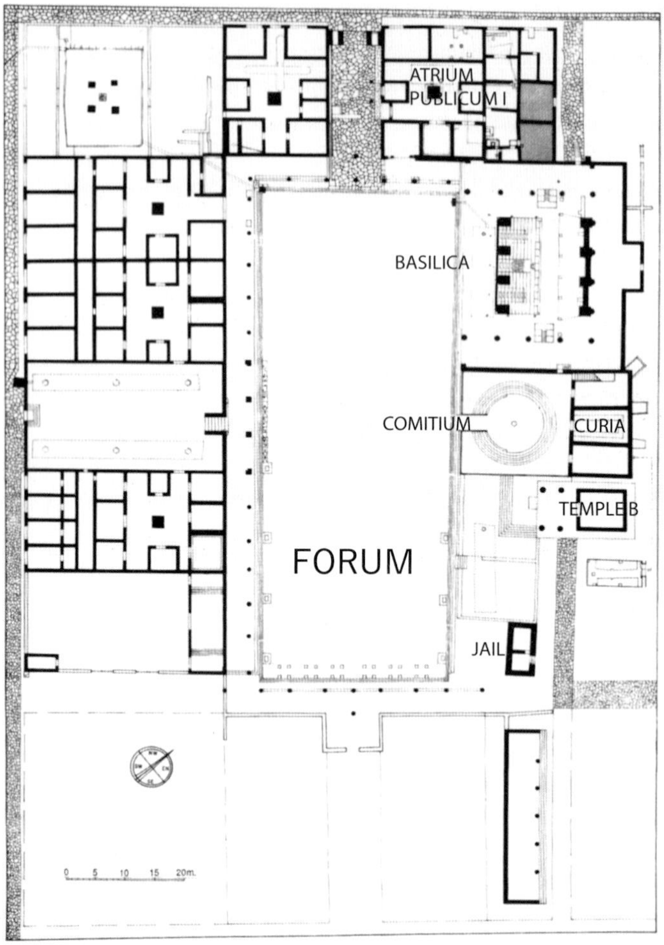

Fig 3. Cosa, Forum. Restored Plan 40/30 B.C.–A.D. 265/275. Atrium Buildings I–V are shown from northwest to southwest and the Basilica, Comitium and Curia, and Temple B from northeast to southeast (adapted from Brown 1993, 239, fig. 76).

of the glass from this deposit may be somewhat earlier than that found in Deposit VII, despite the later terminal date for Deposit VIII as a whole.

The deposit contained four deliberately colored and naturally colored cast ribbed bowls (**Nos. 55**, **76–77**, **86**), three cast linear-cut bowls (**Nos. 97**, **119**, **123**), a cast emerald green rectangular dish (**No. 127**), a purple stemmed bowl (**No. 144**), and a cast opaque blue rectangular dish (**No. 146**). It also yielded a small amount of blown glassware: a blue beaker (**No. 344**), two bluish-green bowls (**Nos. 383–384**), two bluish-green flasks (**Nos. 506–507**), and two small bluish-green unguent bottles (**Nos. 557**, **561**).

Deposit IX: Atrium Publicum Room 25A, Levels II–III and Room 25, Sectors B–D, Level III (late first century A.D. through early third century A.D.)

Level III of sectors B, C, and D of Room 25, and sector A, Levels II and III, represent the occupational fill of a shop built into the Annex of Atrium Publicum I and facing Street "O" (see fig. 2).[14]

[14] Dyson 1976, 139–140, Deposit IX = Deposit 7 (Late Shops) in Dyson 1976.

The debris began to accumulate late in the first century A.D., possibly following repairs necessitated by the partial destruction of the building in 40/45 (see Deposit VII). The coins, sigillata chiara, and amphorae date principally to the second century, although some material is assigned a date as early as the late first or as late as the early third century. The accumulation ceases altogether in the third century, presumably owing to the abandonment of the shop. Although the deposit is not closely dated, it does provide rough chronological evidence for many prominent types of glass that otherwise have no dated context at Cosa.

The glass from the deposit is almost entirely free-blown and colorless. It includes three types of tableware associated with the Imperial period: the well-known colorless cast plates and bowls with broad, overhanging rims (**Nos. 171**, **173**, **191**); a series of unusual plates and bowls with flaring rims and thick edges, triangular in cross section (**Nos. 403**, **406**, **409**, **411**, **412**); and blown colorless plates with flaring sides (**Nos. 415**, **417**, **419–419a**). The deposit also contains a variety of colorless beakers and other vessels with applied threads in the same fabric (**Nos. 269**, **281**, **283–284**), some in a leaf pattern (**Nos. 299–300**), others with a snake-thread motif (**Nos. 286–288**, **292**, **295–296**, **298**). Some of these beakers have beaded stems, as do three examples with no surviving decoration (**Nos. 470a–c**). Colorless and naturally colored jugs, flasks, and bottles (**Nos. 497**, **499**, **502–503**), including a prismatic bottle fragment (**No. 536**), comprise the rest of the glass in the shop. A fragment of a cast mosaic-glass vessel (**No. 23**) is most certainly intrusive in this late context. Glass objects from Deposit IX include an opaque white gaming piece (**No. 588**), a tiny cylindrical tube (**No. 594**), and a turquoise blue tessera (**No. 596**).

Deposit X: Forum Cistern (SW Extension) below Fallen Vaulting (fourth century A.D. through mid-fifth century A.D.)

At the northeastern end of a cistern that bounded the Southwest Extension of the forum (see fig. 3), a deposit of material dating to the fourth through the mid-fifth century was discovered. It appears that the cistern was deliberately filled in during late antiquity.

The sigillata chiara D found in the deposit is said to date to the early to mid-fourth century, and a sigillata plate with a figure holding a Christian cross probably dates to the late fourth or fifth century. Like Deposit IX, this deposit is imprecisely dated, but it includes three fragmentary stemmed cups (**Nos. 480–481**, **486**), so common on late Roman and early Byzantine sites throughout the Mediterranean but rarely published from dated levels of any sort.[15]

ASIDE FROM THESE DEPOSITS, there are additional areas on the site that yielded types of glass with a chronologically homogeneous appearance. Since the coins, pottery, lamps, and other datable objects from these areas also proved to have a chronologically cohesive character, despite the existence of extraneous earlier or later material, it is useful to mention these groups. Although these loci are not sealed deposits in the sense of Deposits I–VIII, they do possess the character of a single period, much like Deposits IX–X. They provide, as well, cumulative and interlocking evidence for conclusions offered in the chapter that follows. Two such loci are described here, but, because of intrusive material, they are not noted under the entries in the catalogue itself.

[15] Dyson 1976, 161. Deposit X = Deposit 8 (Forum Cistern) in Dyson 1976, but the terminal date of the deposit, based on a more recent examination of the pottery evidence, is now considered to extend to the mid-fifth rather than early fifth century A.D. See *Cosa V*, 307–319.

Forum SW, Levels I and Ia (Shops in Atrium Publicum Level III)
(ca. 50/40 B.C. to ca. A.D. 25/50, with both earlier and later material)

Forum SW refers to a trench placed inside the outer wall of several shops located in Atrium Publicum III on the southwest side of the forum (see fig. 3). Levels I and Ia designate a layer of fill that accumulated during the period when the shops were in use.[16] When the shops were abandoned, the fill was left relatively undisturbed, although it was not sealed. The pottery and objects in the fill date mainly from the first century B.C. to the middle of the second century A.D., but with a significant concentration of material from the first century B.C. and early first century A.D.

Finds of glass in the fill closely resemble the cast and early blown tablewares found in Deposits V, VII, and VIII, themselves dating from the decades immediately prior to ca. 25/15 B.C. to as late as ca. A.D. 40/45. Fragments include a cast mosaic-glass dish (**No. 25**), a tooled and blown color-band bottle (**No. 31**), colored and naturally colored cast ribbed bowls (**Nos. 48**, **50**, **90b**), and five cast linear-cut bowls (**Nos. 98**, **99**, **120f**, **120i**, **120k**). Among blown vessels of similar date are blue and green dishes and bowls (**Nos. 317**, **324c**, **324d**, **331**), a purple beaker (**No. 346**), colorless beakers (**Nos. 427**, **438**), bluish-green flasks (**Nos. 505**, **513**), but also an intrusive late antique stemmed cup (**No. 477**).

Atrium Publicum Rooms 6, 11, and 21, Level I
(ca. 25/15 B.C. to ca. A.D. 40/45)

These rooms are part of the Augustan house in Atrium Publicum I described in Deposits III and V (see fig. 2).[17] Unlike the rooms mentioned in those deposits, rooms 6, 11, and 21 were disturbed following their destruction in A.D. 40/45. Nonetheless, much of the material in Level I so closely resembles Deposit V that, with certain exceptions, it too must date to the period when the house was occupied.

The glass includes two cast ribbed bowls, one colored (**No. 65**) and one naturally colored (**No. 118**); a cast colorless plate with broad, overhanging rim (**No. 168a**); two examples of colored blown tableware (**Nos. 328**, **361**); and a naturally colored blown beaker (**No. 429**).

FIGURE 4 summarizes the glass in dated deposits discussed in this chapter.

[16] Moevs 1980a, 175–177.

[17] Moevs 1973, 28–31, with additional information furnished by Frank E. Brown.

Deposit and Dating	*Catalogue Numbers*
Deposit I: Basilica ENE 6, Level III and ENE 7, Level IV (ca. 150/145 B.C. to ca. 72/70 B.C.)	No. 1
Deposit II: SUNY House Room 3, Cesspool, Sediment Level (prior to ca. 25/15 B.C.)	No. 92
Deposit III: Atrium Publicum Room 16, Level III (prior to ca. 25/15 B.C.)	No. 100
Deposit IV: SUNY House Room 4, Levels 0-B and I (prior to ca. 25/15 B.C.)	Nos. 12, 22, 135, 145, 147, 364
Deposit V: Atrium Publicum Rooms 14 and 16, Level I (ca. 25/15 B.C. to ca. A.D. 40/45)	Nos. 5, 6, 7, 8, 9, 10, 11, 12, 13, 14, 14a, 30, 32, 57, 65, 83, 84, 90f, 101, 104, 121, 136, 137, 161, 204, 251, 360, 371, 372, 382, 426, 430, 436, 512, 514, 549, 555, 565, 592, 595
Deposit VI: Atrium Publicum Room 11, Cesspool (ca. 25/15 B.C. to ca. A.D. 40/45)	No. 154
Deposit VII: Atrium Publicum Room 22, Level II (decade prior to ca. A.D. 40/45, with some earlier material to ca. 25/15 B.C.)	Nos. 21, 70, 85, 107, 125, 134, 139, 207, 257, 308, 308a, 309, 310, 311a-e, 315, 316, 317, 318, 319a, 320, 323, 333, 337, 342, 351, 353, 354, 358, 373, 374, 375, 376, 377, 378a–d, 379, 380a–n, 381, 445, 446, 446a, 450, 452, 454, 463, 509, 510, 517, 534, 535, 537, 538, 545, 546, 547, 550, 551, 558, 559, 562, 563, 564
Deposit VIII: Basilica ENE 1, Level III; ENE 3, Level III; ENE 4, Level II (prior to ca. A.D. 50/55 but mainly Tiberian	Nos. 55, 76, 77, 86,96, 97, 119, 123, 127, 144, 146, 344, 383, 384, 506, 507, 557, 561
Deposit IX: Atrium Publicum Room 25A, Levels II–III and Room 25, Sectors B–D, Level III (late first century A.D. through early third century A.D.)	Nos. 23, 171, 173, 191, 269, 281, 283, 284, 286, 287, 288, 292, 295, 296, 298, 299, 300, 403, 406, 409, 410, 411, 412, 415, 417, 419, 419a, 420, 470a–e, 497, 499, 502, 503, 536, 588, 594, 596
Deposit X: Forum Cistern (SW Extension), below Fallen Vaulting (fourth century A.D. through mid-fifth century A.D.)	Nos. 480, 481, 486

Fig. 4. Summary of glass in dated deposits.

3 ♦ Classification, Arrangement, and Description of Excavated Glass

General Remarks

Perhaps the most effective method for classifying and describing excavated glass is that devised by Donald B. Harden and Jennifer Price for publication of the material from Fishbourne Roman Villa (1971). This system is derived largely from Harden's classification of the glass from Karanis, Camulodunum (1947), and numerous other sites. In it, the glass is first divided into broad categories based on technique of manufacture (core-formed, cast, mold-blown, free-blown decorated, and free-blown undecorated), then further subdivided according to *types* (see discussion below) within each of these categories. The major exception is free-blown undecorated glass, which is arranged by shape, moving from open to closed forms. This scheme was intended primarily for use with fragmentary glass from Roman sites but is also applicable to pre-Classical, Hellenistic, medieval, or Islamic glass, either fragmentary or intact. The present catalogue has made use of a similar system.

The reason for adopting such a system of classification, however, has never been explained in full. Nor have there been recent discussions of the methodology involved in cataloguing glass or the special problems inherent in preparing a lengthy study of such excavated material. Accordingly, I have written this chapter in the hope that it will not only explain the rationale behind this catalogue but also serve as a guide for others who intend to publish glass.

Organization of the Catalogue by Types

The catalogue aims at describing the individual fragments of glass from Cosa in an accurate and consistent manner and, wherever possible, arranging them according to types (or classes) based on sets of distinctive characteristics common to all fragments of that type. These sets of shared characteristics may involve similarities of shape, decoration, color, material, fabric, function, manufacturing and finishing techniques, or any combination of the above. It is also true in most instances that vessels of a single type are presumed to have a common set of inclusive dates and to share a common provenance(s) and pattern of distribution—matters of general concern to ancient historians, art historians, museum curators, and collectors, as well as archaeologists and glass specialists.

For core-formed, cast, decorated mold-blown, and decorated blown glass, the determination of clearly defined types is usually not a complicated matter, even when only small fragments survive. For example, a late Hellenistic core-formed alabastron (**No. 1**) or a Roman mold-blown gladiatorial beaker (**No. 205**) is immediately identifiable from only a tiny piece of the entire vessel. In both cases a highly visible set of characteristics makes the identification secure, and once the identification is made, all fragments of that type or group can be gathered together for study and publication.

For reasons of clarity and convenience, each type usually has been given a name, generally a short descriptive title. The names invariably are modern since ancient terms, except for certain shapes, are imprecise or lacking altogether. Most titles are derived from the most obvious traits of the type, such as the ribbing on ribbed bowls. Unhappily, some titles are inevitably cumbersome (colorless vessels with broad, overhanging rims), yet they too are necessary for consistency and ease of reference.

As noted above, the types themselves are marshaled within five or six categories based upon technique of manufacture and the presence of decoration. Since the major techniques are, for the most part, in rough chronological sequence, they bring a dating dimension into the overall classification. Core-formed vessels appear first, then cast, both polychrome (i.e., mosaic) and monochrome, followed by decorated mold-blown glass, and finally free-blown glassware, decorated and plain. Unusual, unrelated, unidentifiable, and very late material, generally medieval, follows the ancient vessels. Window glass and glass objects are also placed at the end of the volume. Within each of these broad categories the relevant types are arranged at the discretion of the cataloguer, but if possible, they are placed in an approximate chronological order or arranged so that related types are together. Thus marbled mosaic-glass ribbed bowls are the last category of mosaic wares to be discussed so that they can be placed next to monochrome colored and naturally colored ribbed bowls.

Inevitably, there are inconsistencies. Free-blown glass beakers, whether plain or decorated with cut lines and horizontal grooves (**Nos. 424–467**) are grouped with undecorated blown glass, not with decorated vessels. Common sense argues that it would be unwise to classify basically identical beakers into two or three separate sections, especially given their fragmentary state and the possibility that some examples do not preserve their original decoration. Similarly, mold-blown prismatic bottles (**Nos. 525–548**) are classified with free-blown bottles and flasks. Despite the presence of simple molded designs or inscriptions on some examples, prismatic bottles are classified in almost all publications with unadorned storage wares, and not with decorated or figural mold-blown vessels. By the same token, individual pieces such as **Nos. 193** and **354**, a facet-cut bowl and purple two-handled jar with applied decoration, are catalogued with other examples of their overall type or class, even though most other examples lack decoration. In such cases, practical considerations dictate the choices.

Within a given type or class, there can be further divisions based on shape, date, provenance, decoration, nuance of style, or sheer quantity of material. Cast linear-cut bowls in this catalogue, for example, are divided into two subclasses based on the shape of rim and body. Such subclasses are often arbitrary yet useful for purposes of organization.

Equally arbitrary are the considerations that determine the order of individual fragments within a single type and their arrangement in the final catalogue. Helpful criteria include degree of preservation, nuance of shape, size of vessel, color, and so forth. The criteria vary according to type, again depending on the fragments at hand and the cataloguer's preferences, but in this volume certain guidelines have been followed. For instance, better preserved fragments are placed before more fragmentary examples of the same type; rims are generally catalogued before sides, and sides before bases; rims or base-rings are put in order of decreasing diameters; larger vessels are catalogued before smaller examples of the same shape; and more impressively decorated fragments precede less outstanding pieces. In arranging undecorated vessels by shape, open forms precede closed forms, usually in the following order: plates, dishes, pyxides, bowls, saucepans, beakers, stemmed cups, jars, flasks and jugs, prismatic bottles, unguent bottles, and miscellaneous items such as lamps, inkwells, ladles, spoons, lids, and handles. These criteria, as well as others mentioned in this chapter, are not prescriptive, but they are indicative of the concerns of the cataloguer and of the need for a set of guidelines to facilitate cataloguing and to ensure a degree of consistency.

The main advantages of such a typological system are two-fold: first, it aids the cataloguer in effectively and coherently handling the large number of disparate types encountered on Roman sites; and second, it enables him to cope with great quantities of small fragments whose complete shapes and decoration are no longer preserved. A morphological approach, based solely on shape, proves either to be impractical, given the fragmentary nature of the material, or unsatisfactory in that it disperses examples of a single type (and therefore similar date and provenance) throughout the catalogue according to form. Alternatively, it juxtaposes vessels of entirely different type, decoration, date, and provenance simply because they happen to share a common shape. If this catalogue had relied on such a system, the cast translucent colored fine wares (**Nos. 125–144**) would have been divided into four separate chapters, each devoted to a different shape: circular dishes, rectangular dishes, pyxides, and bowls. Or, stated another way, all plates (or bowls, or beakers, etc.) would be described together in a single chapter, whether they were cast examples of the Hellenistic era or free- and mold-blown examples of the early, middle, or late Roman periods. As a result, any feeling for the steady development of the industry over several centuries would be lost, as would any understanding of the glassware available in a single period.

This is not to suggest that catalogues employing a typological classification necessarily omit information, only that such an arrangement often makes it difficult to find it. Most users of archaeological catalogues seek specific comparanda or information concerning the provenance, distribution, style, or date of a vessel whose type is known or discernible. They look for discussions of a given vessel within the context of its *type* as a whole and are little concerned with the individual shape as such.

Unfortunately, one of the most prominent categories of Roman glass does not fit well into a typological classification, or for that matter into any convenient or precise arrangement. These are the free-blown undecorated (or slightly decorated) vessels of the Roman Imperial period. On most Roman sites they make up the bulk of the fragments, but, like coarse pottery, these utilitarian vessels lack adequate stylistic traits to assign them readily according to types. Accordingly, they are usually classified by shape for lack of better criteria. Occasionally, it is possible to discern related groups, such as the blown monochrome colored tablewares (**Nos. 307–373**) in this catalogue. Such a group presumably was manufactured at one time, in one region, and probably by a limited number of workshops, in this case probably a shop in central Italy during the first half of the first century A.D. At present, however, few similar groups of utilitarian tablewares have been sufficiently studied to determine regional or workshop production. Although an overall classification of plain, blown tablewares, based on such attributions, would be of the greatest value for increasing our knowledge of the Roman industry, it is as yet premature to employ such a system. Far too little utilitarian glass has been assigned securely to specific regions or factories, or placed within close chronological limits.

Given this situation, it remains necessary to rely on a morphological system of classification, and several have been developed to handle undecorated domestic wares. The most satisfactory approach, and certainly the most widely imitated, is that of Clasina Isings in *Roman Glass from Dated Finds* (1957). In this publication, Isings defines 134 shapes, arranged from open to closed forms and divided into five sections, each treating the forms of a century or half century. Within sections, cast vessels precede blown. By introducing chronological divisions and distinguishing between two of the major methods of manufacture, Isings usually avoids juxtaposition of vessels with similar shapes but of different type or from vastly different periods. At the same time, vessels of approximately the same date are often placed in close proximity. But like all morphological schemes, it presumes intact or mostly intact examples and assumes a limited number of shapes, ones that can easily be defined, named, and placed in a known chronological order. Unfortunately, most archaeological sites yield only fragments, and these, more often than not, cannot be assigned Ising's forms with any certainty. In such instances, references to the forms are perhaps best omitted, at least for the purposes of classification. Too many cataloguers, overly eager to equate their fragments with these forms, have done so with little regard to the type and date of the glass involved, so long as the shape appeared to be comparable. As a result, many of these identifications are misleading or even incorrect. In this catalogue Isings's forms are used primarily to direct the reader to additional references for those fragments whose shapes are considered characteristic of the type involved (e.g., Isings Form 17, small blown bowls with pinched ribs, **Nos. 249–258**). Finally, it is well to remember, too, that even when Imperial tablewares are intact, they can represent an almost unmanageable array of forms and variations. The publication of 1,570 vessels in the Rheinischer Landesmuseums Trier, almost entirely blown and of local manufacture, defined no fewer than 165 shapes, including forty-four subgroups.

At Cosa, the undecorated blown glass is so fragmentary and indistinct that little of it can be positively or even tentatively identified. Consequently, relatively few examples have been selected for publication, and those that are included have been arranged according to shape, with subclasses sometimes identified. On occasion, vessels of the same basic form but of different date are placed in chronological order. Among plates and bowls, for example, the first-century bluish-green wares (**Nos. 374–391**) are listed first, then plates and bowls with applied fillets (**Nos. 392–402**), dating from the late first or second century, and finally the bowls with triangular rims (**Nos. 403–413**), as well as several other kinds of colorless vessels (**Nos. 414–424**), all belonging to the second or third century A.D. The remaining fragments of utilitarian blown glass are ordered simply by shape, moving from open to closed forms as described on page 17 above. Dated subgroups are infrequent, either because the fragments were not found in datable contexts or because they could not be positively associated with specific forms. Indeed, certain fragments may have been incorrectly assigned. Rims here attributed to prismatic bottles (**Nos. 525–535**) may in fact belong to free-blown flasks, jars, or jugs. In part for this reason, the headings employed in the catalogue are purposely broad, encompassing more than one shape ("Plates, Dishes, and Bowls" or "Flasks and Jugs"). It is also probable, given the fragmentary state of the collection, that many forms once present in antiquity cannot now be detected at all.

Yet another method for handling utilitarian glass seeks only to publish material from dated deposits and to marshal these deposits in chronological order, generally with vessels arranged by shape within each group. Such a scheme provides an overview of the glass available at a given site at specific points in time and illustrates change over the course of the site's history. The drawback of this system, however, is that it excludes unstratified and undated material, and on most excavations such material forms the bulk of the glassware.

From the above discussion, it should be clear that no single method of classification is perfect or fully capable of encompassing all the glass from a given site, especially when dealing with Roman Imperial wares. Unlike ceramic studies, which before cataloguing separate the pottery into well-known "types" (Arretine, Gaulish sigillata, green-glazed, etc.), glass publications regularly include all the finds in a single continuous catalogue, regardless of how many types are present or how diverse their techniques of manufacture, decoration, provenance, or date. Even pieces whose identification is uncertain are included, and these at times form a significant portion of the manuscript. In short, until more is known about ancient glass, modern classification systems will never be wholly successful in organizing the material, nor should this be expected. After all, such arrangements were never intended by the craftsmen who manufactured the articles we study. Nor should any classification scheme be imposed *a priori* on a collection of fragments or vessels. Every site yields different sorts of glass, in varying quantities and states of preservation, and each site will pose unique cataloguing problems, ones requiring special solutions. It is also likely that as more catalogues appear, and as more is learned about ancient glass, the classification system advocated here will require refinements, even major alterations. Just as new catalogues cause us to reassess our knowledge of the subject by making additional material available, so too will they cause us to change the methods by which we classify and present fresh material.

Section Introductions in the Catalogue

Each type or class of glass is discussed in an introductory section preceding the catalogued entries belonging to that type. The general characteristics of the type are described, and the idiosyncrasies of individual Cosan fragments are noted. Evidence regarding the chronology, distribution, and provenance of the type is discussed insofar as it is known. Comparable material from other excavations or museums may also be cited, but only if the pieces are truly similar or shed light on the date or provenance of the type in question. At other times parallels are cited or illustrated if they provide intact examples of forms or decoration that at Cosa are known only from fragments. In an archaeological catalogue such as this, no attempt has been made to list systematically every known parallel. Rather, emphasis has been placed on dated parallels from Italian and, where important, European or eastern Mediterranean sites.

Although the introductory sections describe a great many types or groups spanning several centuries, they do not form a continuous history of the industry in ancient Italy. Not only does Cosa lack examples of certain types of fine wares known from other Italian sites (cameo glass, for example), but the utilitarian vessels are too fragmentary to reflect the rich variety once in use by the residents of Cosa.

Description of Catalogued Entries

General Remarks

Individual fragments or vessels are catalogued in a uniform manner, each description containing as many as fifteen elements, some of which are optional, others necessary (unless the fragment does not admit to that type of information). These elements, in the order in which they usually appear, are as follows: state of preservation, name of vessel or object, color, weathering, material, fabric, quality of craftsmanship, shape in detail, decoration in detail, techniques of manufacturing and finishing, dimensions, disposition, bibliography, and parallels. In this catalogue each description is prefaced by a catalogue number, inventory number, find-spot, and date (if from a datable context). Plate numbers are also given. When fragments are too small, too numerous, or identical with others already described, they may be listed in a condensed format, as is true of sixteen tiny but recognizable fragments of cast, linear-cut bowls (**Nos. 120a–p**).

State of Preservation

Preservation refers to the number of extant fragments and the parts of the vessel they represent (rim, side, bottom, and so forth); whether they join or not; and whether they provide a complete or nearly complete profile. If they do, “profile restorable” is stated. The word “intact” is used to describe vessels with no breaks or losses; "complete" means that the entire vessel is accounted for but is broken and repaired.

Name of Vessel or Object

The generic name assigned to a vessel or object (plate, bowl, beaker) is listed with its state of preservation, occasionally prefaced by a more precise indication of shape or type (fusiform alabastron or shallow ribbed bowl), or a comment on its size (small, medium, large). In many instances, when the glass is fragmentary, it is difficult to determine its exact shape or to distinguish readily between one of two possible shapes (a plate versus a dish, for example).

Color

A description of color may be as complex as the polychrome pattern of a striped mosaic-glass bowl (see **No. 12**) or as simple as the single word "colorless," describing a great many Roman beakers. Unfortunately, the color variations of ancient glass are altogether too numerous and caused by too many diverse factors for ease of classification. Moreover, specialists have found no uniform or mutually acceptable method for describing color in glass. The Munsell Color Charts, for example, are best used with opaque objects such as pottery and have only limited application to glass, which is usually translucent. Scientific descriptions of color, obtained by spectrum analysis and represented graphically, are physically exact but prohibitively costly.

This catalogue has sought to use as few terms as are consistent with the material at hand, keeping the following observations in mind. First, the ancient glassmaker had less control, and often less interest, in producing a precise color or tint than does his modern counterpart, who has access to modern chemistry and technology. Second, glass batches in antiquity were often small, and one factory over a brief period could manufacture products noticeably varied in color while otherwise similar. Third, accidental or intended changes in the oxidation/reduction state of the batch, or in the process of making a vessel, can dramatically alter the color. Fourth, the thickness of a glass, being transparent or translucent, decidedly affects our perception of the value of its color. The thick handle of a purple jug will appear almost black, whereas a thin fragment of the side seems very pale by comparison. Fifth, the effects of time, weathering, and devitrification, especially after the glass lies buried for centuries, can distort, disguise, or wholly change the original color. And finally, the importance of coloration in cataloguing ancient glass is often exaggerated. Rarely is color the primary factor in attributing a fragment to a particular type of glassware.

Because of this, descriptions of color in this catalogue are brief, meant to indicate, where possible, the *original intention* of the glassmaker. Accordingly, vessels are divided into three broad categories: naturally colored, decolorized, and "intentionally" colored. The first includes translucent and transparent bluish-green, light green, yellowish-green, pale olive, and pale yellow glasses, all colors associated with the free-blown and mold-blown utilitarian wares of the Roman Empire. These colors result from metallic oxides, often iron, present in the raw materials and cullet used to make the glass. Theoretically, additional chemical colorants were not added by the glassmaker (although I must admit that it is often difficult to determine whether a color is due to the natural composition of the batch, the introduction of cullet, or the intentional addition of chemical compounds).

Colorless or decolorized glass is produced by mixing manganese or antimony, or sometimes both, with the batch, thereby neutralizing the metallic oxides that otherwise tint the glass. In antiquity, colorless glassware was much prized, especially by pre-Classical and Hellenistic glassmakers because it introduced the novelty of transparency, otherwise only associated with costly rock crystal.

It again achieved popularity during the mid-first through third centuries A.D., when almost all fine and much ordinary tableware of the Roman Empire was purposely decolorized.

It should be noted, however, that most colorless glass retains a bluish or greenish tinge, especially visible along a fracture or edge, as is true of modern window glass. If pronounced, this hue is mentioned in the description (colorless with bluish-green tinge). Also, if too much manganese is added or incompletely mixed in the batch, purple streaks or even purple glass may result. In describing colorless vessels, the term "clear" is too imprecise to be used, and the word "white" is employed only to describe *opaque white* vessels.

Intentionally or decoratively colored glass is the result of added chemical substances (metallic oxides) that produce a specific color in the fused batch, in much the same way as food coloring is held in suspension in water. These purposely colored glasses may be opaque or translucent. Opaque colors are achieved through the use of chemical opacifiers, such as tin or titanium oxides for opaque white, in addition to colorants. The most common opaque colors associated with the Hellenistic and Roman industries are white, yellow, red to brownish-red, light blue, and a variety of greens, including olive green and jade green. Opaque orange, "flesh-color," and brown exist, but black on close inspection is, in fact, a translucent dark green, blue, brown, or purple that only seems to be opaque.

Other decoratively colored glasses are translucent or, rarely, transparent. The most commonly employed by the Hellenistic and Roman glasshouses are shades of blue, green, purple, and yellow to golden-brown, or brown. Two striking variants of green and blue, employed for both cast and blown vessels of the Augustan Age, are a dark blue-green (aquamarine) and a vibrant blue, here termed peacock blue. Turquoise blue and dark blue were exceedingly popular in the later Roman Empire, especially for trailed threads and blobs applied to colorless or lightly colored vessels. In this catalogue these intentional colors are occasionally described as being dark, light, or pale. In some instances the degree of their translucency or opacity is also noted. When two mixed colors are in evidence, such as golden-brown or bluish-green, the predominant color is stated last, the modifier first. For polychrome or mosaic-glass vessels, the ground color or overall colored effect is cited to enable the reader more easily to visualize the vessel.

Weathering

Weathering refers to all the detrimental effects of age, moisture, and chemical attack causing decomposition on or under the surface of the glass. If weathering has been especially virulent, much of the glass may have disintegrated, and the original thickness of the vessel may be markedly diminished. Cataloguers must be careful to distinguish between the color of the weathered surface and the actual color of the glass beneath, and to distinguish clearly between weathering of the surface and bubbles or impurities within the glass (see "Material"). They should also be aware that some glass from archaeological excavations or museum collections has been so completely cleaned, or even repolished in modern times, that only the most persistent forms of weathering are in evidence. The principal terms for weathering, as described in Harden 1936, 9–11, and Buechner 1960, 114–117, with illustrations, are as follows:

dulling The loss of translucency and clarity, owing in part to surface weathering or devitrification, in part to scratches and physical damage to the surface.

iridescence A surface with rainbow-like color caused by thin layers of seemingly transparent or whitish films that have flaked away. Iridescence is the most common form of weathering, and it appears on almost all ancient and medieval glasses, even those preserved in very dry climates.

pitting Small but distinct hollows caused by decomposition of the surface. Like iridescence, it affects most ancient glass.

corrosion Pitting sometimes has progressed to the point where the surface is destroyed. Corrosion can take the form of irregular grooves (sometimes called "ploughing"), not unlike worm holes in old wood. These can be mistaken for intentional decoration.

silvery weathering Numerous layers of silver-colored iridescent films on the surface. They flake away with ease once the glass is in a dry environment.

milky weathering Layers of opaque white films, often granular or gritty to the touch and associated with considerable pitting. Milky weathering often cannot be removed and permanently destroys the surface of the glass.

enamel weathering A dark but iridescent layer of weathering resembling multicolored enamels. Like milky weathering, it adheres to the surface.

black scaly weathering A hard, dark encrustation, usually with numerous layers of iridescent films below it.

The terms defined above are descriptive, and not scientific definitions of decomposition or weathering. Such definitions would probably be difficult to formulate and their significance negligible for those who study ancient glass. Two fragments of the same vessel, lying only a few centimeters apart, can weather in distinctly different ways, as is true of a colorless plate found in Atrium Publicum Room 25C, Level II (**No. 169**).

At Cosa the glass was washed and thoroughly cleaned following excavation so that only the most persistent weathering products remained on the fragments. Nonetheless, it is evident that all the glass was affected to some degree by decomposition, usually by iridescence, slight pitting, and surface dulling. The phrase "iridescence and pitting" occurs with great frequency in the catalogue. Also, at Cosa, certain varieties of glass consistently were affected by specific forms of weathering. Purple pieces, for example, are associated with enamel weathering, opaque white and yellow by milky weathering, and opaque red by a characteristic hard, green surface. Certain colorless fragments, such as those from snake-thread vessels, are usually coated with whitish, milky weathering, granular to the touch and irreparably obscuring the clarity of the glass.

Material

Material refers to the physical aspect or appearance of the glass at the time of its manufacture, and it involves such factors as the presence and disposition of bubbles, blowing spirals, or accidental inclusions in the glass. Bubbles and blowing spirals are of interest because they can provide clues concerning the manufacturing process. Elliptical bubbles, for example, indicate stress as the vessel is being rotated and manipulated on the blow pipe. Modern glassmakers recognize several types of inclusions. These include "seed," groupings of tiny bubbles; "stone," usually unmelted particles of the batch, bits of crucible, or sand; and "scale," small flakes of metal tools that find their way into the glass, such as the flecks of iron that still adhere to a pontil wad on the underside of a stemmed cup at Cosa, **No. 481**, pl. 30. Reference to crazing (a network of hairline cracks) on the surface or to strain cracking under the surface also comes under this heading. The latter condition is the result of internal stress in the glass, often due to improper annealing or sudden changes in temperature during manufacture.

Fabric

Fabric is used with two meanings, one conventional, the other technical. In the first instance it refers to the glass *per se*, as in the sentence, "The decoration is in the same fabric as the vessel." Otherwise, fabric refers to the actual chemical composition of the glass, and this can be determined only by spectroscopy, neutron activation, or other scientific means. Visible differences in color, material, and weathering may enable a cataloguer to segregate different fabrics, but only chemical analyses can verify these suggestions. Since all ancient and medieval glasses are of the silica-soda-lime variety (see "glass" in "Glossary"), it is the presence, absence, or proportions of trace elements (especially lead, uranium, or tin) that are potentially the most useful in obtaining information on the provenance, date, or manufacturing processes of a particular vessel or type of glassware. At present, however, there are an insufficient number of analyzed and dated samples from known sites to provide the controls against which new samples can be compared.

Quality of Craftsmanship

Quality of craftsmanship is noted only when a fragment or vessel is of exceptionally fine or exceptional workmanship.

Form in Detail

Form refers to the shape of the vessel and its parts. Descriptions of form proceed from the top of the vessel to the bottom, from the inside to the outside. In this catalogue rims, in relation to the side, may be outsplayed, offset, upright, vertical, horizontal, or inturned. With respect to their shape and method of finishing, rims may be tubular, double-tubular, knocked-off, or solid and thickened. The edge of the rim may be rounded, pointed, flattened, fire-polished, or left unfinished. The term "lip" is used at times to describe a distinct but short protrusion outward from the rim.

Necks may be cylindrical, tapering upward or downward, or they may be funnel-shaped. Some shoulders are at right angles to the body, others simply flare or curve outward. Sides usually are convex curving, but they may be straight and tapering downward or vertical, concave, inturned, indented, or carinated with two or more convex curves in succession. The body of a vessel refers to everything below the rim (or below the neck and shoulder) but above the base. The term "bottom" means the resting surface of the vessel; it may be flat, convex, concave, or kicked. Bases refer to any projecting element upon which the vessel rests. As such, a base may be stemmed or toed, or consist of a tubular, cut-out, or solid base-ring. Plates and dishes also have floors that have an upper and underside. Handles are described with such conventional phrases as strap handle, hip handle, or stirrup handle; they may also be fitted with thumb pieces.

Although these terms do not exhaust the descriptive vocabulary used in this catalogue, they do indicate those that appear most often. They also suggest how few terms are needed to describe the shapes of fragments normally encountered in Roman glass. Since this catalogue deals almost exclusively with fragmentary vessels, terms to describe the intact forms of vessels are not discussed here. The "Glossary" should be consulted whenever questions concerning terminology arise.

Decoration in Detail

Descriptions of decoration are as varied as the multiple methods and motifs employed by ancient glassmakers to enhance the attractiveness of their products. In the case of core-made bottles (**Nos. 1–2**), threads of glass in colors contrasting with those of the body are applied and combed with a tool, producing zigzag and festoon patterns. With cast mosaic-glass (**Nos. 3–44**), the decoration is inherent in the manufacturing technique, which makes use of polychrome sections of cane(s) fused into colorful, repetitive patterns. Cast monochrome vessels (**Nos. 45–204**) may be decorated with cut grooves and ridges, either singly or in bands (**Nos. 92–123**). They also may derive their decoration from molds or tools used during the manufacturing process, as is true of the ribbing on ribbed bowls (**Nos. 27–91**).

Mold-blown vessels also make use of molds or forms to impart floral, figural, or geometric designs that are raised in relief on the surface (**Nos. 205–229**). The subject matter can vary from classical myth (**No. 220**) to gladiatorial combat (**No. 205**); and even Latin and Greek inscriptions occur (**Nos. 206–207, 539–540**). Most free-blown glass is plain and unadorned; in antiquity it served as serviceable household wares, especially after glass became exceedingly common in the mid-first century A.D. Nonetheless, certain types or groups, chiefly finer imperial tablewares, are decorated. Decoration may take the form of applied trails, coils, blobs, or prunts of glass (**Nos. 266–300**) or of simple tooled indentations (**Nos. 301–306**), pinched ribs or protrusions (**Nos. 249–265**), or patterns impressed on the surface while the glass is hot. Finer glassware after cooling may be facet-cut, engraved, or abraded with geometric patterns or even figural scenes (**Nos. 232–248**). Still other vessels were painted or enameled once the glass was cold. Perhaps the most elaborate forms of decoration include cameo glass, sandwich gold-glass, and cage-cups, all luxury goods not represented at Cosa.

When present, the kind of decoration, together with a full description of its color and motifs, as well as its location on the vessel, are all incorporated into the catalogue entry.

4 ♦ The Catalogue

HELLENISTIC CORE-FORMED GLASS

*Mediterranean Group III (**Nos. 1–2**)*

The excavations at Cosa yielded two examples of core-formed bottles, a fusifom alabastron (**No. 1**) and a tiny fragment of a second vessel (**No. 2**), either another alabastron or, more likely, a large-bodied amphoriskos. Both are the typical and predominant products of a prolific core-forming industry, now referred to as Mediterranean Group III, that emerged in the second century B.C. This was the last core-forming industry (within a tradition beginning in the Bronze Age) to manufacture small glass containers by this relatively cumbersome method (see"sand-core technique" in "Glossary"). Although most examples of Group III are known from published sites in the Levant or eastern Mediterranean, the distribution of the group is exceptionally wide in the Mediterranean region, ranging from the Levant, Aegean, and Black Sea westward to Italy and the coastal areas of the western Mediterranean. In the late Hellenistic period these bottles and their prized contents of perfumed oils or cosmetics were among the few categories of glass vessels to reach the Roman communities of central Italy. Production of Group III bottles appears to have ceased by the close of the first century B.C. although certain examples have been discovered in contexts of the first century A.D. The principal studies of this group are Fossing 1940, British Museum Coll. 1981, McClellan 1984, and Grose 1989.

At Cosa the fusiform alabastron (**No. 1**) was found in a datable trench (Deposit I) associated with the construction of the forum basilica and dated to the second half of the second century B.C., most probably around 150/140 B.C. Other dated alabastra from Italian sites include one recovered from Grave 57 at Ornavasso, along with denarii attributed by the excavator Bianchetti to 88 and 84 B.C. (but with the Arretine bowl with the signature L. Sarius L(uci) l(ibertus) Surus: Ornavasso 1895, 228, pl. 15.15; Graue 1974, 225, pl. 64.3; see Oxé and Comfort 1968, 400 no. 1660); another from the Roman cemetery at Ancona attributed generally to the second and first centuries B.C. (Marconi 1934, 16); and four others from Morgantina in Sicily (unpublished), also belonging to the second and first centuries B.C.

Undated fusiform alabastra from central or northern Italy (but lacking definite provenances and dates) include ones from Ticino and Siena (Calvi 1973, 55–56, pl.1, and G. Pellegrini, "Appendiche Museografia," in *Studi e materiali di archeologia e numismatica* 3 [1905] 314–315, nos. 422–423); and yet another from Tharros on Sardinia (British Museum Coll. 1981, 127, no. 335). Three others, known only from a photograph, were in the hands of a Roman antiquities dealer early in this century (DAI neg. no. 4507).

In the eastern Mediterranean comparable fusiform alabastra, accompanied by large-bodied amphoriskoi with strap and S-shaped handles, are especially plentiful, as the lists in Fossing 1940, 111–114; British Museum Coll. 1981, 122–125; and McClellan 1984 demonstrate. Several specimens have been found in Greece, one from a grave on Crete not far from Herakleion (Alexiou 1965, 555, pl. 698) and another from Amphipolis in Thrace (Lazarides 1957, 71, pl. 23). Both were recovered from contexts that are regarded as late Hellenistic.

MEDITERRANEAN GROUP III

1. CD 963 and CD 1019 — Pl. 1
Bas. ENE 6, III; ENE 7, IV — Deposit I
ca. 150/145 B.C. to ca. 72/70 B.C.

One non-joining and two joining fragments of the neck and body of a fusiform alabastron (profile mostly restorable); light blue (translucent but appearing opaque) with opaque white decoration. Iridescence and pitting. Upper part of body is decorated in a feather pattern of opaque white threads; lower part is encircled by a spiraling white thread. Core-formed with threads of glass combed and marvered flush to the surface. Max D body ca. 3.7 cm; A Th .25 cm.
Bibliography: Grose 1973, 177, pl. 189, no. 1.

2. CF 2374 — Pl. 1
A.P. Room 15 Exterior, I

Small fragment of neck and shoulder of unguent bottle; dark green appearing opaque black, with opaque yellow decoration. Iridescence and much pitting. Shoulder flares outward from cylindrical neck; a translucent light green scar of lug or handle at junction of neck and shoulder. Shape uncertain, but probably an amphoriskos. Opaque yellow thread encircles neck and shoulder. Core-formed with threads of glass marvered flush to the surface. D neck ca. 1.7 cm; A Th shoulder .5 cm.
Remarks: In all likelihood, this fragment was part of an amphoriskos of Mediterranean Group III: see Grose 1989, 123–125 and 169–174, Class III:E and Amphoriskos Forms III:2A–III:3.

HELLENISTIC AND EARLY IMPERIAL ROMAN CAST MOSAIC GLASS

Forty-four fragments of cast mosaic glass were found at Cosa. Although few in number, they represent almost every prominent category of mosaic glass manufactured by the later Hellenistic or early Roman workshops.[1] Included are examples of network, striped, short-strip, marbled, gold-band, color-band, and an assortment of composite mosaic glass patterns. Most belong to types characteristic of the early Roman casting industry founded in Italy late in the first century B.C. (see discussion in Grose 1983, 40–44). However, many of these styles and patterns were originally devised in the Hellenistic era (among them, network, striped, and gold-band glass) and only later adopted by the Roman-Italian industry. As a group, these vessels were among the first kinds of glassware to appear on Roman sites in Italy.

Virtually all pieces of cast polychrome glass retrieved from Cosa represent small bowls or beakers (twenty-two examples). The most common form is the marbled ribbed bowl (fourteen pieces), followed by network mosaic-glass beakers (nine examples). Other shapes include dishes, jars, and small bottles. With few exceptions the fragments were discovered in the domestic quarters of Cosa and constitute vessels used as serviceable household objects. Not surprisingly, the inhabitants at

[1] Only one fragment at Cosa (**No. 23**) was discovered in a context that might indicate that it belonged to a group of mosaic vessels manufactured by a revived late Roman mosaic industry of the third and fourth centuries A.D. On the existence of such an industry, see the brief account in Sardis 1980, 88–89, and Grose 1983b, 285.

Cosa left behind no examples of the finer mosaic glasses manufactured by the earliest Hellenistic industry of the late third and second century B.C. (See Grose 1981, 63, 70–71). Nor are there examples of the exceptionally large or elaborate mosaic vessels produced by the later Hellenistic or Roman factories and known from finds at Rome and elsewhere.

Fifteen pieces are from dated deposits, and these help to clarify the chronological development of mosaic glass, including the relationship between types of Hellenistic and Roman date. Indeed, it is now possible to trace the historic evolution of certain cane patterns from the time they were first devised in the Hellenistic period until the techniques fell from fashion in the early or mid-first century A.D. Here, this has been done for network, striped, gold-band, and marbled vessels. The finds from Cosa further demonstrate that many varieties of Roman mosaic glass first appeared a decade or two earlier than has previously been suspected based on the finds from provincial sites in central and northwestern Europe.

The method used here to classify mosaic glass according to pattern is meant to facilitate cataloguing and not to suggest mutually exclusive groups. Although some workshops may have concentrated on making vessels of a single pattern, others clearly manufactured vessels in various cane designs but often in identical shapes, as is true of the factory that produced the network, striped, and composite mosaic bowls of the Antikythera group (see below and Grose 1983, 43). It is also apparent from the study of the fragments at Cosa that a correlation usually exists between a specific pattern and one or more characteristic shapes (for example, marbled ribbed bowls or network beakers), but unusual forms should also be expected. Far more shapes and variants exist than is evident from the relatively small corpus of published examples. The same holds true of color combinations. Although certain colors are routinely associated with particular cane configurations (and others exceedingly rare), exceptions do occur. Finally, there are unique pieces, such as the white millefiori dish (**No. 22**), that, as yet, do not fall into any recognizable type or group.

In order to appreciate the complete range of Hellenistic and Roman mosaic glass and to observe the diversity within individual types, it is useful to examine the extensive collections of fragments housed in major European and American museums. Of special importance are the collections at Rome in the Vatican Museums (Vatican Coll. 1975), the Museo Nazionale Romano (unpublished and being recatalogued), the Antiquarium Comunale (unpublished and recently recatalogued). Equally important are the fragmentary vessels found at The Metropolitan Museum of Art, formerly in the Gréau and Morgan Collections (Gréau Coll. 1903), the Corning Museum of Glass (Corning Coll. 1979), the Toledo Museum of Art (unpublished), and the Victoria and Albert Museum (unpublished). Although the sources of these fragments are seldom known with certainty, most pieces, such as those in the Toledo Museum of Art (but originally part of the Coleman-Curtis Collections; Sturgis 1894), were acquired in Rome following the unification of Italy in 1870. At that time the city was transformed by building projects throughout the area of ancient habitation, and considerable archaeological material was unearthed. Minor objects such as fragments of glass made their way into the hands of local antiquities dealers, who often polished one surface, then encased the fragments in cardboard mounts with gilded edges. In fact, many of the fragments in the museums noted above are still exhibited in this distinctive manner.

Network Mosaic-Glass Vessels (***Nos. 3–11***)

Hellenistic and Roman glassmakers alike manufactured mosaic vessels using preformed canes that exhibited a network pattern (also called "*reticella*" or "lace-mosaic glass" in other publications).

The canes were formed by twisting fine colored threads (usually opaque white or opaque yellow but also other colors) around a larger colorless cane or, occasionally, around a central cane that was colored, most often blue, purple, or golden-brown. At times the glassmaker spirally wound only a single thread about the central cane; at other times, two parallel spiraling threads were used. In some cases two threads were spirally wound in opposite directions, forming reverse spirals (Grose 1983, 42, ill.). Once made, the canes were cut to requisite lengths and cast in or over molds to produce vessels with a "lacy" appearance. A single network cane whose central element (and twisted threads as well) was in a contrasting color to those of the body was applied as a rim.

Examination of vessels demonstrates that the use of network canes has a long history, beginning in the Hellenistic period, probably late in the third century B.C., and continuing down through the Augustan/Tiberian era, when the Roman industry adopted the style for its own purposes (Grose 1983, 42–43). Although it is not yet possible to determine with certainty the exact evolution of the technique, or to integrate all known examples into a single scheme,[2] study of existing vessels allows the identification of at least three major phases, each representing a different period and presumably a different center of production (see Grose 1984, 30, fig. 5).

The earliest class of network vessels, all bowls, is associated with the finds from the well-known tomb groups at Canosa in southern Italy, said to belong to the later third and second centuries B.C. and customarily attributed to the famed but elusive factories at Alexandria (Harden 1968, esp. 27 and 42; Oliver 1968, 49–50, 52–54, 57–58; von Saldern 1975, 41–42, 44; Grose 1981, 63–65; and Grose 1983, 42–43, 45).[3] Bowls of this group are typified by a hemispherical or, more rarely, a parabolic shape with upright rims and convex bottoms. The canes, usually in pairs, spiral upward from the bottom to the rim, forming almost horizontal rows,[4] and are colorless, with either opaque yellow or opaque white threads. The coil at the rim is often colorless with a spiraling blue thread, but other combinations exist, such as golden-brown or purple with an opaque white thread. At least one example uses a colorless strip with a distinct pink tinge. These bowls were cast, probably by winding softened canes around a mold, then fitting an outer mold over it to hold the canes in place until they were fused in the kiln. At least twelve examples are known (see vessels cited above and Damascus Coll. 1969, 100, fig. 38, a previously ignored bowl probably from a Syro-Palestinian context; another excavated at Nimrud, Dillon 1907, 39–40;[5] and two unpublished pieces of uncertain provenance in a private collection).[6] Of the others, four are thought to have come from Canosa, a fifth from Ancona, with two more acquired in Greece, and yet another ascribed possibly to Turkey. Not surprisingly, Cosa, founded in 273 B.C. as a colony for impoverished farmers, yielded no examples of this early class.

The second group of network bowls is represented by glassware recovered from the celebrated shipwreck off the island of Antikythera, located on the east–west shipping route between the Aegean

[2] Examples of vessels that do not readily fall into any of the groups mentioned here include an unusual pyxis in the British Museum (acc. no. 96.6-30.3) and three broad, shallow dishes (D 12–13 cm) with slightly convex bottoms, housed in the British Museum and the Victoria and Albert Museum (cited in Oliver 1968, 67). These latter vessels, however, are closely related to certain mosaic dishes (see p. 43 below) that are Hellenistic in date.

[3] It should be noted that no example of this class has been recovered from a securely dated context. Consequently, their dating is inexact, as is the dating of the Canosa group in general.

[4] Not all bowls thought to belong to the Canosa group have canes laid in a spiraling fashion. At least one, now in the Corning Museum of Glass but from a tomb group discovered in Italy, possibly at Canosa, has canes laid in parallel rows across the vessel (Corning Coll. 1979, 177–178, no. 462, color pl. 22; cf. Parlasca 1976, 150, pl. 18).

[5] I wish to thank Professor Dan Barag, Hebrew University, for bringing this bowl to my attention.

[6] These two bowls, part of a larger hoard of vessels belonging to the Canosa group, are to be published in an upcoming issue of the *Journal of Glass Studies*.

and Italy. The artifacts in the cargo indicate that the ship sank between ca. 80 and 50 B.C., and, consequently, the eleven glass vessels, including two network bowls, are assigned to the first half of the first century (Weinberg 1965, 38–39, nos. 10–11, figs. 20–23; De Solla Price 1974, 5–9; Grose 1981, 65–66). These bowls are characterized by their small hemispherical shapes and high trailed base-rings. Again laid in pairs, the canes spiral from the bottom to the rim, forming almost horizontal concentric rows.[7] Additional network vessels of this sort are cited in Oliver 1968, 55–57, along with other bowls of the same profile but in striped or composite mosaic-glass patterns. Not mentioned is a fragmentary example, presumably belonging to this group, from Delos (Delos 1970, 234, no. 308, pl. XXVII), which in all likelihood dates to the period before the sack of the island in 88 (or 69) B.C.[8] A unique piece from the western Roman provinces was found in Severinstrasse, Cologne, and dated by Fremersdorf to the first century A.D. (Fremersdorf 1958, 21, color pl. 5; cf. LaBaume n.d., no. H3, pl. 44.3). Since this vessel is apparently not from a sealed deposit, the exact date is actually uncertain. In any event, the bowl must have been an heirloom considering the date assigned to other examples of this group.

At Cosa, **No. 3** and possibly **No. 4** are related to the Antikythera group. The shape and dimensions of **No. 3**, as well as its uniform horizontal rows of canes, make such an identification reasonably secure. **No. 4** has a similar appearance, but its diameter is larger than is customary, and its attribution to this group is correspondingly less certain.

The last identifiable group of network vessels was manufactured by the Roman industry, presumably in Italy, during the final decades of the first century B.C. and early decades of the succeeding century. Vessels in various shapes were produced, but the small bowl or beaker, such as **No. 5**, was preeminent. It is characterized by an upright rim and straight, almost vertical, side tapering downward to a slightly concave bottom. Base-rings are never employed, in contrast to the vessels of the Antikythera group. Usually the body is made of colorless canes twisted with opaque white threads (**Nos. 5–10**), less commonly with opaque yellow (**No. 11**). On some examples colored canes (blue, purple, golden-brown, twisted with opaque colored threads) are employed in the body (see Corning Coll. 1979, 194, no, 527, color pl. 29). A blue cane twisted with white customarily forms the rim (**Nos. 5-10**), but many other color combinations occur, including two colored threads (opaque yellow and opaque red, for example) twisted around a central colorless cane (see Gréau Coll. 1903, II, pls. 102–103, and Corning Coll. cited above).

Small bowls or beakers similar to **No. 5** at Cosa have also been found at Vindonissa 1960, 13, nos. 4–5, pls. 1, 17 (incorrectly restored with convex bottoms). Other examples are cited in Vatican Coll. 1975, 38–39, nos. 190–191, pl. 6, and Corning Coll. 1979, 194, no. 524, pls. 29, 41. A partly intact example, in the Museo Nazionale Romano in Rome, is illustrated in Grose 1984, 30, fig. 6.

Other shapes include small cylindrical bowls with vertical sides (Corning Coll. 1979, 179, no. 525, pls. 29, 41); large bowls (D 18.5 cm) with straight outsplayed sides (Corning Coll. 1979, 193–194, no. 523, color pl. 27 and pl. 41), said to be from the Roman cemetery at Agro Adriese (Sangiorgi Coll. 1914, 62, no. 221, pl. XLII); and lidded pyxides (Greifenhagen 1962, no. 4, allegedly from Cumae and probably belonging to the Roman production). The lid of a small pyxis, almost certainly of the Roman group, is preserved in the Toledo Museum of Art (acc. no. 23.1876, unpublished).

[7] At least one example, presumably related to this group, has parallel network canes laid side by side across the vessel (Gréau Coll. 1903, II, pl. 108.2; cf. Oliver 1968, 57, fig. 10).

[8] Because the fragment is quite small and lacks a preserved section of rim, positive identification of the vessel is difficult. Another example from the Aegean, albeit in a composite cane pattern, was found in a tomb at Agios Thomas near Eraklion on Crete. In profile it is identical to the network bowls from the Antikythera shipwreck. Although the tomb is characterized as "Roman" by the excavator, it must date to the late Hellenistic period. With it was found a fusiform alabastron identical to **No. 1** at Cosa (Alexiou 1965, 555, pl. 698).

Technically, the vessels of the Roman group were mostly made by sagging a preformed disk of network canes into, or more probably over, a "former-mold," as illustrated in Grose 1984, 29, fig. 9. This caused the canes to stretch and widen progressively from the rim to the body and bottom while simultaneously causing the thickness of the sides to taper from the rim to bottom as well. Because only one surface was in contact with the mold, only the interior and the edge of the rim required rotary-polishing.

Parallels and Dated Examples of Network Mosaic-Glass Vessels

Cosa: **Nos. 5–10**. Deposit V, ca. 25/15 B.C. to ca. A.D. 40/45.

Rome: Museo Nazionale Romano. Inv. no. 294080. H 6.5 cm; D 8.8 cm. Small bowl similar to **No. 5** at Cosa. Illustrated in Grose 1984, 30, fig. 6.

Vatican Coll. 1975, 38–39, nos. 190–191, pl. 6. Find-spots uncertain but presumably from Rome. Fragments not drawn, but the shapes are probably similar to **No. 5** at Cosa.

Rome: Antiquarium Comunale. Fragments belonging to small bowls. Unpublished.

Sangiorgi Coll. 1914, 62, no. 221, pl. XLII. Large bowl, cited in the text, said to date to the early Imperial period. From Agro Adriese.

Bibracte 1899, pl. LVI, 50. Fragment from ca. 5 B.C. Shape not identified.

Haltern 1901, 172, pl. XXXIII, fig. 3; 1903, 96, fig. 20; 1909, 373, pl. XXXVIII. Fragments from ca. 11 B.C. to A.D. 16, presumably belonging to the Roman group.

Vindonissa 1960, 12–13, nos. 4–5, pls. 1.4 and 1.5, 17. Three dated fragments from the Tiberian/ early Claudian or Claudian/Neronian levels. Additional parallels from other sites listed on p. 16.

Camulodunum 1947, 293, no. 1, pl. LXXXVIII. Period IV, A.D. 49–60.

Staré Hradisko 1961, 55, no. 602-2111, pl. 50.6. Late La Tène burials, of which one contains a fragment of network glass of unstated shape but possibly of the Roman group.

Based on current evidence, the earliest known examples of the Roman group are those found at Haltern, Bibracte (if indeed this fragment belongs to the Roman type of network glass), and possibly at Staré Hradisko, an eastern European site described as "late La Tène." These finds suggest that such vessels may have been in production as early as the last decade(s) of the first century B.C. On the other hand, the examples from Vindonissa and Camulodunum come from Tiberian, Claudian, and Neronian levels and indicate that most pieces are later and belong to the first quarter of the first century A.D. The examples at Cosa (**Nos. 5–10**, representing as many as six different vessels) are dated between ca. 25/15 B.C. and ca. A.D. 40/45 and confirm such a chronology. The absence of network glass from Flavian contexts further suggests that the type was not available beyond the middle of the century. In fact, the chronological and geographical distribution of the group is so limited that it can be argued that Roman network glass enjoyed only a brief period of popularity and rarely circulated beyond Italy, where production presumably was centered.

Network Mosaic-Glass Vessels

3. C69.382 Pl. 1

Arx W. Slope, Sounding 1, 0

Fragment of rim and side of small bowl; network pattern composed of twisted canes of colorless and opaque white threads. Two different canes were used: one composed of two parallel white threads twisted spirally around a central colorless cane; the other, a single white thread twisted spirally around a central colorless cane. On the fragment, the two canes alternate in almost horizontal rows parallel to the rim; presumably the two canes are paired and spiraled upward from the center of the bottom to the rim. A cane of golden-brown twisted with an opaque white thread forms the rim. Pitting and dulling. Upright rim with rounded edge; convex curving side. Cast in canes, then rotary-polished on the interior and outside of rim; rest of exterior fire-polished. D ca. 9.5 cm; A Th .3 cm.

4. C66.801 Pl. 1
V-D, NW Balk, Sounding 1, I

Fragment of rim and side of small bowl; network pattern composed of twisted canes of colorless and opaque white threads. Each cane consists of a single white thread twisted spirally around a central colorless cane. On the fragment the canes are laid in almost horizontal rows and probably spiraled upward from the center of the bottom. A cane of blue twisted with an opaque white thread forms the rim. Iridescence and much pitting. Upright rim with rounded edge; straight side, tapered downward. Cast in canes, then rotary-polished on the interior and outside of rim; rest of exterior fire-polished. D ca. 12 cm; A Th .25 cm.

5. CE 2016 Pl. 1
A.P. Room 16, I Deposit V
ca. 25/15 B.C. to ca. A.D. 40/45

Three non-joining fragments of rim and side, and of side and bottom, of small bowl (profile restorable); pattern composed of twisted canes of colorless and opaque white threads. Each cane composed of a central colorless cane with two opaque white threads in a reverse spiraling pattern. A cane of blue twisted with two opaque white spiraling threads forms the rim. Iridescence and pitting. Upright rim with rounded edge; nearly straight side, tapering downward; slightly concave bottom. Cast in canes laid in parallel strips across a mold, then rotary-polished on the interior; exterior fire-polished. Restored H 5.2 cm; D ca. 9.5 cm; Th tapers from .3 cm at rim to .1 cm at bottom.
Bibliography: Grose 1973, 177, pl. 189, no. 2.

6. CE 2182A Pl. 1
A.P. Room 16, I Deposit V
ca. 25/15 B.C. to ca. A.D. 40/45

Fragment of rim and side of small bowl; network pattern as in **No. 5**. Weathering, shape, and technique of manufacturing also the same. D ca. 9–10 cm; A Th .2–.15 cm.

7. CE 2183A
A.P. Room 16, I Deposit V
ca. 25/15 B.C. to ca. A.D. 40/45

Fragment of rim and side of small bowl; network pattern as in **No. 5**. Weathering, shape, and technique of manufacturing also the same. Possibly from the same vessel as **No. 6**. D ca. 9–10 cm; A Th .2–.13 cm.

8. CE 2109
A.P. Room 16, I Deposit V
ca. 25/15 B.C. to ca. A.D. 40/45

Tiny fragment of rim and side of small bowl; network pattern as in **No. 5**. Weathering, shape, and technique of manufacturing also the same. Possibly from the same vessel as **No. 5, 6,** or **7**. D uncertain.

9. CE 840
A.P. Room 14, I Deposit V
ca. 25/15 B.C. to ca. A.D. 40/45

Tiny fragment of rim and side of small bowl; network pattern as in **No. 5**. Weathering, shape, and technique of manufacturing also the same. Possibly from the same vessel as **No. 5**, **6**, **7**, or **8**. D ca. 9–10 cm; A Th .25 cm.

10. CE 2110
A.P. Room 16, I Deposit V
ca. 25/15 B.C. to ca. A.D. 40/45

Tiny fragment of side of small bowl; network pattern composed of a central colorless cane twisted with opaque white threads. Some canes consist of a colorless cane twisted with opaque white threads. Some canes consist of a colorless cane twisted with two spiraling threads in reverse directions; others have two parallel white threads spiraling around a central colorless cane. Iridescence and pitting. Curving side. Shape probably similar to **No. 5**. Technique of manufacturing also the same as **No. 5**. A Th .12 cm.

11. C69.121 Pl. 1
S.H. under NW wall, Rooms 2 and 6, 0-B

Fragment of rim and side of small bowl; network pattern composed of twisted, colorless canes with opaque yellow threads. Each cane consists of a central colorless cane twisted with two yellow threads in a reverse spiraling pattern. A cane of blue twisted with an opaque white thread forms the rim. Iridescence and pitting. Smaller than **No. 5**, but similar in shape. Technique of manufacturing also the same. D ca. 8.5 cm; Th tapers from .3 cm at rim to .1 cm on side.

*Striped Mosaic-Glass Vessels (**Nos. 12–14**)*

On archaeological sites yielding examples of network mosaic wares, it is not uncommon to find fragments of striped or ribbon mosaic glass as well. In fact, the two varieties closely resemble each other. Both usually employ network canes in the body and at the rim. Both share similar shapes and manufacturing techniques within each period of production, and both enjoy comparable patterns of distribution. Even their historical evolution in the late Hellenistic and Roman periods appears to be almost identical. Indeed, the affinities between contemporary vessels of the two types are so striking that they must have been made either in the same factories or in close communication with each other. The fundamental difference between the two types is that striped wares contain colored monochrome strips of glass, not network canes alone, and these strips are laid side by side in parallel rows across the vessel rather than in a spiraling fashion.

As is true of network glass, the striped pattern first emerged in the Hellenistic period, presumably in factories located somewhere in the eastern Mediterranean. The earliest known example is from the Antikythera shipwreck, where a small hemispherical bowl with an upright rim and a high, applied base-ring was discovered. The body is composed of alternating colored strips cut from two different canes and separated from one another by single network canes (Weinberg 1965, 37–38, no. 9, figs. 18–19). One of the strips is opaque yellow; the other is composed of squarish sections of composite mosaic glass set end-on-end. Based on the date of the shipwreck, ca. 80–50 B.C., this vessel and others like it must belong to the first half of the first century B.C. (for additional examples, see Oliver 1968, 55–57). Cosa yielded no example of this group, but Morgantina in Sicily, a mixed Greek and native settlement more involved in trade with the eastern Mediterranean, produced a fragmentary bowl considered to be of this group (Grose 1982, 25–26, fig. 6a). The fragment was found in a context containing late Hellenistic fill to ca. 20 B.C.

Following the invention of the striped pattern, it was adopted late in the first century B.C. by the Romano-Italian casting industry, presumably active at Rome or elsewhere in Italy. This industry manufactured a variety of striped vessels, which can be readily distinguished from their Hellenistic predecessors (Grose 1984, 31, fig. 7). Moreover, just as the striped example from Antikythera closely resembles its contemporary network and composite mosaic counterparts, the striped vessels of the newly devised Roman group also resemble the shapes of contemporary network glass and, indeed, are found in the same loci at Cosa and at most other Italian or European sites with Augustan or Julio-Claudian levels.

The principal shapes associated with the Roman group are small bowls with upright rims and nearly vertical sides, tapering downward to a slightly concave bottom (for an intact example, see Grose 1978, 76, fig. 11; others appear in the list that follows); somewhat larger bowls with flaring rims and sides (Mariacher 1959, pl. V opposite p. 12; Corning Coll. 1979, 199, no. 540, color pl. 28); and broad shallow bowls with upright rims and slightly concave bottoms, such as **Nos. 12–13** at Cosa, or an example from Brescia (Tamassia 1964, 38, no. 11, pl. 3; also see the list below). Other forms such as footed bowls, carinated bowls, and pyxides are known from fragments, as in the Corning Coll. 1979, 201–202, color pl. 30; Gréau Coll. 1903, II, pls. 104–106; and the collection of the Toledo Museum of Art (unpublished).

In respect to technique, the composition of the colored canes on vessels of the Roman group is considerably more complex than is initially apparent. On close examination one sees that individual colored canes are *not* laid down one at a time but are prefabricated in groups of two or three to form wider composition canes. This, of course, facilitated the assembling of the canes when making large numbers of vessels. **No. 12** at Cosa is typical. It consists of two composite canes, one composed of three colored strips, the other of two. Separating each composite strip is a single network cane. Thus, despite the presence of forty or fifty narrow strips, the body is actually formed from far fewer composite canes (see color ill. in Grose 1983, 42). Especially common among vessels of the Roman group are composite canes employing strips of green, opaque yellow, green; or strips of blue juxtaposed with opaque red (Grose 1983, 42–43).

Equally complicated is the vertical cross section of each colored strip. With few exceptions each color consists of three (occasionally two) of these superimposed layers of glass (again, Grose 1983, 42). The color blue in **No. 12**, for example, is in reality a layer of opaque white laminated between two layers of translucent blue; green is a thin layer of opaque yellow with translucent green on either side. More novel is the green of **No. 13**. It is achieved by the passage of light through two layers of blue with opaque yellow in between. The use of opaque white and yellow in the green and blue strips may well have been motivated by a desire to increase the luminosity of these otherwise dark colors.

Although most strips are continuous lengths of a single color, some strips are composed of alternating sections of monochrome glass and squarish, polychrome mosaic glass insets fused end-to-end to form a cane. Such a strip occurs in **No. 14**. Similar strips are also employed in striped vessels of the late Hellenistic period (see the bowl in Weinberg 1965, cited above).

With respect to casting methods, a two-stage sagging technique seems most probable for striped vessels of the Roman group (Grose 1984, 32, fig. 9). First, composite canes were cut to appropriate lengths, then fused together to form a circular disk. Afterwards the disks were placed in or over former molds and heated until they sagged to create a finished vessel. After annealing, the interior surface and outside of the rim were rotary-polished.

Parallels and Dated Examples of Striped Mosaic Glass Vessels

Cosa: **No. 12**, Deposit IV, prior to ca. 25/15 B.C.; **Nos. 13–14a**, Deposit V, ca. 25/15 B.C. to ca. A.D. 40/45.

Castiglione della Pescaia: Museo Civico, Grosseto, Room XV, Case I. Small hemispherical bowl with upright rim and slightly concave bottom. Unpublished.

Ostia II 1970, 82, pl. XVI, figs. 273–274. Fragments from Flavian levels.

Ostia III 1973, 391. Four fragments in levels dated from the Flavian period to early second century A.D.

Luni II 1977, 271 and 274, pl. 158.16. Four fragments from strata of the second century A.D. or later. One fragment of a hemispherical bowl, said to be like Isings Form 1; another example is described as a carinated bowl identical to Isings Form 2.

Vatican Coll. 1975, 36–38, nos. 151–152, 154–155, 157, 179, 181–182, pl. 7 and color pl. opposite p. 24. Fragments without find-spots but presumably from Rome or Italy. Shapes unidentified. Fremersdorf, unfortunately, includes striped mosaic glass under the heading *Bandgläser*, a term that also includes such unrelated wares as color-band bottles.

Rome: Antiquarium Comunale. Five fragments of small hemispherical bowls and broad shallow bowls without find-spots but presumably from Rome. Unpublished.

Rome: Excavation near the Porta San Giovanni in which numerous intact pieces (!) and fragments of striped vessels were said to have been found; present whereabouts not stated (Bologna Cat. 1964, 373).

Mariacher 1959, pl. V opposite p. 12. Small bowl with flaring rim and side. Said to be from the first century A.D. and now housed in the Museo Civico, Adria.

Neuburg 1949, 11, no. 29, pl. IX. A small hemispherical bowl said to be from Italy and now in the Metropolitan Museum of Art, inv. no. 17, 194.560.

Novaesium 1904, 416, fig. 23. Fragment considered "post-Haltern" but earlier than Flavian period.

Vindonissa 1960, 9–13, nos. 1–3, pls. 1, 1–3 and 17, 11. Fragments dated before A.D. 60/75. Parallels from additional sites listed on p. 16.

Camulodunum 1947, 293, no. 2, pl. 87. Fragment from period IV, A.D. 41 to 69.

Magdalensberg 1979, 19, nos. 1 and 4, pl. 17. Augustan, until ca. A.D. 15.

Cherchel, Algeria: Broad shallow bowl, possibly from an Augustan burial (Leveau 1970, 111, figs. 8 and 11).

Zara, Dalmatia: Small hemispherical bowl with slightly concave bottom (Marconi 1932, color pl. opposite p. 36).

Among dated examples listed here, **No. 12** at Cosa assumes special importance because it provides unambiguous evidence that such bowls were being manufactured by the last quarter of the first century B.C. The other three examples at Cosa (**Nos. 13–14a**) are also dated but only indicate that striped wares were in use during the first half of the first century A.D. This evidence is in accord with other dated specimens, both from Italy and from the provinces of northwestern Europe and North Africa, where they are also attributed to Augustan Julio-Claudian contexts. Fragments found at Ostia and elsewhere in Flavian or second-century levels may indicate that such vessels remained in circulation after midcentury, or even later, but in all probability these fragments simply represent pieces redeposited from earlier levels.

Yet another means of assembling canes on Roman striped vessels of this period is seen in a broad, shallow bowl from a cemetery at Agro Adriese in northern Italy (Sangiorgi Coll. 1914, 61, no. 214, pl. 40; republished in Corning Coll. 1979, 200–201, no. 545, color pl. 27). In this case the vessel is divided into four wedge-shaped quadrants, and the canes in each quadrant are arranged to fit symmetrically into a triangular wedge. This "quadripartite pattern" is associated with broad, shallow bowls with upright rims and slightly concave bottoms, not dissimilar to **No. 12** at Cosa. Such bowls almost always contain lengths of purple with two or three thin, parallel opaque white lines. Although few intact vessels have survived, fragments are numerous. When a given provenance is known, it almost always is Italian, or occasionally one of the provinces of North Africa or Europe. In date, such pieces appear in Augustan or Julio-Claudian contexts. **No. 14** at Cosa is probably a fragment from such a vessel.

Parallels and Examples of Quadripartite Pattern of Striped Mosaic Glass

Cosa: **No. 14**, Deposit V, ca. 25/15 B.C. to ca. A.D. 40/45.
Corning Coll. 1979, cited above. Said to be from Agro Adriese in an early Imperial context.
Bologna Cat. 1964, 372–373, no. 539, pl. 137. Bowl from Cavarzere (Rovigo), now in the Museo Nazionale Atestino at Este.
Pompeii: Museo Nazionale, Naples, inv. no. 13558; Alinari Photo neg. no. 12509. Unpublished.
Morgantina: Antiquarium, Aidone, inv. no. 60.1305. From a level of late Hellenistic fill ending early in the first century A.D.; Grose 1982, 25–26, fig. 6b.
Gudiol y Cunhill 1925, 12, no. 40. Bowl from Spain in the Instituto Amatller in Barcelona.
Thill 1975, 67–79. Bowl from Hellange, Luxembourg.
Magdalensberg 1979, 19, no. 1, pl. 17. A fragment presumably dating before A.D. 45.
Toledo Museum of Art: A small number of fragments in diverse shapes, not just broad, shallow bowls, most likely from Italy. Unpublished.

A quadripartite striped bowl of a different sort was discovered in a Roman building at Bocca di Magra in Liguria (Frova 1976, 56, fig. 71). In profile it resembles **Nos. 17–18** at Cosa and has a flat or slightly convex bottom. Unlike vessels of the more standard type described above, it makes use of few, if any, network canes in the body, and the predominant colors are blue, opaque yellow, and opaque white. Its existence emphasizes the diversity of glass production in the early Empire, even within an otherwise uniform mosaic tradition.

Striped Mosaic Glass Vessels

12. C68.519 — Pl. 1
S.H. Room 4, I — Deposit IV
Prior to ca. 25/15 B.C.

Nine joining and six non-joining fragments of rim, side, and bottom of broad, shallow bowl (profile restorable); striped pattern composed of two alternating composite stripes separated from each other by a cane of network mosaic glass (a colorless cane wound with two opaque yellow threads in a reverse spiraling pattern). Each stripe is in fact a broad composite stripe composed of three colored canes: one composite stripe is green, opaque yellow, green; the other is blue, opaque red, blue (see color fig. in Grose 1983, 42). Thus the overall pattern consists of eight colors (green, yellow, green; twisted colorless/opaque yellow; blue, opaque red/blue; twisted colorless/opaque yellow), a pattern repeated over the entire surface. Moreover, in cross section each colored stripe is composed of three thin, superimposed layers of glass. Blue, for example, is actually comprised of blue, opaque white, blue; green is green, opaque yellow, green; red is colorless, opaque red, colorless; and yellow is colorless, opaque yellow, colorless. A twisted cane of network glass identical to that in the body forms the rim. Iridescence and pitting. Upright rim; convex curving side; concave bottom. Cast in composite canes laid in parallel rows across a mold, then rotary-polished on interior, also on exterior to junction of side and bottom; rest of exterior fire-polished. H 3.5 cm; D ca. 14 cm; A Th .2 cm. *Bibliography:* Grose 1983, 42–44; Grose 1984, 31, fig. 8a.

13. CE 2013 Pl. 1
A.P. Room 16, I Deposit V
ca. 25/15 B.C. to ca. A.D. 40/45

Fragment of rim and side of small bowl; striped pattern composed of two alternating stripes of glass separated by a twisted cane of network mosaic glass (blue wound with opaque white spiraling threads). Each stripe is itself a broad composite stripe: one composite stripe is opaque yellow, green, opaque yellow; the other is blue and opaque red. The pattern thus formed consists of seven colors: yellow, green, yellow; twisted blue/opaque white; blue, opaque red; twisted blue/opaque white. Also, in cross section each colored stripe is layered: blue, yellow, and red are identical in composition to those described in **No. 12**. Green, however, consists of three superimposed layers of blue, opaque yellow, and blue, thereby producing the effect of green. A cane of network glass (colorless wound with two parallel spiraling threads of opaque yellow) forms the rim. Iridescence and pitting. Upright rim; convex curving side. Shallow, hemispherical shape. Technique of manufacturing as in **No. 12**. D ca. 11 cm; Th tapers from .3 cm at rim to .2 cm on side.

14. CE 2012A
A.P. Room 16, I Deposit V
ca. 25/15 B.C. to ca. A.D. 40/45

Three non joining fragments of side and bottom of broad, shallow bowl; striped pattern with mosaic-glass insets. On the fragments, a composite stripe of green, blue, green alternates with a stripe composed of rectangular sections of mosaic glass, thereby forming a stripe. Each mosaic section consists of a central element in green ground containing two concentric circles of opaque white and four corner elements in purple ground containing two concentric circles of opaque white. In cross section the colored stripes are layered as in **No. 12**. Convex curving side, slightly concave bottom. Shape probably similar to **No. 12**. Weathering and technique of manufacturing probably also the same. It is possible, however, that the cane pattern was laid out in four wedge-shaped units, dividing the vessel into quadrants rather than in a pattern of parallel stripes laid side by side across a mold. A Th .15-.2 cm.

a. CE 2012B, A.P. Room 16, I, Deposit V, ca. 25/15 B.C. to ca. A.D. 40/45. Tiny fragment of striped vessel; green alternating with blue stripes. Unlike **No. 14**, the green in cross section is layered blue, opaque yellow, blue. Shape uncertain. A Th .2 cm.

Gold-Band Mosaic-Glass Vessels (***Nos. 15–16***)

Perhaps the most handsome examples of mosaic glass found at Cosa are three small fragments of gold-band vessels, all found in the same area of housing (V-D, Street M) but not necessarily from the same vessel. One fragment (**No. 15b**) preserves a portion of the side of a large container (D 14–18 cm), possibly a bowl or pyxis but seemingly too large to be restored with **No. 16**, a section of a small base-ring. **No. 15a** is too tiny to aid in any identification. It is certain, however, that whatever the complete shape(s) was, it was not one of the three most frequently encountered gold-band forms: small globular or carinated bottles, small lidded pyxides, or alabastra of two main types (see Grose 1983, 41, maps). Nor can these fragments easily be restored as parts of such lesser-known shapes as skyphoi, saucepans, patellae, and plates or bowls of diverse sizes and dimensions. Nonetheless, they clearly emphasize the little appreciated fact that the gold-band technique, at least in Roman times, was used to manufacture a wide variety of vessels. This diversity can only be appreciated when examining the many fragments housed at the Corning Museum of Glass (Corning Coll. 1979, 203–208) and the Metropolitan Museum of Art (unpublished). For descriptions of intact gold-band vessels, see Oliver 1967, 20–25, and more recently, Berger 1980, 9–13, esp. p. 12, and Grose 1983, 41, 43–44; and also the list below.

In color, the fragments at Cosa closely resemble the small bottles, bowls, and pyxides mentioned above. In all instances bands of royal blue, purple, emerald green, and golden-yellow to golden-brown, along with a strip of gold-foil laminated between two layers of colorless glass, make up a serpentine pattern repeated several times (usually in three identical sections) over the body of the vessel. The Cosa fragments also resemble these bottles and pyxides in thickness and in manufacturing technique, at least insofar as they are carefully rotary-polished on both surfaces and feature horizontal cut lines to highlight the exterior profile.

Although most previous studies of gold-band glass have usually treated all shapes as part of a single type, the diversity of forms, colors, pattern configurations, and even manufacturing methods indicates the existence of several distinct groups that presumably are the products of different workshops and times. This hypothesis receives preliminary corroboration from distribution patterns now beginning to emerge from a study of examples with known find-spots (see distribution maps in Grose 1983, 41, 43–44). The long cylindrical alabastra, for example, comprise two or more uniform series of vessels. Most pieces share one of two shapes (see Oliver 1967, 20–12: his types A and B), a vertically set zigzag cane configuration, the use of turquoise or medium blue glass (as well as other colors), and a common manufacturing technique, which is discussed in Cummings 1980, 40–42. Often the colored bands are embedded in a colorless or pale green matrix, and most intact examples have detachable stoppers. These alabastra are said to have come from a great many locations throughout the Mediterranean: Izmir, Samothrace, Amrit (Syria), Egypt, Cyprus, Cumae, Palestrina, and Magna Graecia—all sites or areas within the framework of the Greek trading orbit of the later Hellenistic period. This suggests an eastern Mediterranean production site.

In date, these alabastra are said to belong to the early first century A.D. or to the last decades of the preceding century (Oliver 1967, 33). In fact, only one example has been recovered from a meaningful context, an alabastron from a grave on Samothrace dated to the Augustan period (Dusenbery 1967, 38, no. 8, fig. 9). More recently a fragmentary vessel of uncertain form, possibly a bowl but clearly not an alabastron, was found in a La Téne III context at Béthisy-Saint-Martin (Oise, France) and dated around 35/30 B.C. (Berger 1980, 11). Owing to the inclusion of a turquoise blue band, this vessel may relate to the alabastra and provide an additional dated example of this group. Given present uncertainties, however, it is best only to conclude that the alabastra represent one, and possibly two or more, group(s) that were made in the late Hellenistic epoch but are absent from Roman contexts in Italy and northwestern Europe.

To the contrary, the small bottles, bowls, and pyxides, along with other related vessels, including the fragments from Cosa, are almost exclusively found on Roman sites belonging to the early Imperial epoch. Most of these sites are located within Italy, extending from the Bay of Naples north to the

Po valley and the Veneto, or in the Roman provinces of central and northwestern Europe (again, see the maps in Grose 1983, 41). Find-spots include Cumae, Pompeii, Herculaneum, Naples, Rome, Cosa, Volterra, Terni, Bologna, Aquileia (Italy), Salona and Zgoraj (Slovenia), Cologne (Germany), Magdalensberg (Austria), Fishbourne and Yorkshire (Britain), Balsa (Portugal), and Ampurias (Spain). From the eastern Mediterranean come only a few examples: Paphos on Cyprus, Kertch on the Black Sea, and, surprisingly, Karanis in Egypt. Among dated pieces are globular bottles, one from an Augustan context at Magdalensberg 1979, 118–120, 163, and another from a Claudian or Neronian tomb at Cologne (Fremersdorf 1932, 285). The vessels from Pompeii and Herculaneum antedate the eruption of Mount Vesuvius in A.D. 79, and the three tiny fragments from Fishbourne belong to levels dating respectively from ca. A.D. 43/75 and A.D. 75/100 (Fishbourne 1971, 326–327, nos. 8–10, fig. 137).

Based upon distribution and dated contexts, the Roman group of gold-band vessels appears to have been manufactured in Italy during the first half (or three-quarters) of the first century A.D. Suggestions that these vessels were made at centers such as Alexandria (Fremersdorf 1932, 285) or Aquileia (Calvi 1968, 95) lack any supporting documentation. By virtue of common characteristics the undated examples at Cosa can also be attributed to the first-century Roman production of Italy.

Parallels and Dated Examples of Gold-Band Mosaic-Glass Vessels

Rome: Antiquarium Comunale. One fragment. Unpublished.

Rome: Museo Nazionale Romano. Numerous fragments in several shapes. Unpublished.

Vatican Coll. 1975, 38, nos. 186–189, pls. 4 and 6. No. 186 is a cylindrical alabastron. Nos. 197–189 are too fragmentary to determine their shapes but probably belong to the Roman group. Presumably all four fragments came from Rome or Italy.

Naples: Museo Nazionale Archeologico, inv. no. 109.634. Small lidded pyxis from Pompeii, cited in Calvi 1968, 95.

Herculaneum: Antiquarium. Saucepan, unusual in shape and cane configuration.

Magdalensberg 1979, 118–120, 163, nos. 918 and 1200, pls. 14 and 17. Augustan and post-Augustan to ca. A.D. 45, respectively.

Oppenländer Coll. 1974, 104–105, no. 276. Small bowl with straight sides, tapering diagonally to base-ring. D 5.7 cm.

La Baume n.d., no. M, 1. Rim fragment of bowl of uncertain shape but with convex curving sides, found at Cologne. D 8 cm.

Corning Coll. 1979, 203–208, nos. 556–583, color pls. 29 and 31, for small bottles, pyxides, bowls, and other vessels, most of which belong to the Roman group described here. Nos. 584–585, color pl. 31, are alabastra that belong to the late Hellenistic group.

Toledo Museum of Art: Three fragmentary and intact late Hellenistic alabastra and several fragments of pyxides, bottles, and shallow bowls of the Roman group. Unpublished.

Seemingly unrelated to either the standard types of late Hellenistic alabastra or the Roman group described above are several vessels with unusual or unique shapes, color combinations, or pattern configurations. A biconal alabastron and a wing-handled skyphos in the Metropolitan Museum of Art (Oliver 1967, 26–30) are graphic examples, as is a saucepan in the Herculaneum Antiquarium (see list above). The same is true of a broad, shallow dish with a flaring side, comparable to **Nos. 17–18** at Cosa, found at Morgantina in a level containing late Hellenistic fill antedating ca. 30 B.C. (Grose 1982, 24–25, fig. 5A). The precise relationship of these vessels and the others described here cannot be determined at this time, but once again they underscore the diversity of cast mosaic glass of the first centuries B.C. and A.D.

Gold-Band Mosaic-Glass Vessels

15. C71.195A and B Pl. 1
H.T. Room 10, 0 and Street M, Drain Sediment

Two small, non-joining fragments of vessel or vessels; gold-band pattern composed of serpentine stripes of purple, green, blue, golden-yellow, and sandwich gold-glass. Separating each color is a fine opaque white line, possibly owing to weathering. In cross section the purple and yellow stripes consist of superimposed layers of colorless glass with a colored layer in between them; blue is a layer of opaque white glass with a layer of translucent blue on either side; green is a layer of opaque yellow with a layer of green on either side; and the gold-band consists of two layers of colorless glass with "crushed" gold foil between them. Slight iridescence and pitting. Excellent craftsmanship. Fragment A preserves part of a straight side. Fragment B preserves a portion of side and curved angle. The curve of the fragment and the direction of the polishing marks and cut line (mentioned below) suggest the fragment is from the lower portion of a vessel, and also establish the diameter of the piece. Shape uncertain. On the exterior of fragment B, a horizontal cut line at junction of the side and the basal angle. Cast, rotary-polished, presumably on a lathe, on both surfaces. D body ca. 14–18 cm; A Th .2 cm. Only fragment B appears on pl. 1.

16. C71.201 Pl. 1
H.T. Room 9, 0

Tiny fragment of base-ring (?) of vessel; gold-band pattern composed of bands of green, blue, and sandwich gold-glass. See **No. 15** for description of the layering of each color. Slight iridescence and pitting. Excellent craftsmanship. Fragment proceeds diagonally outward and terminates in a flat edge. Shape uncertain. On the exterior and interior, beneath the edge, are narrow horizontal cut lines. Shape uncertain, but fragment may belong to same vessel as **No. 15**. Cast, rotary-polished, and cut on both surfaces. D base-ring ca. 5 cm; Max Th .3 cm.

Composite Mosaic-Glass Dishes with Applied Rims (***Nos. 17–19***)

Among vessels at Cosa with composite mosaic patterns are three fragmentary examples (**Nos. 17–19**) of a homogenous series of broad, shallow dishes with outsplayed sides and applied rims. These dishes are generally uniform in shape and overall appearance, but diameters vary from 15 to 18 cm, with some examples as small as 11 to 12 cm and a few as large as 21 cm (see parallels listed below). On all pieces the sides are sharply outsplayed and terminate in an applied rim composed of two twisted threads of roughly similar size. In most instances one is opaque white and the other colored, often purple but also blue or golden-brown. The bottoms are either flat (**No. 17**) or slightly convex, causing the vessel to wobble slightly. These dishes have relatively thick walls that seldom taper from the rim to the bottom, a feature characteristic of much Hellenistic mosaic glass.[9]

[9] Although these dishes bear some resemblance to the much larger mosaic plates associated with the Canosa hoards, such as the examples illustrated in Harden 1968, 25–26, nos. 3–4, figs. 10–15, they differ sufficiently in size, thickness, rim treatment, and the absence of cut horizontal grooves to argue that they are not directly related to the Canosa glasses. Moreover, examples of both types have never been found together in the same archaeological contexts. Based on present evidence, the smaller mosaic dishes discussed here seem to be a later and possibly separate development.

In the examples at Cosa, as elsewhere, the colored pattern is customarily formed from sections of one or two (sometimes more) principal canes, each in a dark ground (purple, blue, golden-brown, or aquamarine) and containing an opaque white or opaque yellow spiral, at times with a tiny rod in yet another color at the center of the section.[10] Interspersed among these sections, at least on many pieces, are a small number of rectangular or squarish insets in opaque white, yellow, or light blue. When such insets are included, the opaque colors usually face only the upper surface because their undersides are masked in yet another color. A few pieces also contain rectangular insets of purple flanked on either side by opaque white strips. As is true of **Nos. 18–19** at Cosa, the row of sections nearest the rim often makes use of only one of the main canes, while the second row contains both.

These vessels were probably cast over former molds, then rotary-polished on the interior and on the outside of the rim. The rest of the exterior, as seen on intact museum examples, was fire-polished. Overall, these dishes give the appearance of careful craftsmanship, as is characteristic of so much Hellenistic glassware.

Parallels and Dated Examples of Composite Mosaic-Glass Dishes with Applied Rims

Cosa: **Nos. 17–19**. D 15–16 cm.

Ager Cosanus: Rim fragment, D ca. 15 cm. Unpublished.

Rome: Museo Nazionale Romano, inv. no. 10851. D 18 cm. Provenance uncertain but presumably Rome or Italy. Unpublished.

Rome: Antiquarium Comunale. Several fragments of rims. Unpublished.

Vatican Coll. 1975, 31, no. 26, pl. 4 and color pl. opposite p. 80. D 17.3 cm. Provenance uncertain but presumably Rome or Italy.

Pompeii: Museo Nazionale Archeologico, Naples, as illustrated in Barovier 1974, 115, fig. 6. Estimated D 11–12 cm.

Cumae 1913, 740, pl. 118, no. 18. D 12.5 cm.

Morgantina: Antiquarium Aidone, inv. no. 57.646. D ca. 16–17 cm. From a context containing late Hellenistic fill to the close of the first century B.C. Grose 1982, 25, fig. 5b.

Paestum: Museo Archeologico. Fragment from an undated context in the Bona Dea sanctuary, University of Michigan–University of Perugia excavations. D ca. 15 cm. Unpublished.

Florence: Museo Archeologico. D 15 cm. Unpublished.

Corning Coll. 1979, 181, no. 475, color pl. 24 and pl. 41. D 12.8 cm. From Sangiorgi Collection, said to have an Etruscan provenance.

BMFA Coll. 1968, nos. 13–14. These two vessels are similar in shape but use cane patterns that are noticeably different from the others cited here. No. 14, purchased in Rome, is unusually small, D 11.6 cm. No. 13 has a diameter of 17 cm.

Gréau Coll. 1903, II, pl. 92, nos. 3–4 and 6. Fragments of three comparable dishes. Provenance uncertain.

Toledo Museum of Art: One intact dish (acc. no. 23.1464) and twelve fragmentary examples, including a dish with a D of 21 cm (acc. no. 80.424). The fragments are thought to have been collected in Rome. Unpublished.

Whenever a provenance is known or alleged, it is invariably Italian, extending from Etruria in the north to Campania and Magna Graecia in the south. Dated examples are so few as to be

[10] Yet another cane pattern associated with these vessels is an opaque white or yellow "star" with five to ten rays centered in a blue, purple, golden-brown, or aquamarine ground (see Oliver 1968, 61–62, 67, for examples and discussion).

inconsequential, but the growing preponderance of evidence, both positive and negative, points to the second and first centuries B.C. The fragment at Morgantina, for example, is from a mixed Hellenistic level containing no material that is later than the close of the first century. The example from Cumae is said to be from a tomb that also antedates the Christian era. Finally, the complete absence of such vessels from the Roman provinces of central and northwestern Europe is indicative of the fact that these dishes had ceased production by the time the Romans became active in those areas late in the first century B.C. and early in the first century A.D.

Presumably related in date, if not in place of manufacture, is a second series of broad, shallow dishes with composite patterns and applied rims. These vessels, first published as a group in Oliver 1968, 65–68, closely resemble the dishes described above but differ in that the rims are upright, the bottoms slightly convex, and the diameters somewhat smaller (11.5 to 14.5 cm). Again, cane patterns vary, but most examples employ one, two, or more principal canes, usually with an opaque white spiral set in a darker colored ground. Many also contain opaque yellow or opaque white rectangular insets, visible only on the upper surface of the vessel, or rectangular sections of opaque white and purple strips. Many examples also have two small suspension holes drilled below the rim.

Of the twenty-two examples published by Oliver, only two have provenances, both Italian. One is attributed to the Roman Campagna, another to Nervi, southeast of Genoa. Several others from the Durand Collection, according to Oliver 1968, 67, may have been acquired in Italy, and the examples in the Museo di Antichita at Turin (Mariacher 1959, pl. 4), in the Vatican Collection (Vatican Coll. 1975, 31, no. 27, pl. 4 and color pl. 67), and the Museo Nazionale Romano in Rome (inv. no. 294058, D 13.2 cm, unpublished), must also have come from Italian sites. No example is known from a dated context, but Oliver 1968, 68, argues for a date in the latter part of the first century B.C.

Probably related to both groups of dishes are certain hemispherical and conical bowls also with applied rims and comparable composite mosaic patterns (see examples in Oliver 1968, 58–65). Although none has yet been found with a dish in a controlled excavation, stylistic criteria argue for a common provenance(s), at least in particular instances. Several of the bowls have been recovered from dated contexts—notably one from a burial considered to belong to the late second century B.C. (Oliver 1968, 58, 63–64, who correctly revises downward the date first assigned by the excavator). This would suggest that the dishes, customarily assigned to the first century B.C., may in fact commence in the preceding century.

Composite Mosaic-Glass Dishes with Applied Rims

17. C69.381A and B Pl. 1
Arx W. Slope, Sounding 1, 0

Two non-joining fragments of rim and side, and of side and bottom of broad, shallow dish (profile restorable); composite mosaic pattern composed of polygonal sections of colorless ground containing opaque white spirals, green ground containing opaque yellow spirals, purple ground containing opaque white spirals, and rectangular insets that are opaque yellow or opaque white on the upper surface but colorless on the underside. A twisted cane of opaque white and purple threads forms the rim. Predominant color is purple. Pitting and dulling. Good craftsmanship. Outsplayed rim and side with rounded edge; flat bottom. Cast in sections of canes; rotary-polished on the interior; fire-polished on exterior. Restored H 2.1 cm; D ca. 16 cm; A Th .2 cm.
Bibliography: Grose 1973, 177, pl. 189, no. 3.

18. C68.271 Pl. 1
Forum SE, 0

Two joining fragments of rim and side of broad, shallow dish; composite mosaic pattern composed of polygonal sections of golden-brown ground containing opaque white spirals and rectangular insets of blue containing opaque white spirals. The row nearest the rim consists entirely of golden-brown sections; the second row alternates between golden-brown and blue sections. A twisted cane of purple and opaque white threads forms the rim. Predominant color is golden-brown. Iridescence and pitting. Outsplayed rim and side with rounded edge. Cast in sections of canes, then rotary-polished on both surfaces. D ca. 15–16 cm; A Th .35 cm.

19. C70.570 Pl. 1
Forum SE, Trench 70C, 0

Fragment of rim and side of broad, shallow dish or bowl; composite mosaic pattern composed of polygonal sections of blue ground containing opaque white spirals and pentagonal sections of golden-brown containing opaque white spirals. The row nearest the rim consists entirely of blue sections; the second row alternated between blue and golden-brown sections. A twisted cane of blue and opaque white threads forms the rim. Predominant color is blue. Pitting and dulling. Outsplayed rim and side with rounded edge. Cast in sections with canes, then rotary-polished on both surfaces. D ca. 16 cm; A Th .3 cm.

Composite Mosaic-Glass Vessels: Miscellaneous Patterns and Shapes (***Nos. 20–28***)

Fragments of nine mosaic vessels in diverse polychrome patterns were also discovered at Cosa during the course of excavations. Only two, however, are sufficiently preserved to determine a complete or almost complete profile. **No. 20** is part of a small hemispherical bowl with a twisted coil applied as a rim. In form, it probably resembled the network bowls (**Nos. 5–11**) and, likc thcm, probably had a slightly concave bottom. **No. 22** is a unique flat-bottomed dish with a pointed projecting lip. It is composed of segments of a single cane with a true millefiori pattern set in a white ground. No exact parallel is known for either the bowl or dish.

No. 24 is too tiny to determine the shape of the vessel, but it employs a recognizable pattern of short rectangular sections of canes in blue, opaque yellow, and opaque red, which are set at right angles to one another. This "short-strip" configuration, which is closely associated with striped mosaic glass, is also known from a great many fragments in the Gréau Coll. 1903, II, pl. 117, no. 2, no. 555, color pl. 30, and the Toledo Museum of Art, with additional pieces in the Antiquarium Comunale (Rome) and Museo Nazionale Romano (Rome). The only intact vessels employing such a pattern are a handsome, footed ribbed bowl now in the Victoria and Albert Museum (Honey 1946, 22, pl. 2c) and a small lens-shaped dish in the Museo Archeologico in Florence (unpublished). In collections of fragments footed ribbed bowls are quite frequently associated with this particular cane configuration. As is also true of striped mosaic glass, each colored strip is composed of two or three thin, superimposed colored layers (see ill. in Grose 1983, 42).

Considerably more common than short-strip mosaic glass is another polychrome pattern represented by two fragments at Cosa, **Nos. 26 and 26a**. This pattern employs canes in a green ground enclosing concentric circles or rods of opaque yellow and opaque red glass (occasionally other colors). Plates, dishes, bowls, and patellae of all sorts made use of this color combination. An intact bowl, dating between A.D. 5 and 20, was uncovered in a grave in Tessin (Simonett 1941, 136, no. 32, fig. 116, pl. 9.1). From Magdalensberg come additional fragments that are thought to date between 25 B.C. and A.D. 45 (Magdalensberg 1979, 24, no. 25, pl. 19; 71, no. 259, pl. 19). Other bowls in this pattern are housed in the Museo di Antichità at Turin (Mariacher 1959, pl. 4) and the Corning Museum of Glass (Corning Coll. 1979, 180, no. 467, color pl. 24). Fragments are found in the Vatican Coll. 1975, 35, no. 116, pl. 7; Gréau Coll. 1903, II, pl. 87, nos. 9–11; pl. 95, nos. 2–4, 7; pl. 98, nos. 1–4; pl. 100, nos. 1–4, 6–9; Corning Coll. 1979, 180–181, nos. 468–470, color pl. 25; with unpublished fragments in the Toledo Museum of Art, the Antiquarium Comunale, and the Museo Nazionale Romano in Rome.

The number of surviving vessels and fragments in these colors suggests that this pattern was exceedingly popular in the late first century B.C./early first century A.D. Complicating the picture, however, is a growing body of evidence that the pattern—as well as the mosaic technique—either continued in production into later Imperial times or was revived in the third or fourth century A.D. Excavations at Sardis, for example, yielded two fragmentary vessels and eight plaques in this fabric (Sardis 1980, 88–89), and the Metropolitan Museum of Art's excavations at Khargeh Oasis in Egypt produced many more examples, presumably from late antique contexts (cf. Grose 1983b, 285). Since so little of the later material has been identified or studied, it is not yet possible to distinguish easily between the earlier and the later production.

At Cosa, only the opaque white millefiori dish (**No. 22**) is from a meaningfully dated context. It was located in Deposit V, belonging to the decades preceding ca. 25/15 B.C. This dish, along with the other cast wares from that deposit, are among the first significant types of glass tablewares to be used by residents of the town. Whether they were the products of the Romano-Italian casting industry or imports from the eastern Mediterranean, however, cannot now be determined.

The other datable fragments are from loci too late to be instructive. **No. 21** was found in Deposit VII and was already fragmentary and trodden into the earthen floor of a shop when the building was destroyed, ca. A.D. 40/45. **No. 23**, discovered in Deposit IX, dating to the late first to early third century A.D., must have been redeposited.

Composite Mosaic-Glass Vessels: Miscellaneous Patterns and Shapes

20. C69.155 Pl. 2
S.H. Garden, I

Fragment of rim and side of small bowl; millefiori and other cane patterns, composed of purple flowers with eight white petals surrounding a center of seven blue rods ringed in opaque white; interspersed among the flowers are rectangular mosaic insets of opaque white glass with green sections at either end. A twisted cane of two blue threads forms the rim. Predominant color is purple. Iridescence and pitting. Upright rim with rounded edge; nearly vertical side, tapering downward. Probably a hemispherical bowl with a slightly concave bottom similar to **No. 5**. Cast in sections of cane, then rotary-polished on interior and on outside of rim; rest of exterior probably fire-polished. H ca. 4.5 cm; D ca. 9 cm; Th tapers from .3 cm at rim to .1 cm on side.

21. CE 2039 Pl. 2
A.P. Room 22, II Deposit VII
Decades preceding ca. A.D. 40/45

Three non-joining fragments of side and bottom of bowl; millefiori and other cane patterns, flowers with seven opaque white petals around a center of blue and opaque white concentric circles; also insets of purple ground containing two concentric circles of opaque white; insets of green ground

containing three concentric circles of opaque yellow; and hexagonal sections of purple ground containing opaque yellow flecks. Predominant color is purple. Much pitting. Straight side, tapering diagonally downward; slightly concave bottom. Cast in sections of cane, then polished on both surfaces. A Th .2 cm.
Bibliography: Grose 1975, 36, no. 1.

22. C68.505 Pl. 2
S.H. Room 4, I Deposit IV
Prior to ca. 25/15 B.C.
Five joining and two non-joining fragments of shallow dish (profile restorable); millefiori pattern composed of opaque white flowers with ten blue petals surrounding a center of purple, opaque white, and opaque yellow concentric circles. Predominant color is opaque white. Milky weathering and pitting, reddish weathering on opaque yellow glass. Nearly horizontal rim with pointed edge sloping obliquely to the inside; outsplayed side, tapering diagonally downward; flat bottom. Cast in sections of cane, then polished on both surfaces. H 1.6 cm; D ca. 15.5 cm; A Th .2 cm.

23. CE2111 Pl. 2
A.P. Room 25A, II Deposit IX
Late first through early third century A.D.
Two tiny, non-joining fragments of rim and side and of side and bottom of shallow bowl or other vessel; composite mosaic pattern composed of sections of green ground with centers of opaque yellow and opaque white concentric circles or, alternatively, with centers of opaque yellow and opaque red circles. Heavily corroded. Upright rim with rounded edge; nearly straight side, tapering diagonally downward; flat bottom. Cast in sections of canes set diagonally, then polished on both surfaces. D ca. 12–14 cm; A Th .25 cm. Fragment of bottom not illustrated.

24. C69.233
S.H. Room 17, I
Tiny fragment of vessel; short strip mosaic pattern composed of rectangular sections of blue glass alternating with rectangular sections of opaque red glass bordered by bands of opaque yellow. In cross section the blue glass is in three layers: blue, opaque white, blue; the red is colorless, opaque red, colorless. Iridescence and pitting. Shape uncertain. Cast in rectangular sections of cane laid at right angles to one another, then polished on both surfaces. A Th .2 cm.

25. C69.322
Forum SW, I
Tiny fragment of basal angle and bottom of dish or bowl; composite mosaic pattern composed of sections of purple ground containing opaque white spirals and sections of green ground containing opaque yellow spirals. On upper surface of fragment, there are rectangular insets of opaque yellow glass. Iridescence and much pitting. Flat bottom with indications of a convex curving side. Shape uncertain. Cast in sections of cane, then polished on both surfaces. A Th .2 cm.

26. C68.657
S.H. Room 5, 0-B
Six joining fragments of side of vessel; millefiori pattern composed of green flowers with multiple opaque yellow petals surrounding a center of opaque white, opaque red, and green concentric circles.

Pitting and dulling. Almost straight side. Shape uncertain. On exterior, three horizontal cut lines in a band around the body. Cast in sections of cane, then polished on both surfaces. A Th .2 cm.

27. C68.673
S.H., 0-A
Tiny fragment of side of vessel; composite mosaic pattern composed of polygonal sections of opaque white and dark blue glass adjacent to circular sections of pale aquamarine streaked with opaque white and opaque yellow. In cross section the dark blue is two layers of blue with opaque white in between; the white is opaque white backed with a layer of colorless glass. Iridescence and pitting. Convex curving side. Shape uncertain. Cast in sections of cane, then rotary-polished on interior; exterior fire-polished. A Th .18 cm.

28. C68. 674
Forum ENW, 0
Tiny fragment of side and bottom of vessel; composite mosaic pattern composed of blue sections containing yellow spirals. Pitting. Shape uncertain. Cast in sections of cane; rotary-polished on interior; exterior possibly fire-polished. A Th .2 cm.

Marbled Mosaic-Glass Ribbed Bowls and Miscellaneous Marbled Vessels (***Nos. 29–44***)

Ribbed Bowls

The most common mosaic pattern encountered on archaeological sites of the late first century B.C./early first century A.D., Cosa included, is marbled mosaic glass. The shape most frequently associated with this pattern is the ribbed bowl. At Cosa seventeen of forty-four cast mosaic fragments are marbled, and fourteen of these are ribbed bowls. At Morgantina six out of sixteen fragments are marbled, at Vindonissa 194 fragments from a total of 263 ribbed bowls are marbled (Vindonissa 1960, 12), and at Camulodunum seventeen out of thirty-three mosaic examples are marbled ribbed bowls (Camulodunum 1947, 288). At Cosa other shapes with a marbled design include a dish (**No. 29**) and a blown color-band bottle (**No. 31**), both in golden-brown with opaque white veining, as well as a jar (**No. 30**) in a purple ground, also with white streaks.

The origin of marbled mosaic glass is uncertain, but surely it was inspired by the variegated designs of colored stones such as onyx or agate. True marbled examples do not occur among Hellenistic vessels, such as those of the Canosa or Antikythera groups, and in fact no piece known to the author antedates the establishment of the Roman industry in Italy late in the first century B.C. It is probable, however, that the marbled pattern is related to Hellenistic "onyx glass," produced by a technique employing extremely large sections of preformed canes, often golden-brown with streaks or spirals of opaque white glass, laid side by side to form a repeating pattern (as illustrated by a bowl in Oliver 1967, 15, fig. 3, and discussed in Cummings 1980, 38 and 42). Such a pattern, however, differs from marbling that is characterized by its random veining, presumably the result of folding together two batches of different colored glass in one crucible (Cummings 1980, 14–15). The earliest known examples of onyx glass are three highly unusual lidded piriform jars, two in the National Museum at Athens (unpublished) and a third in Museo Archeologico in Florence (inv. no. 70802, unpublished). The two from Greece were discovered near the village of Palaiokastron, not far from ancient Metropolis, in Thessaly, and are thought to be from a grave belonging to the second or early first century B.C. (Oliver 1967, 16–17; cf. Hoffmann 1965, 278). A fourth vessel, found in a grave at Cumae, is a biconical alabastron, also dated to the second or early first century B.C. (Cumae 1925, 87–90, fig. 5; cf. British Museum Coll. 1981, 125, where an earlier date is suggested).

Given the close association between marbled designs and ribbed bowls, it is possible that the marbled configuration emerged simultaneously with the introduction of the ribbed bowl. This occurred sometime in the second half of the first century before Christ (for discussion, see pp. 54–56 below). However, the earliest securely dated examples of marbled ribbed bowls are from Haltern and are no earlier than the last decade of the first century B.C. Others are also known from Augustan contexts, such as those cited in Magdalensberg 1979, 26–31. The two dated examples at Cosa (**Nos. 30** and **32**) are from levels as late as ca. A.D. 40/45 and consequently provide no evidence concerning the earliest chronological development of marbled bowls but demonstrate their popularity in the first half of the first century A.D.

In color, these vessels, as well as other marbled forms, are either blue, golden-brown, or purple (but never green), usually with opaque white veining. Occasionally, white veining is combined with opaque yellow (**Nos. 42–43a**) or is solely yellow (**No. 37**). At other times several colors are included, as in **No. 44**, which employs blue, purple, and white to create a colorful effect. In most instances the streaking is random, but some vessels display a repetitious pattern of cane sections (often with spirals) akin to onyx glass (see Corning Coll. 1979, 188, nos. 501–502, color pl. 27) or use circular flecks of color in a monochrome colored matrix, such as **No. 39**. In some cases rectangular insets also appear with the veining (**No. 37**).

The marbled ribbed bowls exhibit the same variations in shape and ribbing as do their monochrome counterparts (see pp. 54–56 below). Some are small hemispherical bowls, such as **No. 34**, which in form and dimensions resemble not only monochrome ribbed bowls but even mold-blown ribbed bowls (**Nos. 211–212**) and free-blown colored bowls with pinched ribs (**Nos. 249–259**). Others are broad and shallow (Corning Coll. 1979, 188, no. 501) or large and deep (**No. 33**). Still others are footed with high dome-shaped bases, as was probably true of **No. 32**. When intact, **No. 32** doubtless resembled better-preserved examples now in several museum collections (Gréau Coll. 1903, II, pl. 141, nos. 1–2, or Corning Coll. 1979, 190, nos. 507 508, pls. 26 and 41).

Rims are either upright or slightly outsplayed, giving the profile a gentle S-curve, and bottoms, unless footed, are either flat or slightly concave, as is typical of most products of the early Roman casting industry. Decoration at times takes the form of internal horizontal grooves below the rim or around the body, but this occurs less commonly than on cast monochrome ribbed bowls. In some cases the ribs are numerous, quite short, and closely spaced around the shoulder; in other cases they are shallow and undulating (**No. 32**). Finally, there are numerous marbled examples of the so-called "pillar-molded" variety with pronounced, evenly spaced ribs set around the body. The ribs at times decorate only the side, terminating at the junction with the bottom (**No. 34**), but occasionally extend to the center of the bottom (**No. 39**), forming a resting platform.

Technically, these bowls must have been cast by several methods. The two-part molding technique is usually cited (Corning Coll. 1979, 33–34), but a sagging method employing former-molds is perhaps more likely in many cases (see p. 32 above). The ribs in the latter case would have been created through use of a tool or the imprint of a special device prior to sagging (Cummings 1980, 26–29). Not all specialists, however, would agree with Cummings either on the use of former-molds or on the mixing technique to create the marbled effect.[11]

[11] The suggestion that marbled mosaic patterns are the result of folding together two batches of different colored glass (Cummings 1980, 14–15) may in a few instances be true. However, having examined several hundred cast ribbed bowls, the author has found without exception that they were formed from sections of preformed canes placed side by side.

PARALLELS AND DATED EXAMPLES OF MARBLED MOSAIC-GLASS RIBBED BOWLS

Cosa: **No. 32**, Deposit V, ca. 25/15 B.C. to ca. A.D. 40/45.
Luni I 1973, 465, inv. nos. CM 489, CM 2678, C 291.
Luni II 1977, 271 and 274, inv. nos. CM 6526 and 4918, pl. 158.13. Later example is a marbled ribbed bowl from a level dated to A.D. 40–50.
Arezzo: Museo Civico. Rim fragment similar to **No. 32**. Unpublished.
Vatican Coll. 1975, 35–36, nos. 119–138, pl. 6. Small fragments of marbled and onyx mosaic ribbed bowls, without find-spots but presumably from Rome or Italy.
Rome: Antiquarium Comunale, provisional inv. no. 146, a fragmentary deep marbled bowl similar to **No. 32**; five additional fragments of smaller bowls in process of being catalogued. Unpublished.
Ostia II 1970, 82, pl. XVI.276. Fragment from a Flavian level.
Aquileia 1968, 171, no. 161, pl. 10, plus additional fragments in the Museum at Aquileia. All unpublished.
Simonett 1941, 134, no. 27, pl. 9.4. Marbled ribbed bowl from an Augustan or early Tiberian burial.
Florence: Museo Archeologico; uncatalogued rim fragment from a vessel similar to **No. 32** at Cosa.
Morgantina: Antiquarium, Aidone. Two rim fragments, uncatalogued, from contexts containing late Hellenistic fill to the early first century A.D. Grose 1982, 24, fig. 4a. Haltern 1901, 173. Fragment from a context dating to ca. 11 B.C. to A.D. 16.
Magdalensberg 1979, 26–31, with parallels listed. No. 64, p. 32, is from an early Augustan context, until 10 B.C.
Camulodunum 1947, 294, no. 9, fragment from early Claudian level.

For additional parallels from Italy and northwestern European sites, see listings in Isings 1957, 17–21, Vindonissa 1960, 12, 17–18, and Magdalensberg 1979, 26–31.

In date, the marbled ribbed bowls and other marbled vessels appear most plentifully in Augustan or Julio-Claudian contexts. Their general absence from northern European sites occupied after the mid-first century A.D., or from the finds at Pompeii and Herculaneum, suggests they were no longer available after midcentury. The argument that the marbled ribbed bowls are earlier than their monochrome counterparts is not borne out by the evidence. At Cosa, as elsewhere, mosaic ribbed bowls are contemporary with monochrome examples of similar appearance, even if they disappear from production at an earlier point in time.

Marbled vessels, especially ribbed bowls, while constituting the most popular mosaic pattern in Italy and in Europe, are relatively rare on sites in the eastern Mediterranean, but exceptions occur (Dura-Europos 1963, 26, no. 101). This provides a strong argument for an Italian provenance, at least for most examples. Multiple centers of production, however, cannot be ruled out.

Miscellaneous Marbled Mosaic-Glass Vessels

Aside from ribbed bowls in marbled patterns, other forms also make use of similar colored configurations: plain bowls, dishes, pyxides, jars, and small unguent bottles. At Cosa, **No. 29** is a golden-yellow dish streaked with opaque white for which there is no immediate parallel. **No. 30**, a purple jar with white veining, is similar but not identical in shape to a mosaic jar found in a burial at Locarno and dated between ca. A.D. 5 and 20 (Simonett 1941, 140, no. 25, fig. 119, pl. 9.3). This same shape (Isings Form 4) also reappears in contemporary blown glass in opaque white and opaque blue fabrics, as illustrated in Magdalensberg 1979, 94–95, no. 729, pls. 5 and 14, and dated to the Augustan period; also p. 95, nos. 730–732. At Cosa, the pattern appears to be randomly marbled, whereas the jar from Locarno is actually comprised of very large cane sections in a purple ground with spirals of opaque white. The vessel from Cosa was retrieved from Deposit V, belonging to the years spanning ca. 25/15 B.C. to ca. A.D. 40/45.

No. 31 at Cosa preserves the rim and neck of a small, blown color-band bottle, typical of scores of comparable vessels known from other archaeological sites and museum collections. In terms of technique of manufacture, size, shape, and color combinations and designs, these bottles exhibit considerable variation. Some appear to be assembled from preformed canes or strips of glass that were combined in some fashion, then inflated on a blow pipe.[12] Others seem to be free-blown using a gather of marbled glass. Still others are free-blown with festoons or streaks of colored glass applied to the surface. Some may even have been cast in molds or forms of some sort. Since many examples, including **No. 31**, made use of preformed cane elements, it is catalogued with cast mosaic-glass and not with free-blown unguent bottles.

Because it is not uncommon to find a variety of color-band bottles in the same cemetery—or even in the same grave—it appears that such bottles share, for the most part, a common chronological evolution in the late first century B.C. and early first century A.D. It may also suggest, despite differences of technique, that many or most were the products of the same centers of production, or of different centers in operation simultaneously but knowledgeable of each other's wares. When dated, most examples are from Augustan or Julio-Claudian contexts. A fair number of bottles are known from archaeological sites in Italy, the Aegean, and the Syro-Palestinian region, but a great many now in the museum collections lack provenances. Only a few color-band bottles are listed below.

Parallels and Dated Examples of Miscellaneous Marbled Mosaic-Glass Vessels

Cosa: **No. 30**, Deposit V, ca. 25/15 B.C. to ca. A.D. 40/45.

Cosa: **No. 31**, Forum SW, I, with material mostly of the late first century B.C./early first century A.D.

Arezzo: Museo Civico. Three bottles. Unpublished.

Pavia Coll. 1983, 124–125, no. 110. Small ribbed bottle from Tomb 5 at Gropello Cairoli (Pavia). Attributed to the end of the first century B.C./early first century A.D., with discussion on pp. 106–107.

Aquileia 1968, 28–51; among others, nos. 85–88. 90, 92, 115–116, color pls. 5–6. With discussion and parallels cited.

Cumae 1913, col. 740, nos. 7 and 10, pl. 118. Two bottles.

Sangiorgi Coll. 1914, 60, nos. 211–212, pl. 39. Provenance of no. 212 is said to be central Italy; provenance of the other bottle is uncertain.

Vatican Coll. 1975, 36, no. 149, pl. 4 and color pl. opposite p. 80. Presumably from Italy.

Magdalensberg 1979, 119–121, no. 922, pl. 14. Bottle dated before ca. A.D. 45, when occupation at the site ended. Additional parallels cited on p. 119.

Samothrace 1967, 40–41, nos. 14–15, figs. 15–16. Bottles attributed to first half of first century A.D.

[12] Cummings 1980, 44, in discussing gold-band bottles, suggests such a manufacturing method but emphasizes the difficulty in determining the precise procedures once involved.

Alexandra 1972, 115–118, pl. 76. Three bottles from tombs in Athens said to belong to the first century A.D.

ROM Coll. 1975, 23 and 28, nos. 79–81, pl. 6, illustrates three additional examples with references to others, including several from the eastern Mediterranean.

Miscellaneous Marbled Mosaic-Glass Vessels

29. C71.193 Pl. 2
V-E, 0

Small fragment of rim and side of bowl; marbled mosaic pattern, golden-yellow ground with streaks of opaque white. Pitting and dulling. Upright rim with rounded edge; nearly straight side, tapering downward. Cast, rotary-polished on interior and on outside of rim; rest of exterior fire-polished. D ca. 14 cm; A Th .4 cm.

30. CE 2015 Pl. 2
A.P. Room 16, I Deposit V
ca. 25/15 B.C. to ca. A.D. 40/45

Fragment of rim and side of jar; marbled mosaic pattern, purple ground with streaks of opaque white and colorless glass. Iridescence and pitting. Outsplayed rim with rounded edge; neck curved inward, then outward at shoulder. Cast, rotary-polished on interior; exterior fire-polished. D ca. 8 cm; A Th .4 cm.

31. C69.403 Pl. 2
Forum SW, I

Rim, neck, and indications of shoulder of small color-band bottle; marbled mosaic pattern, golden-brown ground with streaks of opaque white. Silvery weathering and corrosion. Horizontal rim with pointed edge sloping obliquely to the outside; cylindrical neck tapers to flaring shoulder. Slight constriction at junction of neck and shoulder. Blown and tooled, using preformed cane elements; rim rotary-polished. H neck 2.5 cm; D neck 2.1 cm; A Th .2 cm.

Marbled Mosaic-Glass Ribbed Bowls

32. CE 2014 Pl. 2
A.P. Room 16, I
ca. 25/15 B.C. to ca. A.D. 40/45

Seven joining and nine non-joining fragments of rim and side of large ribbed bowl; marbled mosaic pattern, golden-brown with opaque white streaks. Silvery iridescence and pitting. Outsplayed uneven rim with thickened rounded edge; nearly vertical, straight side, with slight curve downward. Probably a footed ribbed bowl. On exterior, rounded and undulating shallow ribs set vertically on body. Cast, rotary-polished on interior, exterior fire-polished. Tooling marks on outside of rim. D ca. 19 cm; Max Th rim .4 cm; Min Th side .17 cm.

33. CD 1079 Pl. 2
Bas. ENW 3, I

Fragment of rim (edge missing) and side of large ribbed bowl; marbled mosaic pattern, purple ground with opaque white streaks. Iridescence and pitting. Slightly outsplayed rim; nearly straight side, tapering downward. On exterior, flat broad ribs set almost vertically on fragment. Cast, rotary-polished on interior; exterior fire-polished. Tooling marks on outside of rim. D shoulder ca. 18 cm; Max Th rim .3 cm.

34. CD 1078 Pl. 2
Bas. ENW 3, I
Five non-joining fragments of rim and side of bowl; marbled mosaic pattern, golden-brown ground with opaque white streaks. Milky weathering, iridescence, and pitting. Slightly outsplayed rim with rounded edge; convex curving side. On exterior, rounded ribs extend to junction of side and bottom.

35–37. There are no descriptions for these ribbed bowls, but **No. 35** is illustrated on pl. 2.

38. C72.100
House of the Birds, Rooms 21 and 22, 0
Fragment of side at bottom of ribbed bowl; marbled mosaic pattern, blue ground with opaque white streaks. Iridescence and pitting. Convex curving side. On exterior, pronounced uniformly spaced ribs; ribs extend beneath bottom. Cast, rotary-polished on interior; exterior fire-polished. D uncertain; Max Th .4 cm.

39. CD 345 Pl. 2
Bas. Nave 7, I
Four joining fragments of side and bottom of ribbed bowl; marbled and composite pattern, purple ground with opaque white flecks and streaks. Iridescence and pitting. Fine craftsmanship. Convex curving side; slightly concave bottom, upper side of bottom flat. On exterior, pronounced uniformly spaced ribs extending to center of bottom. Cast, rotary-polished on interior; exterior fire-polished. D uncertain; Max Th .3 cm.

40. C66.802
V-D, Street M, 0-I
Fragment of bottom of ribbed bowl; marbled mosaic pattern, golden-yellow with opaque white streaks. Iridescence and pitting. Slightly concave bottom, upper side of bottom slightly convex. On exterior, ribs extend to junction of side and bottom. Cast, rotary-polished on interior; exterior fire-polished. Max Th .3 cm.

41. C70.569 Pl. 2
VI-D, E, I
Small fragment of side and bottom of ribbed bowl or other vessel; marbled mosaic pattern, golden-brown with opaque white streaks. Iridescence and pitting. Convex curving side; slightly concave bottom. Precise shape uncertain. On exterior, rounded ribs terminate above the junction of the side and bottom. Cast, rotary-polished on interior; exterior fire-polished. A Th .2 cm.

42. C71.200
V-E, 0
Three tiny non-joining fragments of side of ribbed bowl or other vessel; composite mosaic pattern composed of purple, blue, opaque white, and opaque yellow streaked elements. Pattern obscured

by corrosion. Nearly straight side, tapering downward. On exterior of fragment, a shallow rounded rib. Cast and polished. Max Th .2 cm.

43. CE 2117
A.P. Room 1, I
Small fragment of side near rim of ribbed bowl; marbled mosaic pattern, blue ground with opaque yellow and opaque white strips. Iridescence and pitting. Convex curving side. On exterior of fragment, a pronounced vertical rib. Cast, rotary-polished on interior; exterior fire-polished. A Th .25 cm.
a. C73.71, A.P. VI, South Corner, 0. Description identical to **No. 43**.

44. C70.571
S.H. Street 5, Trench 10–12, 0
Small fragment of side of ribbed bowl or other vessel; marbled mosaic pattern, blue ground with purple and opaque white streaks. Slight iridescence and pitting. Curved fragment with raised section suggesting a rib. Shape uncertain. Cast and polished. A Th .2 cm.

CAST HELLENISTIC AND EARLY IMPERIAL ROMAN MONOCHROME GLASS

Like the cast mosaic-glass at Cosa, the fragments of cast monochrome vessels also represent virtually every prominent type or group of tablewares customarily associated with the late Hellenistic or early Roman glass industries. The earliest wares (with the possible exception of the bowls in the appendix) are grooved bowls (pp. 65–67 below). In all likelihood these pieces were imported into Italy from the Syro-Palestinian region during the course of the mid-second to mid-first century B.C. They are the first type of glass drinking vessels to gain wide currency in classical antiquity, and their distribution encompasses the entire Mediterranean basin. In the first century B.C., they were supplanted by the many kinds of mosaic wares already described and by new types of monochrome vessels, again tablewares, which were manufactured in both the eastern Mediterranean and Italy. Chief among them were ribbed bowls, of which the earliest were devised in the Syro-Palestinian region (Tel Anafa) by the beginning of the first century B.C., then followed in the Augustan Age by other varieties (see pp. 54–58 below), and by the introduction as well of linear-cut bowls (pp. 68–70 below). At approximately the same time, coincident with the reigns of Augustus and the Julio-Claudians, a diverse production of cast translucent and opaque fine wares came into vogue, presumably the innovations of the Romano-Italian workshops (pp. 76–82 below).

The last prominent category of cast vessels to be produced in quantity by the Romans was developed under the Flavians but achieved great popularity in the early to mid-second century A.D. These wares comprise an elegant group of colorless circular plates, dishes, and bowls characterized by angular shapes, broad, overhanging rims, and in some instances by fine facet-cut decoration (pp. 88–91 below). This type represents one of the most successful and spectacular efforts of the Roman casting industry.

As is true of the mosaic wares, the classification system employed here is based on easily recognizable traits in order to define distinctive types or related groups. Some types, such as the grooved bowls, are remarkably homogeneous and can be assigned inclusive dates and regional provenance with some certainty. Others, like the ribbed bowls, represent the products of different workshops and prolonged periods of production in several areas and are more difficult to classify in a simple manner. Moreover, in all categories, there are variants and exceptional pieces that do not conform to any known group. Although it is presently impossible to understand their place in the history of classical glassmaking, they are important in that they reflect the striking diversity and bold experimentation of the late Hellenistic and early Roman epochs.

*Ribbed Bowls: Colored and Naturally Colored (**Nos. 45–91**)*

Monochrome ribbed bowls of various kinds constitute the largest and geographically the most widespread group of all cast tableware fabricated in Classical antiquity. They also represent one of the most long-lived, spanning all of the first-century B.C. and much of the succeeding century. The earliest examples occur on Hellenistic sites, such as Delos and Tel Anafa in the eastern Mediterranean, where they are already in evidence by the first decade of the first century B.C. (see references in list below). The latest examples, such as eight pieces found in their packing crate at Herculaneum, are abundantly present on most Roman sites until the early second century A.D. Not unexpectedly, they do not comprise a single homogeneous type but rather represent several distinct classes whose shapes, ribbing, and coloration suggest diverse times and places of manufacture. Although one such group, the marbled mosaic ribbed bowls, has been discussed earlier (pp. 47–49 above), the vast majority of examples are in fact monochrome, either deliberately colored or naturally colored (with, admittedly, a few pieces in decolorized fabrics).

At Cosa there are twenty-seven examples of translucent colored bowls. Twelve are in shades of blue, six of golden-brown, two of golden-yellow, and one in lime green. Six others are greenish-yellow or yellowish-green, not unlike Hellenistic grooved bowls, here **Nos. 92–93**, and these colors are probably the result of the oxidation/reduction state of the glass while hot and not due to the addition of colorants (for discussion of spectroscopic analyses of such colors, see Grose 1979, 57 and 62). Of these, **No. 46** stands out by virtue of its brilliant cobalt blue color and careful craftsmanship. Bowls of purple, emerald green, aquamarine, and peacock blue are lacking at Cosa and are rare or nonexistent anywhere. This is surprising since these colors were often used for the closely related linear-cut bowls (pp. 68–70 below) or the translucent fine wares (pp. 76–82 below), which are contemporaneous. Also in a decided minority are ribbed bowls in opaque fabrics, although some examples in opaque white, light blue, and several greens (but oddly never opaque red) are known to exist (e.g., Aquileia 1968, 72, no. 163, pl. 9; Oppenländer Coll. 1974, 97 and 99, nos. 258–259).

In form and style of ribbing the bowls at Cosa exhibit the same diversity as do their marbled counterparts described in the preceding section. Some are large and deep (**Nos. 45–47**); others are smaller with gentle S-curve profiles (**Nos. 48–51**) and closely resemble one series of the linear-cut bowls (here, **Nos. 94–101**); and still others enjoy an almost hemispherical body with an upright rim (**Nos. 56–64**). At least one specimen (**No. 50**) possesses a broad, shallow shape with short, multiple ribs concentrated at the shoulder. Otherwise, ribs can be gentle and undulating (**No. 57**), boss-like (**Nos. 62–63**), or pronounced and of the pillar-molded kind (**Nos. 45–47** and **58**).[13] At times the ribs extend to the junction of the side and bottom; alternatively, they continue almost to the center of the bottom. In all instances the bottoms are slightly concave or flat.

On half the examples single or multiple horizontal grooves are cut on the interior (but never the exterior) surface. On bowls such as **Nos. 53** and **62–64**, which may be contemporary with the grooved bowls noted above (see dating below), grooves tend to be broad and deep. This contrasts

[13] The term "pillar-molded," derived from English glassmaking of the nineteenth century (see "Glossary"), is used here to describe only those ribbed bowls with deep, projecting ribs that are symmetrically and uniformly spaced on the vessel.

with the grooved decoration on the specimens of the Augustan and early Imperial epoch, which tend, like linear-cut bowls, to have one narrow groove below the rim and a band of two at the center of the body (alternatively, two at the center and a small circular groove at the middle of the bottom). Surprisingly, both the colored and naturally colored bowls exhibit grooves, whereas their contemporary mosaic counterparts often lack them.

In regard to method of manufacture the most attractive explanation for ribbed bowls is that they were produced in two stages using former-molds, a technique discussed in chapter 3 (see also Grose 1984, 28–29, fig. 4). Prior to sagging the blank, however, the ribs were created by the impress of a special tool with regularly spaced gaps. Once annealed, the bowls were rotary-polished on the interior and on the top and outside of the rim. Although this procedure may not have been responsible for all classes of ribbed bowls (see discussion in Weinberg 1973, 36–39), it does seem to provide an adequate solution for understanding most characteristics of these vessels.[14]

Most of the intentionally colored ribbed bowls at Cosa belong to the Romano-Italian production of the first four decades of the first century A.D., with the probability that bowls of this sort were devised late in the preceding century. To this period can be assigned the small bowls with S-curve profile and most of the brilliantly colored examples with pillar-molded ribs. This would make them contemporary with mosaic ribbed bowls and linear-cut bowls with identical forms, and even with the free-blown colored bowls with pinched ribs (see discussion on pp. 122–124 below). Dated examples at Cosa appear in Deposits V and VII, respectively assigned from ca. 25/15 B.C. to ca. A.D. 40/45 (but mainly Augustan or Tiberian), and prior to ca. A.D. 50/55 (but again, principally Tiberian). Corroborative evidence comes from Forum SW, I, where two fragmentary bowls (**Nos. 48** and **50**) were recovered with material that had accumulated from the end of the first century B.C. to ca. A.D. 25. On other sites similar bowls, when dated, are also attributed to Augustan through Claudian contexts, as is evident from the comparanda listed below. Among the earliest manifestations of these wares, however, are pieces of Augustan B.C. date, and they indicate that the type emerged sometime in the last quarter of the first century B.C. Examples include bowls from the Regia and the House of Livia in Rome (Grose 1977, 18 and 22, figs. 1–2 and 5–6), Morgantina (Grose 1982, 26, fig. 8), and Albintimilium (see list below).

Probably earlier in time and possibly representing an eastern Mediterranean source are those bowls in greenish-yellow or yellowish-green fabrics, such as **Nos. 53** and **62–64**, especially ones that also possess irregular boss-like ribs (**Nos. 62–63**). These "muddy" colors were first employed by the Syro-Palestinian casting industry of the second and first centuries B.C. (see Grose 1979, 56–57), and the vast majority of ribbed bowls from Cyprus, the Aegean, and the Syro-Palestinian region, unless decolorized, utilize such fabrics. This is true of examples from Delos (unpublished) and elsewhere in the Greek islands, as well as at Tel Anafa (Weinberg 1970, 24–26), Ein-Gedi (Avigad 1962, 183), and Jerusalem (Avigad 1972, 198–200; also, additional unpublished excavations in Jerusalem). In many of these instances the bowls were recovered from levels of the first century B.C., with some examples in contexts of the first half or first quarter of that century. Especially noteworthy are the vessels from Delos, which must antedate the destruction of the town in 88 B.C., or 69 B.C. at the latest.[15] Specimens

[14] One feature not explained by this technique of manufacture is the presence of slightly raised, rounded areas between the upper ends of the ribs. These gentle protrusions are to be seen on most examples but are especially visible on **No. 74**.

[15] Examination of the fragmentary glass from Delos has revealed an unexpectedly large number of colored ribbed bowls, many with boss-like ribs. Their number is so great that they must have been in use at the site before the destruction of 88 and 69 B.C. Supporting this hypothesis is the presence of a great many grooved and fluted bowls (see pp. 65–67 below) of the second and early first century B.C. and, conversely, the absence of all types of glassware attributable to the late first century B.C. or Augustan period (as noted in the text above). In fact, there was only one fragment of a naturally colored pillar-molded bowl among the preserved finds. I wish to thank the École Française d'Archéologie, Athens, and Dr. Jean-Yves Empereur, for permission to see the glass and for their hospitality during my visit.

from Tel Anafa (unpublished) also belong to loci of ca. 125 to 90 B.C.[16] From the west at Morgantina come similar bowls that were discovered in contexts attributable to the late first century B.C. (Grose 1977, 26–27, fig. 8a). Unfortunately, at Cosa none of the pieces in question was found in a dated level, although all came from blocks of housing destroyed ca. 72/70 B.C.

Naturally colored ribbed bowls at Cosa comprise twenty-nine fragmentary vessels, here catalogued separately from their colored counterparts for ease of classification and also because they remained in production after the vogue for color disappeared. The majority are light bluish-green (called aqua or aquamarine in other catalogues), the remainder are in pale to light green. None appear to have been decolorized, as is true of many ribbed bowls on Delos or in the Syro-Palestinian area.[17] In fact, most at Cosa can be attributed to the same Romano-Italian workshops that manufactured their colored counterparts. Among pieces at Cosa, the pillar-molded variety of ribbed bowls is preeminent. Ribs are generally pronounced and uniformly spaced, but rounded, undulating ribs (**No. 79**) or short, densely packed ribs (**No. 70**) also occur. More unusual are vessels like **No. 69** with narrow, widely spaced ribs or **No. 87** with inverted tear-shaped rims. As is argued above, these variations were the result of a manufacturing technique by which the ribs were formed by the impress of a special tool and then sagged one or more times over or in a former-mold. The spacing, depth, and size of the ribs would differ depending on the design of the tool, the force of its application to the blank, and the duration and intensity of heat during the sagging process. Grooves routinely occur on many examples, usually with a single groove beneath the rim and a band of two around the body, but **No. 82** depicts another scheme, with a pair of cuttings at the junction of the side and bottom, and a third, quite small, at the center of the bottom.

In date, the earliest group of naturally colored bowls at Cosa are found in the same contexts as the colored and marbled-mosaic examples, namely Deposits V, VII, and VIII, dating inclusively between ca. 25/15 B.C. to ca. A.D. 40/45. The majority at Cosa, as elsewhere, fall into the Augustan and Julio-Claudian periods and must have been fabricated by the same cluster of shops producing the more colorful pieces. On the other hand, it is clear from other bowls found in Italy (see, for example, Pompeii and Herculaneum in the list below) and from numerous European sites (again, list below) that the manufacture of naturally colored ribbed bowls, especially in broad, shallow shapes (like **No. 70**) continued until late in the first century A.D. In Britain examples are found in strata attributed to the early decades of the second century.[18]

Parallels and Dated Examples of Colored and Naturally Colored Ribbed Bowls

Cosa: **Nos. 65**, **83–84**, and **90f** from Deposit V, ca. 25/15 B.C. to ca. A.D. 40/45; **Nos. 70** and **85** from Deposit VII, decade(s) before ca. A.D. 40/45; and **Nos. 48**, **50**, and **90b** from Deposit VIII, prior to ca. A.D. 50/55.

Albintimilium 1950, 115, nos. 45–46, fig. 57. Two pale green ribbed bowls from Stratum IV, an Augustan B.C. level.

Ostia II 1970, 76, nos. 229–231, pl. 15; 78, nos. 242–243, pl. 15; 81, nos. 268–269, pl. 16.

[16] A reexamination by this author of the strata at Tel Anafa reveals that ribbed bowls are found in levels of the second half of Phase II, dated to ca. 125–95 B.C. This conclusion, together with the evidence cited in the preceding note and the references found in Grose 1979, 67, n. 22, pushes back the invention of the ribbed bowls into the first half of the first century B.C., possibly even somewhat earlier.

[17] Unpublished but seen by the author. At Tel Anafa the percentage of colorless examples is especially striking, suggesting that decolorized ribbed bowls, often in broad shallow forms, were made in large numbers by the Syro-Palestinian industry. The single colorless piece at Cosa (**No. 91**) comes from a vessel of uncertain shape and may have been mold-blown rather than cast.

[18] Examples from Trajanic or Hadrianic contexts are collected in Verulamium 1972, 198–199.

Rome: Antiquarium Comunale. Numerous fragments of colored and naturally colored ribbed bowls. Unpublished.

Rome: Museo Nazionale Romano. Numerous examples of intact and fragmentary colored and naturally colored ribbed bowls. Unpublished.

Grose 1977, 17–25. Two ribbed bowls, one each from the Regia and the House of Livia, Rome, from the contexts of the late first century B.C.

Pompeii 1978, 30, no. 99. Cobalt blue bowl of the pillar-molded kind from Pompeii, before A.D. 79.

Mariacher 1959, pl. 11. Golden-brown ribbed bowl like **No. 57** at Cosa, now in the Museo Civico, Bologna.

Spinazzola 1928, 228. Several examples of colored and naturally colored ribbed bowls from Pompeii and environs. Before A.D. 79.

De Franciscis 1963, 137–139, fig. 2. Eight intact naturally colored bluish-green bowls found in their packing crate in a shop at Herculaneum. Presumably, they were made shortly before A.D. 79.

Aquileia 1968, 72, no. 163; pl. 9. Opaque white, broad ribbed bowl.

Magdalensberg 1979, 26–34, pls. 1, 18–19. The camp yielded 264 fragmentary examples (mosaic ribbed bowls included) from Augustan levels through ca. A.D. 45, when the site was abandoned. Extensive parallels, mostly European, are listed.

Vindonissa 1960, 18–23, nos. 23–27, pls. 2 and 18. Numerous examples with discussion and parallels cited.

Nida-Heddernheim 1974, 18–24, nos. 1-25, pls. 1–2.

Fishbourne 1971, 328–330, nos. 17–21, fig. 137, pl. 25.

Conimbriga 1976, 158–159, 164; nos. 6–13, pls. 34 and 47. Parallels from the Iberian Peninsula are mentioned.

Verulamium 1972, 198–199, nos. 1, 4–9, 11–12, and 14. Naturally colored ribbed bowls from contexts variously dated from A.D. 49–60 to A.D. 110–120. Many of the latest examples, belonging to the early second century A.D., are cited, including a bowl from Benwell, which first was occupied in A.D. 122.

Grose 1982, 26–27, fig. 8. Five fragmentary colored ribbed bowls from Morgantina, found in levels dated to the late first century B.C.

Samothrace 1967, 39, no. 10, fig. 11. A greenish-yellow bossed bowl assigned to an Augustan burial.

Knossos 1971, 253, 262, 272; figs. 6, 12, 19; pl. 41. Numerous examples of colored and naturally colored ribbed bowls, mostly from contexts of the early first century A.D.

Delos: Numerous examples of ribbed bowls with boss-like protrusions for ribs (like **Nos. 62–63** at Cosa); also broader bowls, mostly colored, with multiple, narrow ribs set around the center of the body. These bowls presumably predate the destruction of the site in 88 B.C. or 69 at the latest. Unpublished.

Cyprus 1956, 128, 193–194, figs. 41 and 52.

Sardis 1980, 11–12, nos. 27–41, pls. 2–20. With discussion.

Tel Anafa 1970, 25–27, nos. 31–37. Colorless, colored, and naturally colored bossed and ribbed bowls from Tel Anafa in Upper Galilee, several of which may come from levels before the Herodian period. Many unpublished.

Dura Europos 1963, 25–27, nos. 102–107, pls. 4 and 20.

Isings 1957, 17–21, Form 3. Numerous parallels cited.

Although a detailed understanding of the evolution of the ribbed bowls is far from complete, it now seems certain that the type passed through at least three different stages in its development. If the available evidence from the eastern Mediterranean is corroborated by future finds, the earliest phase can be assigned to Syro-Palestinian (or eastern) craftsmen who invented the tools to fashion ribs, as well as the sagging technology involving former-molds. This occurred by the first decade of the first century B.C. Whether they conceived of the ribs and bosses as mere decorative devices or as a means of strengthening the vessel cannot now be determined, but once invented, ribbed bowls continued in production for almost two hundred years. In all likelihood, these early examples served as the prototypes for the classes that emerged in Palestine, Italy, and elsewhere at the close of the first century B.C. These vessels, mostly in bright colors or mosaic fabrics, also introduced the true pillar-molded variety of rib and the S-curve profile. Somewhat later, the naturally colored ribbed bowls, often in broad shallow shapes, came to dominate the market and continued to be made until late in the first century A.D. or beyond.

Colored Ribbed Bowls, Outsplayed Rims

45. CD 782 Pl. 3
Bas. ENE 1, II

Fragment of rim and side of large ribbed bowl; blue. Silvery iridescence and pitting. Outsplayed rim with flattened edge; side bulged outward beneath shoulder. On exterior, pronounced ribs set diagonally on fragment and polished horizontally across upper ends. Cast, rotary-polished on interior and on outside of rim; rest of exterior fire-polished. D ca. 20 cm; Max Th .4 cm.

46. C66.295 Pl. 3
V-D North, I

Two non-joining fragments of rim and side of large ribbed bowl; cobalt blue. Slight iridescence and pitting. Excellent craftsmanship. Outsplayed rim with flattened edge; side bulges outward beneath shoulder. On exterior, pronounced, uniformly spaced ribs set diagonally and polished horizontally across upper ends. On interior, a thin horizontal cut line below edge of rim. Technique of manufacturing as in **No. 45**. D ca. 17 cm; Max Th shoulder .6 cm; A Th side .3 cm.

47. C72.89 Pl. 3
VI, VII-D, Street N, O

Four joining and four non-joining fragments of side and bottom of large ribbed bowl; blue. Iridescence and pitting. Convex curving side; slightly concave bottom. On exterior, pronounced rounded ribs set vertically on fragments and extending beneath junction of side and bottom. On interior, a broad band of heavily abraded polishing marks above junction of side and bottom. Cast, rotary-polished on interior; exterior fire-polished. D body ca. 17 cm; Max Th side .5 cm.

48. C69.405 Pl. 3
Forum SW, I

Fragment of rim and side of shallow ribbed bowl; golden-brown. Iridescence and pitting. Slightly outsplayed rim with pointed edge sloping obliquely to outside; side bulges slightly outward beneath shoulder. On exterior, multiple short, narrow ribs set almost vertically around center of body. On interior, a wide horizontal groove below edge of rim. Cast, rotary-polished and cut on interior and on outside of rim; rest of exterior fire-polished. D ca. 13 cm; Max Th .4 cm.

49. C68.678 Pl. 3
S.H. Room 2, 0-B

Fragment of rim and side of shallow ribbed bowl; golden-brown. Iridescence and pitting. Slightly outsplayed rim; side bulges slightly outward beneath shoulder. On exterior, multiple short, narrow ribs set vertically around center of body. On interior, a wide horizontal groove below edge of rim. Technique of manufacturing as in **No. 48**. D ca. 13 cm; Max Th .4 cm.

50. C69.406 Pl. 3
Forum SW, I
Fragment of rim and side of shallow ribbed bowl; greenish-yellow. Iridescence and pitting. Slightly outsplayed rim; side bulges slightly outward beneath shoulder. On exterior, multiple short, narrow ribs set irregularly around center of body. On interior, a shallow horizontal groove below edge of rim. Cast, rotary-polished on interior and on top of rim; rest of exterior fire-polished. On exterior, two tooled horizontal depressions at junction of rim and side. D ca. 14 cm; Max Th .5 cm.

51. C68.676A and B Pl. 3
S.H., 0-A
Two non-joining fragments of rim and side of ribbed bowl; yellowish-green. Iridescence and pitting. Slightly outsplayed rim with pointed edge; side bulges slightly outward beneath shoulder. On exterior, multiple short, narrow ribs set diagonally around center of body. On interior, a wide horizontal groove below edge of rim. Technique of manufacturing as in **No. 48**. D ca. 13 cm; Max Th .4 cm.

52. CF 2409
Forum NE Passage, I
Small fragment of rim (edge missing) of ribbed bowl; blue. Iridescence and pitting. Slightly outsplayed rim; convex curving side. On exterior, narrow but pronounced ribs, polished horizontally across upper ends. Technique of manufacture as in **No. 48**. D uncertain; Max Th .25 cm.

53. C68.675 Pl. 3
S.H. Room 2, 0-B
Fragment of rim and side of ribbed bowl; yellowish-green. Iridescence and pitting. Sharply outsplayed rim with rounded edge; convex curving side. On exterior of fragment, a large, pronounced rib set vertically. On interior, two horizontal grooves at junction of rim and side with a corresponding tooled depression on exterior. Technique of manufacturing as in **No. 50**. D ca. 13 cm; Max Th .7 cm.

54. C70.574 Pl. 3
S.H. Street M, 0-I
Fragment of rim and side of ribbed bowl; yellowish-green. Iridescence and pitting. Slightly outsplayed rim; almost straight side, tapering downward. On exterior of fragment, a pronounced rib set diagonally on body and polished horizontally across upper end. On interior, a shallow horizontal line cut below edge of rim. Technique of manufacture as in **No. 48**. D ca. 14 cm; Max Th .3 cm.

55. CD 1112
Bas. ENE 1, III Deposit VIII
Prior to ca. A.D. 50/55
Small fragment of rim (edge missing) and side of ribbed bowl; golden-brown. Iridescence and pitting. Shape and ribbed decoration similar to **No. 49** but lacking a groove. Manufacturing and technique as in **No. 48**. D uncertain; A Th .4 cm.

Colored Ribbed Bowls, Upright Rims

56. C71.176 Pl. 3
V-E, Cross Drain, Street M, I
Four joining and four non-joining fragments of rim and side of ribbed bowl; golden-yellow. Silvery iridescence and pitting. Upright rim; convex curving side. On exterior, rounded ribs set diagonally on body and extending to junction of side and bottom. On interior, a horizontal groove below edge of rim. Technique of manufacturing as in **No. 48**. D ca. 12 cm; Max Th .3 cm.

57. CE 2070 Pl. 3
A.P. Room 16, I Deposit V
ca. 25/15 B.C. to ca. A.D. 40/45
Ten joining fragments of rim and side, and side and bottom of ribbed bowl (profile restorable); golden-brown. Iridescence and pitting. Upright rim; convex curving side; slightly concave bottom. On exterior, shallow rounded ribs set irregularly on body. On interior, a horizontal groove below edge of rim; also a heavily abraded band of polishing marks just above the junction of the side and bottom. Technique of manufacturing as in **No. 48**. H 5.5 cm; D ca. 11 cm; Max Th shoulder .4 cm; Min Th bottom .1 cm.

58. C66.807 Pl. 3
V-D, Street M, 0-I
Fragment of rim (edge missing) and side of ribbed bowl; blue. Silvery iridescence and pitting. Upright rim; convex curving side. On exterior, rounded and uniformly spaced ribs set vertically on body. Technique of manufacturing as in **No. 48**. D uncertain; Max Th .4 cm.

59. C70.578
S.H. Street M, 0-I
Small fragment of rim (edge missing) and side of ribbed bowl; blue. Weathering, shape, and technique of manufacturing as in **No. 48**. D uncertain; Max Th .4 cm.

60. C73.72 Pl. 4
Unknown locus
Fragment of rim and side of small ribbed bowl; golden-brown. Iridescence and pitting. Upright rim; convex curving side. On exterior, shallow uneven ribs set vertically on fragment. On interior, a horizontal groove below edge of rim. Technique of manufacturing as in **No. 48**. D ca. 11 cm; Max Th .5 cm; Min Th .2 cm.

61. C68.677 Pl. 4
S.H. Room 2, 0-B
Small fragment of rim and side of small ribbed bowl; golden-brown. Iridescence and pitting. Upright rim with almost pointed edge, convex curving side. On exterior, short, vertical ribs set low on body. On interior, three horizontal grooves: one below edge of rim; the other two in a band above junction of side and bottom. Technique of manufacturing as in **No. 48**. D ca. 10 cm; Max Th .3 cm.

62. C68.456 Pl. 4
Arx W. Slope, 0

Fragment of rim and side of ribbed bowl; pale lime green. Iridescence and pitting. Upright rim with rounded edge; convex curving side. On exterior, short boss-like ribs, unevenly shaped and spaced, set around center of body. On interior, two horizontal grooves below edge of rim. Cast, rotary-polished and cut on interior and on top of rim; exterior fire-polished. Possible tooling mark on rib. D ca. 13 cm; Max Th rim .6 cm; A Th side .2 cm.

63. C72.41 Pl. 4
H.T. Room 10, O
Fragment of rim and side of ribbed bowl; greenish-yellow. Iridescence and pitting. Upright rim with slightly flattened edge; convex curving side. On exterior, small bulbous ribs set around center of body. On interior, six horizontal grooves or bands of deep abrasion around body. Cast, rotary-polished on interior and on top of rim; exterior fire-polished. D ca. 12 cm; Max Th .6 cm.

64. C66.804 Pl. 4
V-D North, 0
Fragment of rim and side of ribbed bowl; yellowish-green. Iridescence and pitting. Upright rim; straight side, tapering downward. On exterior, narrow but pronounced ribs set diagonally on fragment. On interior, two horizontal grooves below edge of rim. Cast, rotary-polished and cut on interior and on top of rim; exterior fire-polished. Tooling marks on upper ends of ribs. D ca. 12 cm; Max Th 6 cm.

65. CE 836
A.P. Room 11, I Deposit V
ca. 25/15 B.C. to ca. A.D. 40/45
Small fragment of side near bottom of ribbed bowl; blue. Iridescence and pitting. Curved basal angle; slightly concave bottom. On exterior, pointed ribs terminate at junction of side and bottom. Cast, rotary-polished on interior to the bottom; exterior fire-polished. A Th .2 cm.

66. Additional fragments of colored ribbed bowls.

- **a.** CF 2408, IX, Forum East Corner, Cistern Exterior SW, I. Fragment of side and rib of ribbed bowl; blue.
- **b.** C69.404, VII-E, F, 0. Fragment of side near bottom of ribbed bowl; blue.
- **c.** C70.576, S.H. Street 5, Trench 10–12, 0. Fragment of bottom of ribbed bowl; blue.
- **d.** CE 2180, A.P. Room 24, I. Fragment of bottom of ribbed bowl; blue.
- **e.** CE 2179, A.P. Room 20, I. Fragment of bottom of ribbed bowl; blue.
- **f.** CG 376, Curia Forecourt Center, II. Fragment of bottom of ribbed bowl; golden-yellow.

Naturally Colored Ribbed Bowls, Outsplayed Rims

67. CD 1092 Pl. 4
Bas. ENW 2, I
Fragment of rim and side of large ribbed bowl; natural light green. Iridescence and pitting. Outsplayed rim with flattened edge; side bulges slightly outward below shoulder. On exterior, pronounced rounded ribs set almost vertically on body and polished horizontally across upper ends. Technique of manufacturing as in **No. 45**. D ca. 18–19 cm; Max Th shoulder .6 cm; Min Th side .2 cm.

68. C70.579 Pl. 4
S.H. Street 5, 0-I
Fragment of rim and side of large ribbed bowl; natural light green. Iridescence and pitting. Outsplayed rim with rounded edge; side bulges slightly outward beneath rim. On exterior, pronounced rounded ribs set diagonally on body and polished horizontally across upper ends. On interior, a horizontal groove below edge of rim. Technique of manufacturing as in **No. 48**. D ca. 16–18 cm; Max Th .4 cm.

69. C70.575 Pl. 4
S.H. Street M, 0-I
Fragment of rim and side of large ribbed bowl; natural bluish-green. Iridescence and pitting. Slightly outsplayed rim ground very thin on exterior surface; convex curving side. On exterior, narrow, widely spaced ribs set vertically on body and polished horizontally across upper ends. On interior, a horizontal groove immediately below edge of rim. Technique of manufacturing as in **No. 48**. D ca. 17–18 cm; Max Th .3 cm.

70. CE 2047 Pl. 5
A.P. Room 22, II Deposit VII
Decades prior to ca. A.D. 40/45
Fragment of rim and side of shallow ribbed bowl; natural bluish-green. Iridescence and pitting. Slightly outsplayed rim with flattened edge; side bulges outward below shoulder. On exterior, numerous short, pointed ribs set vertically around center of body and polished horizontally across upper ends. Cast, rotary-polished on interior and on outside of rim; rest of exterior fire-polished. D ca. 12 cm; Max Th .4 cm.
Bibliography: Grose 1975, 37, no. 6.

Naturally Colored Ribbed Bowls, Upright Rims

71. C72.126 Pl. 5
VI, VIII-D, Packing, 0
Fragment of rim and side of large ribbed bowl; natural bluish-green. Iridescence and pitting. Upright rim; convex curving side. On exterior, pronounced rounded rib set diagonally on fragment. Technique of manufacture as in **No. 45**. D ca. 17–18 cm; Max Th .3 cm.

72. CD 1091 Pl. 5
Bas. ENE 1, I
Fragment of rim and side of ribbed bowl; natural bluish green. Iridescence and pitting. Upright rim; convex curving side. On exterior, pronounced, widely spaced ribs set vertically on fragment. Technique of manufacturing as in **No. 45**. D ca. 14 cm; Max Th .4 cm.

73. CF 2410
Comitium, 0
Small fragment of rim and side of ribbed bowl; natural bluish green. Iridescence and pitting. Upright rim; convex curving side. On exterior, widely spaced ribs set vertically on fragment. Technique of manufacturing as in **No. 45**. D ca. 14 cm; Max Th .3 cm.

74. C66.805 Pl. 5
V-D, Street M, 0-I

Fragment of rim and side of ribbed bowl; natural bluish-green. Iridescence and pitting. Upright rim; convex curving side. On exterior, pronounced rib with bulbous upper end. Cast, rotary-polished on interior and on top of rim; exterior fire-polished. D ca. 13 cm; max Th .4 cm; Min Th side .1 cm.

75. C66.808 Pl. 5
V-D, Street M, 0-I

Fragment of rim and side of ribbed bowl; natural bluish green. Iridescence and pitting. Upright rim; convex curving side. On exterior, pronounced rounded ribs set almost vertically on body. Technique of manufacturing as in **No. 74**. On exterior, horizontal tooling marks at junction of rim and side. D ca. 12 cm; Max Th .4 cm.

76. CD 1093 Pl. 5
Bas. ENE 3, III Deposit VIII
Prior to ca. A.D. 50/55

Fragment of rim and side of ribbed bowl; natural bluish-green. Iridescence and pitting. Upright rim; convex curving side. On exterior, pronounced, evenly spaced ribs set almost vertically on fragment and polished horizontally across upper ends. On interior, two horizontal grooves in a band around center of body. Technique of manufacture as in **No. 48**. D ca. 12 cm; Max Th .35 cm.

77. CD 1094 Pl. 5
Bas. ENE 3, III Deposit VIII
Prior to ca. A.D. 50/55

Fragment of rim and side of ribbed bowl; natural light green. Iridescence and pitting. Upright rim with flattened edge; convex curving side. On exterior, pronounced, widely spaced ribs set diagonally on fragment and polished across upper ends. On interior, two horizontal grooves in a band around center of body. Technique of manufacturing as in **No. 48**. D ca. 12 cm; Max Th 4 cm.

78. CE 2119
A.P. Room 6, 0

Small fragment of rim and side of ribbed bowl; natural bluish-green. Iridescence and pitting. Upright rim; convex curving side. On exterior, pronounced ribs polished horizontally across upper ends. Technique of manufacturing as in **No. 45**. D uncertain; Max Th rim .2 cm; Min Th side .05 cm.

79. CD 1113
Bas. Exterior W, I

Fragment of rim and side of ribbed bowl; natural light green. Iridescence and pitting. Upright rim with flattened edge; curving convex side. On exterior, rounded ribs polished horizontally across upper ends. On interior, a horizontal groove .1 cm below edge of rim. Technique of manufacturing as in **No. 48**. D uncertain; Max Th .35 cm.

80. CD 1109 Pl. 5
Bas. NEN 1, II

Fragment of rim (edge missing) and side of ribbed bowl; natural bluish-green. Iridescence and pitting. Upright rim; curving convex side. On exterior, small, closely spaced ribs set almost vertically on body. Technique of manufacturing as in **No. 45**. D ca. 12 cm; Max Th .4 cm.

81. C72.96 Pl. 5
IV-B, Market Building, 0
Fragment of side and bottom of ribbed bowl; natural bluish-green. Iridescence and pitting. Convex curving side; slightly concave bottom. On exterior, pronounced pointed ribs extend to just beneath junction of side and bottom. On interior, three horizontal grooves in a band above the junction of the side and bottom. Technique of manufacturing as in **No. 48**. A Th side .25 cm; A Th bottom .4 cm.

82. CE 2006 Pl. 5
A.P. Room 15, I
Fragment of side and bottom of ribbed bowl; natural light green. Iridescence and pitting. Convex curving side; slightly concave bottom. On interior, two horizontal grooves in a band at junction of side and bottom, and a circular groove, D 1.2 cm, at center of floor. Technique of manufacturing as in **No. 48**. A Th bottom .4 cm.

83. CE 2113
A.P. Room 16, I Deposit V
ca. 25/15 B.C. to ca. A.D. 40/45
Two non-joining fragments of side near bottom of ribbed bowl; natural bluish-green. Iridescence and pitting. Convex curving side. On interior, two horizontal grooves in a band at junction of side and bottom. Technique of manufacturing as in **No. 48**. A Th side .2 cm.

84. CE 2121
A.P. Room 16, I Deposit V
ca. 25/15 B.C. to ca. A.D. 40/45
Small fragment of side near bottom of ribbed bowl; natural light green. Iridescence and pitting. Convex curving side. On exterior, vertical ribs. On interior of fragment, a horizontal groove above junction of side and bottom. Technique of manufacturing as in **No. 48**. A Th side .3 cm.

85. CE 2155 Pl. 5
A.P. Room 22, II Deposit VII
Decades preceding ca. A.D. 40/45
Two small non-joining fragments of side of ribbed bowl; natural bluish-green. Iridescence and pitting. Convex curving side. On exterior, rounded ribs. Technique of manufacturing as in **No. 45**. A Th side .1 cm.
Bibliography: Grose 1975, 37, no. 7.

86. CD 1110
Bas. ENE 1, III Deposit VIII
Prior to ca. A.D. 50/55
Small fragment of side near bottom of ribbed bowl; natural light green. Iridescence and pitting. Convex curving side. On exterior, rounded rib extends beneath junction of side and bottom. On interior, three horizontal grooves in a band at junction of side and bottom. Technique of manufacturing as in **No. 48**. A Th .2 cm.

87. CF 2335
A.P. Rooms 16 and 22 Exterior, II and IV
Two joining fragments of side of ribbed bowl; natural bluish-green. Iridescence and pitting. Convex

curving side. On exterior, prominent tear-shaped ribs with bulbous upper ends, set diagonally on body. On interior, two horizontal grooves in a band above the junction of the side and bottom. Technique of manufacturing as in **No. 48**. A Th side .2 cm.

88. CE 2120
A.P. Room 6, 0
Small fragment of side near bottom of ribbed bowl; natural bluish-green. Iridescence and pitting. Convex curving side. On exterior, narrow ribs extend beneath junction of side and bottom. On interior, two horizontal grooves in a band at junction of side and bottom. Very thin; technique of manufacturing as in **No. 48**. A Th side .08 cm.

89. CD 1111
Bas. ENE 2, I
Fragment of side near bottom of ribbed bowl; natural bluish-green. Iridescence and pitting. Convex curving side. On exterior, narrow ribs extend beneath bottom. Technique of manufacturing as in **No. 45**. A Th bottom .3 cm.

90. Additional fragments of naturally colored ribbed bowls.
- **a.** C68.695, S.H. Garden, 0. Tiny fragment of side near rim of ribbed bowl; natural bluish-green.
- **b.** C69.696, Forum SW, I. Tiny fragment of side near rim of ribbed bowl; natural bluish-green.
- **c.** CE 2118, A.P. Room 4, I. Tiny fragment of side near bottom of ribbed bowl; natural light green.
- **d.** C65.675A, V-D, Casupola, I. Tiny fragment of side near bottom of ribbed bowl; natural bluish-green.
- **e.** CE 2177, A.P. Room 6, I. Tiny fragment of side near bottom of ribbed bowl; natural bluish-green.
- **f.** CE 2182, A.P. Room 16, I, Deposit V, ca. 25/15 B.C. to ca. A.D. 40/45. Tiny fragment of bottom of ribbed bowl; natural bluish-green.

Colorless Ribbed Vessel: Variant

91. CE 2058
A.P. Room 25 Exterior, II and III
Three joining fragments of side near bottom of bowl or other vessel; colorless. Iridescence and slight pitting. Convex curving side. On exterior, rounded, widely spaced ribs set around center of body and extending to junction of side and bottom. Shape uncertain. Vessel possibly mold-blown rather than cast and polished. Max Th side .2 cm; Max Th near bottom .6 cm.

*Grooved Bowls (**Nos. 92–93**)*

Two examples of Hellenistic grooved bowls were discovered at Cosa. One is an almost intact conical bowl (**No. 92**) from a context probably dating before ca. 72/70 B.C.[19] The other (**No. 93**) is a fragment of a smaller vessel, also conical, but from an unstratified level. Aside from two core-

[19] On the dating of Deposit II, see chapter 2.

formed bottles (**Nos. 1–2**), these bowls represent the earliest recognizable types of Hellenistic glass imported to the site.

As a group, these bowls (and several thousand comparable examples from other excavations), form a notably homogenous class, presumably manufactured in the Syro-Palestinian region beginning around or before 150 B.C. and continuing in production until the mid-first century B.C. They are described in detail in Grose 1979, 54–67, where the evidence for their date, distribution, and provenance is fully discussed.

Vessels of this type are, for the most part, either conical or hemispherical in shape and exhibit upright rims and convex bottoms. Diameters vary from 7 to 24 cm, but most fall between 13 and 17 cm. The sides are generally quite thick near the rim (.5–.8 cm) but taper considerably toward the bottom (.15–.3 cm). Decoration takes the form of broad horizontal grooves, arranged singly or in bands, on the interior beneath the rim. On some bowls grooves are also found on the exterior, usually around the bottom of the vessel. **No. 92**, for example, lacks external grooves but does have three bands of polishing or usage marks encircling the lowest portion of the body and bottom. Also part of this type is a series of bowls that are ovoid in shape and possess narrow, external grooves arranged around the body in three bands of two grooves each (Grose 1979, 56).

In color, bowls of this type are either decolorized or naturally colored, generally golden-brown, greenish-yellow, or yellowish-green. Although the latter colors are translucent, they have a characteristic "muddy" appearance. Spectroscopic analysis has shown that the chemical composition of the colorless and naturally colored bowls is identical except for the presence of manganese (a decolorizer) in the former. It has also shown that the golden-brown and greenish-yellow colors are most likely caused by different oxidation/reduction states of the glass, not by the addition of intentional colorants (Grose 1979, 57).

In respect to technique, the grooved bowls were cast in all probability over or in former-molds (see chapter 3 and Grose 1984, 28–29, figs. 4 and 9). The latter method not only explains the tapering thickness of the sides and the remarkable thinness of many examples, especially at the bottom, but also explains why, in most instances, only the interior, top, and exterior of the rim are rotary-polished, whereas the rest of the exterior is fire-polished.

Parallels and Dated Examples of Grooved Bowls

Cosa: **Nos. 92–93**. **No. 92** from Deposit II, probably before ca. 72/70 B.C., or before 25/15 B.C. at the latest.

Florence: Museo Archeologico, inv. nos. 70795–70796. Two conical grooved bowls, the first natural bluish-green, the second colorless. From Etruria.

Albintimilium 1950, 75, no. 85, fig. 32. Conical bowl from Stratum VI-A, 50–20 B.C. but with material as early as 90 B.C.

Turin: Museo di Antichità. As illustrated in *FA* II 1947, cols. 190–192. Description and provenance not stated, but presumably a grooved bowl from an Italian source.

Volterra 1975, 22. Conical bowl from a Hellenistic tomb said to date to the early first century B.C.

Vatican Coll. 1975, 43, no. 260, pls. 11 and 12; probably 65, nos. 600–601, pl. 26 (erroneously dated to the third century A.D.); 77, no. 748, pl. 36 (erroneously dated to the fourth century A.D.). None of the pieces has a provenance.

Weinberg 1961, 390, pl. 94. Hemispherical grooved bowl found at Reggio Calabria, reportedly in a second-century B.C. context.

Morgantina 1982, 24, fig. 3. One conical and one hemispherical bowl. From loci containing late Hellenistic fill ending early in the first century A.D.

Grose 1979, 55–59, lists parallels from the eastern Mediterranean, chiefly from Tel Anafa, Jerusalem, and other sites in the Syro-Palestinian area, or from Greece and the Aegean.

Additional examples from the East, not included, are cited below.

Ashdod II–III 1971, 202, pl. 98. Fragmentary bowl in a level not later than the mid-second century B.C., probably first half of that century.

Fitzwilliam Coll. n.d., 22, nos. 30a–c. Bowls from Amathus, Salamis, and Karpassia on Cyprus.

Delos: Antiquarium. Scores of examples, mostly from the residential quarters; all presumably antedate the destruction of Delos in 88 B.C. or in 69 B.C. at the latest.

Parlama 1973, 205–206, pl. 167. Hemispherical bowl with slightly outsplayed rim and external cut lines, presumably related to this type from Kaluvia (Elis) in Greece.

Yalouris 1966, 165, pl. 163. Hemispherical grooved bowl from cemetery at Yialovi (Messenia) near Pylos. Hellenistic date.

Carthage: University of Michigan excavations. Fragments of two grooved bowls. Unpublished.

Ampurias: Several fragmentary examples. Unpublished. Information furnished by Dr. Jennifer Price, University of Durham.

Distribution of these bowls, as presently understood, shows the greatest preponderance of examples in the eastern Mediterranean, specifically in the Syro-Palestinian region. The vast majority of pieces, in fact, are from a single site, Tel Anafa in Upper Galilee. There, over 3,000 examples appeared from less than 5 percent of the tel (Weinberg 1970, 17–27; Grose 1979, 57–58; Grose 1981, 66–68). Elsewhere in the region bowls in varying numbers have been found at Hagoshrim, Khirbet Shema, Ashdod, Samaria, Nessana, a cemetery near Tyre, and especially Jerusalem. Based on a large number of examples from this area, especially the thousands of fragments at Tel Anafa and Jerusalem, it would appear that these bowls were manufactured locally and were among the principal products of the famed Syro-Palestinian glasshouses known to Strabo (*Geog.* 16.2.25) and Pliny the Elder (*HN* 36.190–193). Indeed, these vessels may well have been largely responsible for the reputation of the region as a leading center of glass manufacture.

Outside the Syro-Palistinian region bowls of this type are best known from Greece and the Aegean, areas commercially linked to the Seleucid Kingdom during the Hellenistic period. Examples have been found in the Athenian Agora, Corinth, Messenia, Elis, and Delos, also from Cyprus (Fitzwilliam Coll. n.d., 22), Sardis (Sardis 1980, 7, nos. 6–11, pl. 20), and as far east as Dura-Europos (Dura-Europos 1963, 27–28, nos. 109, 113, pl. 4). In the west, Italy has also yielded grooved bowls, chiefly in Magna Graecia and Etruria but also in Liguria. Farther west are examples from Carthage in North Africa and Ampurias in Spain (see list above).

Whereas most examples of this type are found in contexts belonging to the second half of the first century B.C. through the first half of the following century, the earliest pieces, from Ashdod and the Athenian Agora, are from contexts considered to be no later than ca. 150 B.C. At Tel Anafa the bowls were abruptly introduced in great quantity around 125 B.C., by settlers newly arrived on the site. Although the evidence for the invention of the type is not yet conclusive, there is little question that it enjoyed its greatest popularity in the second half of the century and first quarter (or half) of the next. The finds from Delos confirm this conclusion.

In Italy most examples are discovered in contexts of the first half of the first century B.C. or somewhat later. Given this chronology, it is tempting to suggest that these vessels made their way to central and northern Italy as the result of the Roman military and commercial ventures in the Syro-Palestinian area during the wars of the late Republic.

GROOVED BOWLS

92. C68.640 Pl. 6
S.H. Room 3, Cesspool Bottom Deposit II
Prior to ca. 25/15 B.C., and probably before ca. 72/70 B.C.
Five joining fragments of rim, body, and bottom of large conical bowl (profile restorable); greenish-yellow. Iridescence and pitting. Vertical rim with rounded edge; nearly straight side, tapering diagonally downward; convex bottom. Conical shape. On interior, three horizontal grooves, each .12 cm wide, in a band below edge of rim; three bands of polishing or usage marks around lowest portion of body, rest of exterior fire-polished. H 9.5 cm; D 16.6 cm; Max Th .6 cm tapering to .3 cm near bottom.
Bibliography: Grose 1973, 178, pl. 189, no. 5; Grose 1983, 44.

93. CF 2421 Pl. 6
Temple B, Forecourt Exterior SW, I
Fragment of rim and side of conical bowl; greenish-yellow. Iridescence and pitting. Vertical rim with rounded edge, sloping obliquely to the outside; nearly straight side, tapering diagonally downward. On interior, two shallow grooves in a band below edge of rim. Probably from a conical bowl similar to **No. 92**. Technique of manufacturing as in **No. 92**. D ca. 12 cm; Max Th rim .5 cm.

*Linear-Cut Bowls: Colored and Colorless (**Nos. 94–120**)*

Like the grooved bowls, the cast linear-cut bowls form a homogenous group, both stylistically and chronologically. They appear in large numbers on sites in Italy, the Aegean, the Syro-Palestinian region, and to a lesser extent in North Africa and southern France. In date they are restricted to the last decades of the first century B.C. and first half of the first century A.D., but their greatest popularity coincided with the reigns of Augustus and Tiberius. In the Syro-Palestinian area they are often found on the same sites as those yielding grooved bowls from earlier Hellenistic levels, and, in fact, they may be the lineal descendants of the former. It is even possible that they were made at the same centers, given their similarities of grooved decoration, manufacturing method, and to a lesser extent coloration.

For the most part linear-cut bowls are contemporary with certain classes of cast ribbed bowls and share with them noticeable similarities of form, grooved decoration, color, and manufacturing technique. In many cases the resemblances are so close as to suggest that they may well be the products of the same factories (Grose 1979, 63–65). This is not surprising if these bowls were made by a sagging technique. Then, the only technical difference between the two types would be the application of a tool to impress ribs on the ribbed specimens (see discussion in the sections on ribbed bowls).

The two principal shapes found at Cosa (as well as on most other sites) are a small bowl with a slightly outsplayed rim, gentle S-curve profile, and slightly concave bottom (**Nos. 94–105**); alternatively, a broad shallow bowl with an upright rim, convex curving side, and a slightly concave bottom (**Nos. 108–112**). The pattern of grooves most commonly encountered is that of a single groove below the rim and a band of two around the body, always on the interior (for example, see **No. 94** or **108**). Another common pattern is a band of two grooves at the junction of the side and bottom and a circular groove at the center of the floor (**Nos. 106** and **118**).

Linear-cut bowls are usually fabricated from intentionally colored glass. At Cosa fifteen pieces are blue, ten golden-brown, eight purple, six golden-yellow, and three colorless. Unlike ribbed bowls, none at Cosa are naturally colored bluish-green or light green, but such examples exist (e.g., Toledo Museum of Art, 23.501). Decolorized specimens also occur (**Nos. 115–117**), chiefly in a broad, shallow shape, and are especially popular on Syro-Palestinian sites such as Tel Anafa.

Among examples at Cosa **Nos. 95–105** form a subclass in which all the pieces are virtually identical in shape, dimensions, and decoration. The same holds true of **Nos. 108–112**. **No. 106** is a handsome variant with a rim polished to a point and five horizontal grooves on the side and bottom. **No. 107** is less well made but equally unusual, owing to its relatively thin walls and single groove just below the rim. It is one of the latest known examples of such a bowl, dating to the years immediately before ca. A.D. 40/45. **No. 118** preserves a slightly concave bottom and exhibits a grooved pattern similar to **No. 106**. **No. 119** represents the usual type of undecorated bottom associated with most of these vessels: slightly concave, rotary-polished on the interior, fire-polished on the exterior. For examples from Greece, the Aegean, and the Syro-Palestinian region, see Grose 1979, 63–65, which lists numerous parallels and discusses in detail the date, distribution, and provenance of the type.

Chronologically, the linear-cut bowls are primarily a phenomenon of the Augustan and Tiberian periods. They seem to have been introduced in the last quarter of the first century B.C., as is evident from the context of **No. 100** at Cosa.[20] Certain subclasses were still being produced as late as the decade before ca. A.D. 40/45, the date of **No. 117**. Most examples, however, are confined to levels of the last decades of the first century B.C. or first three decades of the following century. This, at least, is true of most dated examples at Cosa, at the House of Livia and the Regia in Rome, and at Morgantina. It is also true of dated examples from the eastern Mediterranean, principally from Samothrace, Knossos, Kourion, and Tel Anafa (cited in Grose 1979, 63–65), as well as in the recent publication of an early Tiberian deposit of pottery and glass at Corinth (Wright 1980, 163, no. 125, fig. 8). Negative evidence confirms such a chronology. Virtually no linear-cut bowls have been found on sites in central and northern Europe, not even at Magdalensberg or Vindonissa. Their absence from these sites suggests that they were already unavailable after the second quarter of the first century A.D.

As argued in Grose 1979, 63–65, the provenance of the type is thought to be both Syro-Palestinian and Italian, based on the discovery of large numbers in both regions.

Parallels and Examples of Linear-Cut Bowls

Cosa: **No. 100**, Deposit III, dating before ca. 25/15 B.C. **Nos. 101**, **104**, and **118**, Deposit V, dated between ca. 25/15 B.C. and ca. A.D. 40/45. **Nos. 96–97**, and **119**, Deposit VIII, belonging to the decades preceding ca. A.D. 50/55, but principally Tiberian. **No. 117**, Deposit VII, dated to the decade before ca. A.D. 40/45.

Grose 1977, 17–25. Numerous fragmentary bowls from the House of Livia and the Regia found in two dated deposits of the late first century B.C. (or possibly as late as the first decade of the first century A.D.). More than seventy examples appeared in the excavations of the House of Livia. Most examples are unpublished.

Rome: Antiquarium Comunale. Several rim fragments. Provenance unknown but presumably Rome. Unpublished.

[20] Conversely, these bowls do not appear in levels belonging to the first half of the first century B.C. This is true at Tel Anafa, where such bowls are solely associated with Herodian loci. Nor does a single example appear among the finds at Delos, destroyed in 88 B.C. and again in 69 B.C.

Morgantina 1982, 26, fig. 7. Twenty-nine fragments of rims comprising 37 percent of all the cast tablewares on the site. Most are from contexts containing late Hellenistic fill ending early in the first century A.D. Five are from contexts that contain material ending by the close of the first century B.C., and one example, inv. no. 55.220, is from a context dating to the last quarter of the first century B.C.

Luni II 1977, 275, CM 12622, pl. 154.2; also CM 12459 and CM 11831.

Ostia II 1970, 76–77, figs. 226–228 and 232–233. From Flavian levels.

Salona 1975, 150–151, nos. 8–9, pl. 28. These fragments are mistakenly restored as Hellenistic grooved bowls.

Fréjus: Numerous examples brought to my attention by Dr. Jennifer Price, University of Durham. Unpublished.

Colored Linear-Cut Bowls, Outsplayed Rims

94. C71.198 Pl. 6
VI-D, E, Sounding 2, 0

Fragment of rim and side of bowl; golden-brown. Iridescence and pitting. Outsplayed rim; nearly straight side, tapering downward to a rounded basal angle. On interior, three horizontal grooves; one at junction of side and rim, the other two in a band around center of body. Cast, rotary-polished and cut on interior and on outside of rim; rest of exterior fire-polished. D ca. 12 cm; Max Th rim .3 cm; Min Th side .13 cm.

95. C68.684 Pl. 6
Arx W. Slope, 0

Fragment of rim and side of bowl; golden-brown. Iridescence and pitting. Outsplayed rim with pointed edge sloping obliquely to the outside; nearly straight side, tapering downward. On interior, a horizontal groove at junction of rim and side. Technique of manufacturing as in **No. 94**. D ca. 11 cm; Max Th .2 cm; Min Th .1 cm.

96. CD 1119 Pl. 6
Bas. ENE 1, III Deposit VIII
Prior to ca. A.D. 50/55

Fragment of rim and side of bowl; golden-brown. Iridescence and pitting. Slightly outsplayed rim with rounded edge; nearly straight side, tapering downward to a rounded basal angle. On interior, three horizontal wheel-cut grooves; one broad groove at junction of rim and side, the other two in a band around center of body. Technique of manufacturing as in **No. 94**. D ca. 11 cm; Max Th .3 cm; Min Th .1 cm.

97. CD 1121 Pl. 6
Bas. ENE 3, III Deposit VIII
Prior to ca. A.D. 50/55

Fragment of rim and side of bowl; purple. Iridescence and pitting. Slightly outsplayed rim with flattened edge; nearly straight side, tapering downward. On interior, three horizontal grooves; one at junction of rim and side, the other two in a band around center of body. Technique of manufacturing as in **No. 94**. D ca. 12 cm; Max Th .25 cm; Min Th .15 cm.

98. C69.412 Pl. 6
Forum SW, I
Fragment of rim and side of bowl; golden-yellow. Iridescence and pitting. Slightly outsplayed rim with pointed edge; nearly straight side, tapering downward. On interior, two horizontal grooves; one at junction of rim and side, the other one at center of body. Technique of manufacturing as in **No. 94**. D ca. 11 cm; Max Th .2 cm; Min Th .1 cm.

99. C69.434
Forum SW, I
Small fragment of rim and side of bowl; golden-yellow. Iridescence and pitting. Slightly outsplayed rim with a pointed edge; nearly straight side, tapering downward. On interior, one horizontal groove at junction of rim and side. Technique of manufacturing as in **No. 94**. D uncertain; Max Th rim .18 cm.

100. CE 2135 Pl. 6
A.P. Room 16, III Deposit III
Prior to ca. 25/15 B.C.
Small fragment of rim and side of bowl; blue. Iridescence and pitting. Slightly outsplayed rim with flattened edge sloping obliquely to the outside; nearly straight side, tapering downward. On interior of fragment, one horizontal groove at junction of rim and side. Technique of manufacture as in **No. 94**. D ca. 10–12 cm; Max Th rim .4 cm.

101. CE 2137 Pl. 6
A.P. Room 16, I Deposit V
ca. 25/15 B.C. to ca. A.D. 40/45
Small fragment of rim and side of bowl; blue. Iridescence and pitting. Outsplayed rim with rounded edge; nearly straight side, tapering downward. On interior, three horizontal grooves; one at junction of rim and side, the other two in a band around center of floor. Technique of manufacturing as in **No. 94**. D ca. 10–12 cm; Max Th .3 cm; Min Th .15 cm.

102. C68.682
S.H., 0-A
Small fragment of rim and side of bowl; blue. Iridescence and pitting. Slightly outsplayed rim with rounded edge; nearly straight side, tapering downward. On interior, one horizontal groove at junction of rim and side. Technique of manufacturing as in **No. 94**. D ca. 10 cm; Max Th .4 cm; A Th side .3 cm.

103. CD 1123
Bas. SWA, I
Small fragment of rim and side of bowl; blue. Iridescence and pitting. Outsplayed rim with almost pointed edge; nearly straight side, tapering downward. On interior, one horizontal groove at junction of rim and side. Technique of manufacturing as in **No. 94**. D ca. 10 cm; Max Th .3 cm; Min Th .2 cm.

104. CE 2018 Pl. 6
A.P. Room 16, I Deposit V
ca. 25/15 B.C. to ca. A.D. 40/45

Small fragment of rim and side of bowl; purple. Iridescence and pitting. Slightly outsplayed rim with almost pointed edge; nearly straight side, tapering diagonally downward. On interior, two horizontal grooves; one at junction of rim and side, the other around the center of the body. Technique of manufacturing as in **No. 94**. D ca. 10–12 cm; A Th .2 cm.

105. C71.206
V-E, 0
Small fragment of rim and side of bowl; blue. Iridescence and pitting. Slightly outsplayed rim with rounded edge; nearly straight side, tapering downward. On interior, one horizontal groove at junction of rim and side. Technique of manufacturing as in **No. 94**. D ca. 10–12 cm; A Th .2 cm.

Colored and Colorless Linear-Cut Bowls, Upright Rims

106. C70.538 Pl. 7
VI-D, E, 0
Three non-joining fragments of rim and side, and two joining fragments of side and bottom of broad, shallow bowl (profile restored); purple. Iridescence and pitting. Upright rim ground very thin on interior surface and terminates in a pointed edge; convex curving side; slightly concave bottom. On interior, five horizontal grooves: the uppermost is just below the edge of the rim; the second is a broad groove 1.4 cm below the rim; the third and fourth form a band at the junction of the side and bottom; the fifth is a circular groove, D 2 cm at the center of the bottom. Cast, rotary-polished and cut on interior; exterior rotary-polished on upper half of side, fire-polished on lower half of side and bottom. H 5.2 cm; D ca. 18 cm; Max Th side .3 cm; A Th bottom .2 cm.

107. CE 2042 Pl. 7
A.P. Room 22, II Deposit VII
Decade prior to ca. A.D. 40/45
Two joining and six non-joining fragments of rim and side, and side and bottom of broad, shallow bowl (profile restorable); purple. Iridescence, enamel weathering, and pitting. Upright rim tapers to a flattened edge; convex curving side; bottom either flat or very slightly concave. On interior, one horizontal groove just below the edge of rim. Technique of manufacturing as in **No. 94**. H ca. 6 cm; D ca. 16 cm; Max Th .4 cm; Min Th .2 cm.
Bibliography: Grose 1975, 36, no. 3.

108. C69.416 Pl. 7
S.H. Room 17, 0-I
Two joining fragments of rim and side of bowl; golden-yellow. Iridescence and pitting. Slightly inturned rim with pointed edge; convex curving side. On interior, three horizontal grooves; one below edge of rim, the other two in a band around center of body. Technique of manufacturing as in **No. 94**. D ca. 15–16 cm; Max Th .25 cm.

109. C69.410 Pl. 7
S.H. Room 17, 0-I
Two joining fragments of rim and side of broad shallow bowl; blue. Iridescence and pitting. Upright rim with rounded edge; convex curving side. On interior, three horizontal grooves; one below edge of rim, the other two in a band around center of body. Technique of manufacturing as in **No. 94**. D ca. 16 cm; Max Th .3 cm.

110. C68.685 Pl. 7
S.H. Garden, 0-A
Three joining fragments of rim and side of broad shallow bowl; blue. Iridescence and pitting. Upright rim with rounded edge; convex curving side. On interior, three horizontal grooves; one below edge of rim, the other in a band around center of body. Technique of manufacturing as in **No. 94**. D ca. 16 cm; Max Th .3 cm; Min Th .1 cm.

111. CE 2139 Pl. 7
A.P. Room 21, II
Fragment of rim and side of broad, shallow bowl; blue. Iridescence and pitting. Upright rim with rounded edge; convex curving side. On interior, three horizontal grooves; one below edge of rim; the other two in a band around center of body. Technique of manufacturing as in **No. 94**. D ca. 13.5 cm; Max Th .4 cm; Min Th .2 cm.

112. C66.817 Pl. 7
V-D, Street M, 0-I
Fragment of rim and side of shallow bowl; blue. Iridescence and pitting. Upright rim with nearly pointed edge; convex curving side. On interior, three horizontal grooves; one below edge of rim, the other two in a band around center of body. Technique of manufacturing as in **No. 94**. D ca. 14 cm; Max Th .4 cm; Min Th .3 cm.

113. CD 1125 Pl. 7
Bas. Exterior NW, I
Fragment of rim and side of shallow bowl; light purple. Iridescence and pitting. Upright rim with rounded edge; nearly straight side, tapering downward. On interior, three horizontal grooves; a broad groove just below rim, the other two in a band around center of the body (only vestiges of the lowest groove preserved). Technique of manufacturing as in **No. 94**. D ca. 13.5; Max Th .25 cm; Min Th .1 cm.

114. C69.414 Pl. 7
S.H. Garden, 0
Fragment of rim and side of bowl; light purple. Iridescence and pitting. Upright rim with rounded edge; convex curving side. On interior, three horizontal grooves; one below edge of rim, the other two in a band around center of body. Technique of manufacturing as in **No. 94**. D ca. 12.5 cm; Max Th .4 cm; Min Th .2 cm.

115. C69.417 Pl. 7
S.H. Room 17, 0-I
Fragment of rim and side of bowl; colorless with greenish tinge. Iridescence, pitting, and discoloration. Upright rim with rounded edge; nearly straight side, tapering downward. On interior, two horizontal grooves; one below edge of rim, the other around center of body. Technique of manufacturing as in **No. 94**. D ca. 12.5 cm; Max Th .3 cm; Min Th .1 cm.

116. C70.589
Forum SE, Trench 70B, I
Small fragment of rim and side of bowl; colorless with greenish tinge. Iridescence, pitting, and discoloration. Upright rim with almost pointed edge sloping obliquely to the outside; nearly straight

side, tapering downward. On interior, three horizontal grooves; one below edge of rim, the other two in a band around center of body. Technique of manufacturing as in **No. 94**. D ca. 12.5 cm; Max Th .35 cm; Min Th .1 cm.

117. C71.204
V-E, 0

Small fragment of rim and side of bowl; colorless with slight greenish tinge. Iridescence, pitting, and discoloration. Upright rim with almost pointed edge sloping obliquely to the outside; convex curving side. On interior, two horizontal grooves; one below edge of rim, the other one around center of body. Technique of manufacturing as in **No. 94**. D ca. 11–12 cm; Max Th .35; Min Th .2 cm.

118. CE 2020 — Pl. 7
A.P. Rooms 6 and 11, I and II — Deposit V
ca. 25/15 B.C. to ca. A.D. 40/45

Four joining fragments of side and bottom of bowl; purple. Iridescence, pitting, and some enamel weathering. Convex curving side; slightly concave bottom. On interior, three horizontal grooves; two form a band at junction of side and bottom, the third is a circular groove, D 1 cm, at center of floor. The grooved decoration closely resembles that on **No. 106**. Cast, rotary-polished and cut on interior; exterior fire-polished. A Th bottom .1 cm.

119. CD 1148A — Pl. 7
Bas. ENE 3, III — Deposit VIII
Prior to ca. A.D. 50/55

Fragment of side and bottom of bowl; blue. Iridescence and pitting. Convex curving side; slightly concave bottom. No grooves on fragment. Cast, rotary-polished on interior; exterior fire-polished. Max Th .5 cm; A Th side .2 cm.

120. Additional fragments of linear-cut bowls.

- **a.** C71.203, V-E, 0. Tiny fragment of rim and side of bowl; golden-brown. Two grooves on interior.
- **b.** C70.587, S.H. Garden, 0. Tiny fragment of rim and side of bowl; golden-brown. One groove on interior.
- **c.** C68.681, S.H., 0-A. Tiny fragment of rim and side of bowl; golden-brown. One groove on interior.
- **d.** C69.413, S.H. Garden, I. Tiny fragment of rim and side of bowl; golden-brown. One groove on interior.
- **e.** C70.586, Forum SE, Trench 70D, 0. Tiny fragment of rim and side of bowl; blue. One groove on interior.
- **f.** C69.411, Forum SW, I. Tiny fragment of rim and side of bowl; blue. One groove on interior.
- **g.** C72.127, VI–VIII, Street N, 0. Tiny fragment of rim and side of bowl; blue. One groove on interior.
- **h.** C66.843, V-D, Street M, 0-I. Tiny fragment of side of bowl; golden-brown. Two grooves on interior.
- **i.** Missing.
- **j.** C68.698, S.H., 0-A. Tiny fragment of side of bowl; golden-brown. One groove on interior.

k. C69.436, Forum SW, I. Tiny fragment of side of bowl; golden-yellow. Two grooves on interior.
l. C68.699, S.H., 0-B. Tiny fragment of side of bowl; golden-yellow. Two grooves on interior.
m. C70.588, S.H. Room 19, IIA. Tiny fragment of side of bowl; golden-yellow. Two grooves on interior.
n. C68.697, S.H. Room 2, 0-B. Tiny fragment of side of bowl; purple. One groove on interior.
o. C70.629, S.H. Street 5, Sewer. Tiny fragment of side of bowl; light blue. Two grooves on interior.
p. C69.438, S.H., I. Tiny fragment of side of bowl; light blue. Two grooves on interior.

Colored Bowls: Miscellaneous (***Nos. 121–124***)

Four fragmentary bowls differ sufficiently in appearance from the standard series of linear-cut bowls to be placed in a separate section. Three are intentionally colored, one naturally colored. **No. 121** is a part of the rim of a peacock-blue bowl of exceptional workmanship. It exhibits a single horizontal groove on the inside below the rim and a broad area of rotary-polishing on the outside of the rim. **No. 122**, a golden-yellow vessel, is rotary-polished on both surfaces and decorated with two exterior grooves below the rim. Although **No. 123** possesses no signs of polishing on either surface, it has four narrow grooves on the outside and seems to have been cast. It is either conical or ovoid in shape, not unlike the Hellenistic grooved bowls (see pp. 65–67 above), but otherwise bears little resemblance to that type. **No. 124** is the only naturally colored, plain bowl found at Cosa. Its form is reminiscent of some linear-cut bowls, but it differs in color and by the absence of grooved decoration. Close parallels are lacking for all of these vessels.

Three of the above are from dated deposits (V or VIII), both of which contain material belonging principally to the first half of the first century A.D. As such, these bowls once again illustrate the diverse production of the Augustan and Julio-Claudian glasshouses.

Colored Bowls: Miscellaneous

121. CE 2138 — Pl. 8
A.P. Room 16, I — Deposit V
ca. 25/15 B.C. to ca. A.D. 40/45

Two joining and two non-joining fragments of rim and side of bowl; peacock blue. Iridescence and pitting. Fine craftsmanship. Slightly outsplayed rim with pointed edge; convex curving side. On exterior, a broad cut depression below rim. On interior, a horizontal groove below edge of rim. Cast, rotary-polished and cut on interior and on outside of rim; rest of exterior fire-polished. D ca. 16 cm; Max Th side .3 cm.

122. CD 1120 — Pl. 8
Bas. ENE 4, II — Deposit VIII
Prior to ca. A.D. 50/55

Fragment of rim and side of bowl; golden-yellow. Iridescence and pitting. Upright rim with rounded edge; nearly straight side, tapering downward. On exterior, two horizontal grooves in a band below edge of rim. Cast, rotary-polished on interior; exterior rotary-polished and cut. D ca. 11 cm; A Th side .17 cm.

123. CD 1122 and CD 1146 Pl. 8
Bas. ENE 3, III Deposit VIII
Prior to ca. A.D. 50/55

Two joining and three non-joining fragments of side of deep bowl; golden-yellow. Iridescence and pitting. Convex side with slight curve evident on fragments. Shape probably conical. On exterior, four narrow horizontal grooves; three form a band around body, the fourth encircles body near bottom. Cast (?), fire-polished on interior, exterior cut and fire-polished. Uniform Th .3 cm.

124. C72.94 Pl. 8
VI, VII-D, Street N, 0

Two joining fragments of rim and side of broad, shallow bowl; natural bluish-green. Iridescence and pitting. Vertical rim with almost pointed edge; convex curving side. No grooves. Technique of manufacturing as in **No. 121**. D ca. 18 cm; Max Th side .4 cm.

*Fine Wares: Colored, Translucent, and Opaque (**Nos. 125–160**)*

The colored, translucent, and opaque "fine wares," as classified in this section, represent a large number of vessels characterized by rich color, crisp lathe-turned profiles, and a high standard of craftsmanship. The vast majority of examples comprise a fairly homogeneous family of related vessels (**Nos. 125**, **127–134**, **136–143** and **Nos. 146**, **148–158**, **160a–c**), translucent and opaque, respectively, although unique or tangentially related pieces are also in evidence at Cosa (for example, **Nos. 126** and **144**).

These fine wares belong to the first half or first two-thirds of the first century A.D., with the earliest prototypes beginning late in the preceding century (such as **Nos. 135**, **145**, and **147**), presumably as the result of the formation of the Romano-Italian industry (see Grose 1983, 44–45). Because so little glass from Italian sites has been adequately studied, the presence of such fine wares on archaeological sites has previously been overlooked. Likewise, their prominence among the early products of the Roman workshops has never been appreciated or acknowledged.

The principal or "standard" group of these colored fine wares exhibits a pattern of distribution centered on Italy but extending into the provinces of central and northwestern Europe and to a lesser extent North Africa. Substantially fewer examples are published from Greece and the Aegean, Cyprus, Asia Minor, the Syro-Palestinian region, or elsewhere in the eastern Mediterranean. For the type as a whole, an Italian provenance seems assured.

The characteristic translucent colors associated with the type are highly distinctive: emerald green, dark or royal blue, a brilliant deep blue-green (aquamarine), and a peacock or "Persian" blue (Grose 1983, 44–45).[21] No additional colors appear among examples at Cosa, but elsewhere other translucent colors are occasionally found, such as a golden-brown patella in the Toledo Museum of Art, 23.1379 (unpublished), an elegant purple footed bowl in the Museo di Antichità, Turin (Mariacher 1959, pl. 16), or a colorless carinated plate from Pompeii, now exhibited at the Museo Nazionale, Naples, and illustrated in Pompeii 1984. pl. 12c. At Cosa seven pieces are aquamarine, five dark blue, four emerald green, and two peacock blue. The

[21] These four translucent colors characterize the type and are found repeatedly on all sites yielding examples of these wares. Although the author regrets using such imprecise terms as "aquamarine" or "peacock blue," they are essential in differentiating among the shades of blue.

pale blue dish (**No. 126**) and the purple stemmed bowl (**No. 144**), as noted below, stand outside the main series of these fine wares.

Opaque fine wares reveal a somewhat wider range of coloration: brick red (weathering to green), white, light blue, and several shades of green.[22] At Cosa five pieces are red, four light blue, four jade green, three olive green, and two opaque white. Less common, but known from Rome and other locations, are opaque orange (often mottled with opaque red), brown, and "flesh-color." Opaque black on close examination is either a dark translucent purple, brown, or blue. Opaque yellow, so frequently employed in the production of contemporary mosaic glass, is curiously absent from this group.

These same translucent and opaque colors are also found together on virtually all other sites that produce examples of the type. The relative proportion of each color, or the proportion of translucent versus opaque pieces, however, varies from excavation to excavation. At Cosa translucent and opaque vessels occur in almost equal numbers, with translucent aquamarine and opaque red the most frequently encountered. At Camulodunum, Vindonissa, and Magdalensberg, emerald green is the preeminent translucent color and dark blue the next most common. Magdalensberg yielded over twice as many translucent as opaque vessels, whereas Vindonissa produced 144 opaque pieces and only 26 translucent ones (for references, see list below).

With the exception of peacock blue, emerald green, and the opaque greens, most of these colors had been devised by the Hellenistic glass industry in the second and first centuries B.C. Nonetheless, glassmakers of that era seldom employed them to fabricate monochrome vessels.[23] Instead, they preferred to use such colors only in mosaic wares made from polychrome canes. Otherwise, most monochrome glass of the Hellenistic period was decolorized, such as the vessels described in the appendix, or naturally colored like the Syro-Palestinian grooved bowls (**Nos. 92–93**). Only in the Augustan Age did the fashion for strong color emerge, a fact acknowledged by Strabo when discussing the innovative color technology achieved by contemporary glassmakers at Rome (*Geog.* 16.2.25).

The shapes associated with this type are generally as standard as the colors. Although diameters within a given form vary considerably, with many minor variations in profile, most vessels fall into one of a dozen principal shapes (Grose 1983, 43), all closely resembling ceramic or metal prototypes. Represented are circular dishes with convex curving (**No. 125**) or carinated sides (probably **Nos. 131–133**), as well as an unexpectedly large number of translucent and opaque rectangular dishes (**Nos. 127–130**), again a metal shape and one at times exhibiting ornately cut handles (**No. 146**). Pyxides of several styles and forms (**Nos. 134** and **136**) are common, as are medium to small bowls or patellae, some with carinated sides (**Nos. 138**, **141**, **148–151**), others with convex curving sides (**No. 152**). Still others are represented at Cosa only by their bases (**Nos. 142–143**, **154–158**). Not found at Cosa but discovered at Pompeii, Herculaneum, and other sites listed below are large shallow plates (see illustration in Vindonissa 1960, no. 30, pl. 17); circular dishes with almost vertical sides (Mutz 1972, 175, figs. 513–515); broad shallow bowls (Vindonissa 1960, no. 32, pl. 17), and pyxides (?) with concave curving sides (Aquileia 1968, 74, no. 180, pl. 2.6). In many instances, the sides of these vessels are carinated, with two convex curves, one above the other; rims often have an everted lip. Base-rings or basal moldings (as seen on the pyxis, **No. 134**) are always present and mostly are cast as part of the vessel.

[22] Whereas opaque white, red, and light blue are highly uniform in hue, considerable variety exists among the greens. In addition to jade and olive green, found at Cosa, various shades of opaque gray-greens and blue-greens are known from other sites.

[23] Exceptions include a few ovoid or hemispherical bowls in blue or purple; see von Saldern 1975, 42–43, nos. 8–9, both belonging to the Canosa group of early Hellenistic glass.

All of the standard forms cited here are to be found in the translucent colors. In fact, the translucent fine wares appear to constitute entire table services in glass, something quite novel for the Romans of the early first century A.D. The opaque colors, by contrast, are mostly reserved for smaller shapes, especially patellae, bowls, and circular or rectangular dishes. Rarely are they used for the larger carinated forms, such as plates (for an exception, see Corning Coll. 1979, 149, no. 315, pl. 39). At times certain colors are repeatedly associated with specific shapes. Cylindrical pyxides such as **No. 134**, for example, are almost always emerald green.[24] Finally, based on the surviving specimens, vessels in opaque fabrics are more likely to have highly unusual shapes (miniature tables, for example, or matched footed vases from Herculaneum and elsewhere, as listed below). Opaque wares also enjoy more refined and elaborately cut and finished forms, often with carefully executed grooves and ridges (**No. 148**) or detailed basal moldings (**No. 156**). Whether this signifies that the two colored groups were produced (or finished) in different workshops, or in different places or periods, cannot now be easily determined.

Unusual shapes at Cosa include **No. 137**, a peacock blue bowl of unique profile, and **No. 126**, a dish that so closely resembles the colorless dishes in the next section (**Nos. 169–174**) that it may well belong to that type despite its pale blue fabric. Complicating this identification, however, is another dish of identical profile but in emerald green glass. It was discovered in Britain at Kirkby Thore (Charlesworth 1959, 38 and 40, fig. 3.3). Also unusual in form is **No. 139**, a bowl that appears to be a late adaptation of the more standard carinated shape. This piece is closely dated to the decade before ca. A.D. 40/45 but lacks an exact parallel from any other site.

Pieces at Cosa that fall outside the main or standard series of colored fine wares and antedate their development are **No. 135**, a blue lidded pyxis; **No. 145**, an exceptionally large and handsome opaque white plate with a trailed base-ring; and **No. 147**, a jade green rectangular dish with a cut "ear-handle." These three vessels, along with several others, come from Deposit IV and belong to the decades before ca. 25/15 B.C. Whereas the rectangular dish appears to be a direct precursor of comparable dishes that come later (here, **Nos. 127–130**, **146**), the white plate is an extraordinary find. Its closest parallels, if indeed they are similar, are two fragmentary opaque white plates and an opaque (?) blue dish (D 16 cm) now in the Museo Nazionale Romano at Rome (see list below). Probably similar to the pyxis (**No. 135**) is an intact aquamarine lidded pyxis said to have been found near Rome and now in the British Museum (Harden 1968, 44, no. 49).

The fragment of a corroded purple stemmed bowl (**No. 144**) is included in this section only because it seems to have been cast rather than blown. **No. 159** is so fragmentary that it is impossible to determine whether it belongs to an opaque red vessel or to a larger section of inlay.

The immediate inspiration for many of the standard shapes belongs to the Roman ceramic industry, and Berger actually calls the type *Gläser mit keramikähnlichen Profilen* (Vindonissa 1960, 24–30). This relationship is especially evident for the small carinated bowls that possess much the same profile as Haltern Form 12 in Arretine pottery or Dragendorff 27 in Gaulish sigillata. The same holds true for other forms as well (see discussion in Magdalensberg 1979, chap. 12, 65–91). These shapes may also have been influenced by contemporary metalwork, both silver and bronze (see Mutz 1972), if only because such angular profiles are far more typical of metalwork than of ceramics or glass. Contemporary metalworkers also utilized similar methods of polishing and finishing, namely procedures employing lathes (again, Mutz 1972, esp. 174–176).

[24] For other emerald green pyxides, see Trier 1977, 35, no. 90, pl. 33; Spinazzola 1928, 222, with examples from Pompeii and environs; and Magdalensberg 1979, 75, no. 579, pls. 4 and 14, with other parallels cited. Vindonissa 1960, 27, nos. 35–36, pls. 3 and 17, yielded blue examples but lists emerald green pyxides as well. In form the glass pyxides resemble Dragendorff 22 in Gaulish sigillata.

In respect to manufacturing technique, these colored fine wares must have been cast in molds, presumably of two or more pieces.[25] After annealing, they would have been affixed to a lathe and cut, then polished on both surfaces to achieve their final forms. Lathe work is readily discernible on all examples, not only by the presence of cut grooves, ridges, and depressions but also by fine polishing marks on all surfaces. The pattern of concentric depressions and ridges on the bottom of **No. 134**, or the raised circular ridge and central dot of the white plate (**No. 145**), are indicative of lathe work and substantiate Pliny's remark that much glass in the first century was lathe-turned (*aliud torno teritur*; *HN* 36.193). The rectangular dishes are also carefully polished on both surfaces but not in a circular manner. Instead, on preserved pieces the polishing marks are parallel to the sides.

Parallels and Dated Examples of Fine Wares: Colored, Translucent, and Opaque

Cosa: **Nos. 135**, **145**, and **147**, Deposit IV, dated before ca. 25/15 B.C.; **Nos. 136–137**, Deposit V, and **No. 154**, Deposit VI, dating between ca. 25/15 B.C. and ca. A.D. 40/45; **Nos. 125**, **134**, and **139**, Deposit VII, dating to the decade prior to ca. A.D. 40/45; and **Nos. 127**, **144**, and **146**, Deposit VIII, dating to the decades prior to ca. A.D. 50/55.

Luni I 1973, 466–467; CM 187/1, pl. 81.5 (blue patella); CM 536, pl. 81.3 (green patella); CM 826, pl. 81.4 (green vessel); CM 2411/1, pl. 81.6 (green vessel); C 200 (opaque green [?] vessel); CM 1956, pl. 81.9 (green dish); 774, KA 359, pl. 215.14 (blue carinated plate).

Luni II 1977, 276, CM 6928, pl. 154.4 (blue-green cup); CM 8635, pl. 154.5 (opaque white pyxis or dish), dated ca. A.D. 40–50; 276, CM 7292/3, pl. 154.7 (blue rectangular dish).

Ostia II 1970, 77, pl. 15, fig. 235 (green carinated bowl); 80, pl. 16, fig. 255 (opaque blue bowl); 80, pl. 16, fig. 256 (opaque blue bowl).

Ostia III 1973, 41, pl. 14, fig. 11. Opaque red carinated dish.

Vatican Coll. 1975, 44, no. 272, pl. 13 (opaque blue bowl from SS. Marco and Marcellino, Rome); 82, no. 796, pl. 41 (aquamarine saucer from a Roman catacomb); 82, no. 797, pl. 41 (opaque blue cup); 82, no. 799, pls. 41 and 91 (large carinated plate, *meergrün*, probably belonging to this type).

Rome: Museo Nazionale Romano, inv. no. 1098. Emerald green dish with carinated side; inv. nos. 8329 and 60111, two fragmentary opaque white plates not unlike **No. 145** at Cosa, with lathe-cut circular ridges and raised dots at the center of the bottom; also a blue dish with an upright rim similar to **No. 145** but smaller, illustrated in Mutz 1972, 175, pls. 516–517. From Rome or its environs. Except as noted, all are unpublished.

Rome: Antiquarium Comunale. Several fragments of plates, dishes, bowls, rectangular dishes, and patellae. Unpublished.

Pompeii: Museo Nazionale, Naples. Numerous examples of translucent fine wares in all shapes. For illustrations of several pieces, see Spinazzola 1928, 222, bottom shelf. Antedate A.D. 79. Unpublished.

Pompeii: Antiquarium. Several examples on exhibit. Antedate A.D. 79. Unpublished.

Herculaneum: Antiquarium. Inv. no. 1972, small emerald green bowl with hemispherical body like that of **No. 141** (D 6.7 cm); inv. no. 2866, large emerald green carinated plate (D 29 cm); inv. no. 2086, one of two opaque light blue footed vases of unique form (D 14 cm); inv. no. 2044, miniature opaque turquoise blue rectangular "table" with four legs in the shape of paws (H 4.6 cm; L 21 cm; W 14 cm). Antedate A.D. 79. Unpublished.

[25] Harden in Camulodunum 1947, 300, briefly discusses the molds, but since none have been discovered to date, their composition and use cannot be determined.

Maioli 1974, 30, fig. 26. Opaque red carinated cup from Ravenna.

Russi (Ravenna) 1977, 134, no. 1391, fig. 68.6, 84. Green carinated cup.

Aquileia 1968, 74, no. 180, color pl. 2.6. Blue pyxis with concave curving sides. Also numerous unpublished fragments in the museum.

Mariacher 1988, color pl. 16. Lidded and footed purple vase found at Cavour di Cavoretto, near Turin, in a sarcophagus with a small, cast purple beaker and a coin of Geta (cf. Bologna Cat. 1964, 363, pl. 138).

Magdalensberg 1979, chap. 12, 65–91. Numerous examples of translucent and opaque vessels belonging to this type but confusingly classified according to Isings Forms 2, 5, 20, 22, 23, and so forth. Cast mosaic and free-blown vessels in comparable shapes are interspersed among cast translucent and opaque pieces. Numerous parallels cited. At Magdalensberg all examples presumably date before ca. A.D. 45, the time of the abandonment of the site, but certain fragments are from specific contexts attributed from the Augustan to Claudian era, but with most fragments in Tiberian or Claudian levels. Among the earliest vessels are the following: 70–71, no. 544, pls. 3 and 13 (blue-green carinated patella from an Augustan level to ca. A.D. 1); 70–71, no. 537 (opaque red vessel in a late Augustan context to ca. A.D. 15).

Vindonissa 1960, 24–30, nos. 30–48, pls. 3 and 17. Numerous examples of translucent and opaque fine wares, with lengthy discussions and numerous parallels listed. Dated examples are concentrated in Tiberian/early Claudian contexts but continue into Neronian times, with a few examples preserved from Flavian contexts.

Haltern 1901, 173. Opaque light blue fragment of uncertain shape, dated between ca. 11 B.C. and A.D. 16.

Haltern 1909, 371, no. 3, fig. 16. Opaque red cup dated between ca. 11 B.C. and A.D. 16.

Bibracte 1899, pl. 56, nos. 23 and 25. Two opaque white fragments, dated before the abandonment of the site ca. 5 B.C.

Cologne: Römisch-Germanisches Museum. Illustrated in La Baume n.d., K7, pl. 48 (emerald green plate); K5, pl. 48 (emerald green hemispherical patella); M2, pl. 51 (emerald green carinated patella); M3, pl. 51 (emerald green pyxis with concave curving sides). All from Cologne or environs and ascribed to the first century A.D.

Trier 1977, 23, Form 7, nos. 44 and 46, pl. 30 (emerald green and opaque *türkisgrün* carinated dishes); 34–35, Form 20, no. 90, pl. 33 (emerald green pyxis).

Mutz 1972, 174–175. Emerald green carinated dish from Rheinishes Landesmuseum, Bonn; another with a concave curving side from Musée Curtius, Liège.

Camulodunum 1947, 298–301, nos. 38–47 and 53–60, pls. 87–88. Fragments date between A.D. 10 and 43 or A.D. 43 and 65.

Conimbriga 1976, 160 and 164, nos. 14–17, pl. 34. Two opaque blue(?) and two emerald green dishes or pyxides, of which no. 14 is from a Flavian level. Additional examples cited.

Tarragona: Pasaje Cobos. At least thirty-six examples of circular dishes with carinated or almost vertical sides, rectangular dishes, pyxides, carinated bowls and patellae, small hemispherical patellae, both opaque and translucent. Deposit considered to belong to the second quarter of the first century A.D. Unpublished. Information furnished by Dr. Jennifer Price, University of Durham.

Carthage: University of Michigan excavations. Fragment of opaque red bowl or beaker, probably belonging to this type. Unpublished.

Corinth 1952, 98–99, nos. 620–621, fig. 7, pls. 54–55. Small emerald green carinated dish. First century A.D.

Sardis 1980, 9–10, no. 25, pls. 1 and 20. Colorless(!) carinated cup.

Salona 1975, 150–151, nos. 10–15, pls. 28 and 32.
Haevernick 1981, 420–424. Describes translucent and opaque boat-shaped vessels, perhaps the most unusual form that pertains (in all probability) to the fine wares discussed here.
Gréau Coll. 1903, I, pl. 29, 1–9 (opaque red bowls, some incorrectly restored); I, pls. 34, 1–9; 38, 1–9; 40, 2–8 (opaque cast and blown vessels); III, pl. 193, 5 (rectangular dish).
Corning Coll. 1979, 144–153, color pls. 18–19. Numerous vessels and fragments, most of which belong to the standard Roman series of translucent and opaque fine wares.

At Cosa, dated examples of translucent and opaque fine wares come from no fewer than five different deposits. The earliest is Deposit IV (before ca. 25/15 B.C.), associated with the reoccupation of the town in early Augustan times. Presumably the vessels trapped in this deposit were brought to Cosa by new settlers in the last half of the century and were not articles in use when the Republican house was destroyed ca. 72/70 B.C. Vessels from the deposit include a rectangular dish (**No. 147**) that, as mentioned above, closely resembles later examples belonging to the standard series of fine wares. The same cannot be said for the opaque white plate (**No. 145**) or the blue lidded pyxis (**No. 135**), however. Neither resembles the later standard shapes, and both in fact are highly unusual in form. As argued above, these vessels antedate the development of the standard style, probably by several decades. Either they represent some of the earliest products of the newly established Romano-Italian casting industry, or they were among imports still arriving in Italy from the eastern Mediterranean. Whichever is the case, they nonetheless express the shift in taste to intentionally colored monochrome glass, a change that occurred in the last decades of the first century B.C.

Deposits V and VI (ca. 25/15 B.C. to ca. A.D. 40/45) are the earliest strata at Cosa yielding fine wares of unusual kinds. Although these levels contain material as late as the reign of Claudius, most datable objects are somewhat earlier and can be assigned to the late Augustan or Tiberian eras. Thus the peacock blue pyxis lid (**No. 136**), the peacock blue bowl (**No. 137**), and the opaque red carinated bowl (**No. 154**) probably belong to the second or third decades of the first century A.D. This in turn suggests that the fine wares were first manufactured in quantity at about that time.

More precisely dated are the aquamarine dish (**No. 125**), emerald green pyxis (**No. 134**), and blue carinated cup (**No. 139**), all of which come from a shop storeroom destroyed ca. A.D. 40/45. If these pieces were merchandise awaiting sale, they undoubtedly were of recent manufacture and must date to the years just before A.D. 40/45. Roughly contemporary are the more fragmentary vessels of Deposit VIII (prior to ca. A.D. 50/55, see chapter 2), which contained emerald green and opaque light blue rectangular dishes (**Nos. 127** and **146**).

Considered together, the dated pieces discussed above suggest that these fine wares first became available at Cosa either toward the end of Augustus's reign or early in the principate of Tiberius. Their popularity continued into Claudian or even Neronian times. This chronology, for the most part, is also borne out by Berger's analysis of the fine wares found so profusely at Vindonissa (Vindonissa 1960, 24–30), as well as by the discussions of Harden and Price (Fishbourne 1971, 327–328) and Czurda-Ruth (Magdalensberg 1979, 65–91). On all three sites the greatest concentrations of fragments occur in Tiberian or Claudian contexts, with a small amount of material from late Augustan levels and with a few pieces surviving into Neronian or Flavian times. Given the presence of several intact, closely dated vessels in Deposit VII at Cosa, it seems that the type (or a subtype closely related to it) continued in production until midcentury or later. The large number of examples known from Pompeii and Herculaneum, now exhibited in site museums and in the Museo Nazionale in Naples, further suggests that these fine wares were made at least until the middle of the century. On the other hand, there is no evidence to suggest that they were still being manufactured as late as the last quarter of the first century. Two very late pieces—a highly unusual purple beaker and footed bowl with lid—that were found at Cavour di Cavoretto with a coin of Geta (A.D. 211–212) both in shape and color stand outside the standard series of translucent fine wares and consequently do not affect the dating suggested here. Either they were heirlooms from the first century A.D. (and from an unidentified source), or they were products of an otherwise unknown industry specializing in fine cast colored glass and operating in the late second or third century.

As for the precise date when the standard series was introduced, information at present is ambiguous. Although the earliest varieties of opaque and translucent fine wares began to accumulate at Cosa and elsewhere in the last decades of the first century B.C. (see "Deposit IV," chapter 2), the shapes of these vessels differ sufficiently from those of the main group, as noted earlier, to suggest a different tradition. Also, in color the earliest fine wares are predominantly opaque and not translucent, such as **Nos. 145** and **147** at Cosa, and this may indicate that vessels in opaque colors were introduced in large numbers before those in translucent fabrics. Indeed, these opaque vessels are possibly connected in some manner with certain opaque red and opaque light blue vessels fabricated by Hellenistic glassmakers during the second and first centuries B.C.[26]

Further complicating this picture is an extensive series of composite mosaic-glass plates, dishes, and bowls with either carinated or convex curving sides, which in form are identical (or nearly so) to the translucent and opaque fine wares discussed here (for illustrated examples, see Corning Coll. 1979, 184–188, and Oppenländer Coll. 1974, 118). Such mosaic wares are presumably contemporary with their monochromatic counterparts, yet in distribution they are as frequently found on sites in the Aegean and eastern Mediterranean as they are in Italy. Hence, their relationship with the translucent and opaque vessels discussed here is unclear. No examples of such mosaic vessels occurred at Cosa.

Colored Translucent Vessels

125. CE 2115 — Pl. 8
A.P. Room 22, II — Deposit VII
Decade prior to ca. A.D. 40–45

Two joining and four non-joining fragments of rim, side, and base of shallow dish (profile mostly restorable); deep blue-green (aquamarine). Iridescence and pitting. Fine craftsmanship. Nearly horizontal rim with rounded edge; short outsplayed side, tapering diagonally downward; flat floor with low rounded base-ring. Cast, rotary-polished on both surfaces; base-ring cast as part of the vessel. H 1.8 cm; D ca. 16 cm; A Th .22 cm.
Bibliography: Grose 1975, 36, no. 2.

126. CE 1973 — Pl. 8
A.P. Room 25D, II

Fragment of rim, side, and bottom of shallow dish (profile mostly restorable); light blue. Iridescence, pitting, and discoloration. Broad, nearly horizontal rim with rounded, thickened edge; short curving side; nearly flat bottom. No evidence of base-ring, but one probably existed. Shape perhaps

[26] Cast opaque red beakers, bowls, and plates in small numbers are known from eastern Mediterranean sites in contexts of the second and first centuries B.C. and early first century A.D. The shapes and decoration differ entirely from the vessels described here, but they demonstrate the use of opaque red fabrics by Hellenistic glassmakers. For further discussion, see Stern 1981. Opaque light blue dishes and bowls are also known from eastern sites, including Tel Anafa in Upper Galilee (unpublished) and Delos (unpublished). In some instances such vessels are identical in shape to certain Syro-Palestinian translucent fluted and grooved bowls that date to the second and first centuries B.C. (see discussion of **Nos. 92–93** at Cosa; also Grose 1979).

similar to colorless plate, **No. 169**. On upper side of rim, a broad cut depression .7 cm wide. Cast, rotary-polished and cut on both surfaces. H of side .8 cm; D ca. 17–18 cm; A Th .2 cm.

127. CD 1102 — Pl. 8
Bas. ENE 4, III — Deposit VIII
Prior to ca. A.D. 50–55

Fragment of rim, side, and bottom of rectangular dish; emerald green. Iridescence and pitting. Upright rim with flattened edge sloping obliquely inward; straight side, tapering downward to a rounded basal angle; flat bottom. Cast, polished on interior (polishing marks run parallel to side); exterior fire-polished. H 1.4 cm; L and W uncertain; A Th .13 cm.

128. CE 2071 — Pl. 8
A.P. Room 25, I

Small fragment of rim, side, and bottom of rectangular dish; deep blue-green (aquamarine). Weathering, shape, and technique of manufacturing identical to **No. 127**, except the edge of the rim is rounded. H 1.3 cm; L and W uncertain; A Th .3 cm.

129. CD 1114 — Pl. 8
Bas. NW Exterior 3, I

Small fragment of rim, side, and bottom of rectangular dish; deep blue-green (aquamarine). Weathering, shape, and technique of manufacturing identical to **No. 127**. H 1.3 cm; L and W uncertain; A Th .25 cm.

130. CD 1145
Bas. ENW 3, I

Tiny fragment of rim, side, and bottom of rectangular dish; deep blue-green (aquamarine). Weathering, shape, and technique of manufacturing identical to **No. 127**. H 1.3 cm; L and W uncertain; A Th .25 cm.

131. CD 1115
Bas. Nave, 0

Tiny fragment of rim and side of dish or bowl; deep blue-green (aquamarine). Iridescence and pitting. Short outsplayed rim with rounded edge; carinated side with convex curve (probably one of two). Cast, rotary-polished on both surfaces. D ca. 10–12 cm; A Th .3 cm.

132. CF 2438
A.P. Room 26, I–III

Tiny fragment of rim and side of dish or bowl; deep blue-green (aquamarine). Iridescence and pitting. Short outsplayed rim with rounded edge; carinated side with convex curve (probably one of two). Technique of manufacturing as in **No. 131**. D ca. 12 cm; A Th .3 cm.

133. C69.407
S.H., 0

Tiny fragment of rim and side of dish or bowl; dark blue. Iridescence and pitting. Short outsplayed rim with rounded edge; carinated side with convex curve (probably one of two). Technique of manufacturing as in **No. 131**. D ca. 12 cm; A Th .3 cm.

134. CE 1964 Pl. 8
A.P. Room 22, II Deposit VII
Decade prior to ca. A.D. 40–45

One non-joining and eight joining fragments of rim, side, and bottom of pyxis (profile restorable); emerald green. Iridescence and pitting. Nearly horizontal rim with rounded edge sloping obliquely inward; vertical side bulges slightly at center; flat bottom projects beyond the junction of the side, forming a rounded molding. On underside of bottom, a molded and cut pattern of concentric depressions and ridges: a wide shallow depression near the outer edge of the bottom and three shallow concentric grooves near center of bottom. Cast, rotary-polished and cut on both surfaces. H 4.7 cm; D 11 cm; A Th side .3 cm; A Th bottom .2 cm.
Bibliography: Grose 1975, 36, no. 4; Grose 1983, 44.

135. C68.679 and C68.683 Pl. 8
S.H. Room 4, I Deposit IV
Prior to ca. 25/15 B.C.

One fragment of lid (A) and two non-joining fragments of rim and side (B) of lidded pyxis; dark blue. Iridescence, pitting, and corrosion. Fine craftsmanship. Slightly inturned rim with flattened edge; vertical side bulges slightly at center. Shallow convex lid (underside concave) with a pointed projecting molding at edge, designed to fit over rim of pyxis. On exterior of body, a horizontal cut ridge encircles the center of the pyxis. Technique of manufacturing as in **No. 134**. D ca. 13 cm; A Th side .2 cm; D of lid ca. 12.6 cm; A Th lid .2 cm.

136. CE 2140 Pl. 8
A.P. Room 6, I Deposit V
ca. 25/15 B.C. to ca. A.D. 40/45

Fragment of lid of small pyxis (alternatively, the foot of a stemmed cup); peacock blue. Iridescence and pitting. Shallow convex lid (underside concave) with flat edge. On upper side, two horizontal grooves adjacent to edge; on underside, one narrow groove .2 cm from edge of lid. Technique of manufacturing as in **No. 134**. D ca. 7 cm; Max Th .5 cm.

137. CE 2001 Pl. 8
A.P. Room 16, I Deposit V
ca. 25/15 B.C. to ca. A.D. 40/45

Six joining and seven non-joining fragments of rim and side of deep bowl; peacock blue. Iridescence and pitting. Fine craftsmanship. Outsplayed rim with rounded edge; nearly straight side, tapering diagonally downward. On upper edge of rim, a narrow horizontal groove. On exterior, a rounded horizontal ridge at the junction of the rim and side. Technique of manufacturing as in **No. 134**. D ca. 12 cm; Max Th side .25 cm.

138. CF 2411 Pl. 8
A.P. Room 25C Exterior, I

Fragment of rim and side of bowl; emerald green. Iridescence and pitting. Short outsplayed rim with rounded edge; carinated side with one convex curve and vestiges of a second. Cast, rotary-polished on both surfaces. D ca. 12 cm; A Th .25 cm.

139. CE 1963 Pl. 9
A.P. Room 22, II Deposit VII
Decade prior to ca. A.D. 40/45.

Six joining and three non-joining fragments of rim, side, and base of small bowl (profile restorable); dark blue. Iridescence and pitting. Fine craftsmanship. Upright rim with short rounded lip; carinated side with two convex curves; upper side of base concave, underside flat; slightly outsplayed base-ring. On exterior, three thin horizontal grooves encircle body. Cast, rotary-polished and cut on both surfaces; base-ring cast as part of vessel. H 4.8 cm; D ca. 9 cm; A Th .2 cm.
Bibliography: Grose 1975, 36–37, no. 5; Grose 1983, 44.

140. CF 2303 Pl. 9
A.P. Rooms 25A and 6, I–IV
Three joining fragments of rim, side, and base of small bowl (profile mostly restorable); dark blue-green (aquamarine). Iridescence and pitting. Short, horizontal rim with flattened edge; carinated side with two convex curves; indications of a base-ring. Cast, rotary-polished on both surfaces; base-ring cast as part of vessel. Restored H 3.4 cm; D ca. 7 cm; A Th .3 cm.

141. C70.585 Pl. 9
S.H. Street M, 0-I
Lower half of body and base of small bowl; emerald green. Iridescence and pitting. Carinated side with lower convex curve preserved; upper side of floor concave, underside flat; slightly outsplayed base-ring. Technique of manufacturing as in **No. 140**. D base-ring 3 cm; A Th side .35 cm.

142. C70.583 Pl. 9
S.H. Street M, 0-I
Base and base-ring of small bowl; dark blue. Iridescence and pitting. Upper side of base concave, underside flat; slightly outsplayed base-ring. Technique of manufacturing as in **No. 140**. D base-ring 4 cm; A Th base .3 cm.

143. CD 1116
Bas. Nave, 0
Tiny fragment of base and base-ring (edge missing) of small bowl; dark blue. Iridescence and pitting. Flat floor; slightly outsplayed base-ring. Probably from small carinated or hemispherical bowl. Technique of manufacturing as in **No. 140**. D base-ring ca. 3 cm; A Th base .3 cm.

144. CD 1101 Pl. 9
Bas. ENE 1, III Deposit VIII
Prior to ca. A.D. 50/55
Fragment of side and indication of stem of large footed bowl; purple. Iridescence, enamel weathering, and corrosion. Convex curving side; vestiges of a solid beaded stem. Complete shape uncertain. Possibly cast; finishing and surface wholly obscured by corrosion. Max Th side 1.2 cm; Min Th .5 cm.

Colored Opaque Vessels

145. C68.638 Pl. 9
S.H. Room 4, I Deposit IV
Prior to ca. 25/15 B.C.
Sixty-one joining and non-joining fragments of rim, side, and base of large plate (profile restorable); opaque white. Vessel now partially restored in araldite. Milky weathering, dulling, and pitting. Vertical rim with rounded edge sloping obliquely inward; vertical side with rounded basal angle;

flat floor; low base-ring, square in cross section. On upper side at center of floor, a circular ridge, D 2.5 cm, around a raised central dot. Cast, rotary-polished and cut on both surfaces; base-ring applied as a coil, then cut to achieve a squarish cross section. H 3 cm; D 28.4 cm; D base-ring 20.2 cm. A Th .2–.3 cm.
Bibliography: Grose 1973, 178, pl. 189, no. 6; Grose 1983, 44.

146. CD 1108 — Pl. 9
Bas. ENE 3, III — Deposit VIII
Prior to ca. A.D. 50/55
Two joining fragments of rim, side, and bottom of rectangular dish; opaque light blue (very slightly translucent). Iridescence, dulling, and pitting. Upright rim with rounded edge; straight side, tapering diagonally downward to a rounded basal angle; flat bottom. Cast, polished on interior (polishing marks run parallel to side); exterior fire-polished. H 1.8 cm; L and W uncertain; A Th .3 cm.

147. C68.375 — Pl. 9
S.H. Room 4, 0-B — Deposit IV
Prior to ca. 25/15 B.C.
Fragment of rim, side, and bottom and portion of handle of rectangular dish; opaque jade green. Iridescence and pitting. Nearly horizontal rim with flat edge; straight side, tapering diagonally downward to a rounded basal angle; flat bottom; small portion of molded and cut "ear-handle" projects horizontally from rim. Cast, polished on both surfaces (polishing marks run parallel to side); underside of bottom fire-polished; handle cast as part of vessel. H 1.3 cm; L and W uncertain; A Th .27 cm.

148. CD 1076 — Pl. 9
Bas. ENE 1, I
Fragment of rim and side of carinated bowl; opaque olive green. Dulling and pitting. Fine craftsmanship. Outsplayed rim with rounded edge sloping obliquely inward; carinated side with two convex curves. On exterior, two pointed horizontal ridges; one above, the other below the junction of the two convex curves. Cast, rotary-polished and cut on both surfaces. D ca. 14 cm; Max Th .3 cm.

149. C70.228 — Pl. 9
S.H. Street M, 0
Fragment of rim and side of carinated bowl; opaque white. Milky weathering and pitting. Outsplayed rim with rounded edge sloping obliquely inward; carinated side with two convex curves. Cast, rotary-polished on both surfaces. D ca. 10 cm; Max Th .3 cm.

150. C73.73
Locus unknown
Tiny fragment of rim and side of carinated bowl; opaque red. Green weathering and pitting. Nearly horizontal rim with rounded edge sloping obliquely inward; carinated side with two convex curves. Technique of manufacturing as in **No. 149**. D uncertain; Max Th .3 cm.

151. C67.553
Forum SW, 0

Tiny fragment of carinated dish or bowl; opaque olive green. Dulling and pitting. Carinated side with two convex curves. Technique of manufacturing as in **No. 148**. D uncertain; A Th .2 cm.

152. CD 1077 Pl. 9
Bas. ENE 3, I
Small fragment of rim and side of small bowl; opaque light blue. Milky weathering and pitting. Slightly inturned rim with flattened edge; convex curving side. On exterior, a horizontal groove below edge of rim. Technique of manufacturing as in **No. 148**. D ca. 8.5 cm; A Th .2 cm.

153. CD 1107 Pl. 9
Bas. ENE 1, I, II
Three joining fragments of rim and side of small bowl; opaque red. Green weathering and pitting. Upright rim with rounded edge; convex curving side. Technique of manufacturing as in **No. 149**. D ca. 7 cm; A Th .2 cm.

154. CE 1216 Pl. 9
A.P. Room 11, Cesspool Deposit VI
ca. 25/15 B.C. to ca. A.D. 40/45
Two joining fragments of side, base, and base-ring of bowl; opaque red. Green weathering and pitting. Convex curving side; upper side of base concave, underside flat; slightly outsplayed base-ring with flat edge. Cast, rotary-polished on both surfaces; base-ring cast as a part of vessel. D base-ring 4.8 cm; A Th side .25 cm.

155. CF 2439 Pl. 9
A.P. Room 25 Exterior NW, I
Small fragment of base and base-ring of bowl; opaque olive green. Dulling and pitting. Upper side of base concave, underside flat; slightly outsplayed base-ring with flat edge. Technique of manufacturing as in **No. 154**. D base-ring ca. 6 cm; A Th base .4 cm.

156. CE 1996 Pl. 9
A.P. Room 12, Cesspool
Small fragment of side, base, and base-ring of bowl or other vessel; opaque jade green. Dulling and pitting. Fine craftsmanship. Convex curving side; upper side of base concave, underside flat; slightly outsplayed base-ring with flat edge. On exterior, above the junction of side and base-ring, a molded and cut horizontal ridge. Technique of manufacturing as in **No. 154**. D base-ring ca. 6 cm; A Th side .3 cm.

157. CD 251 Pl. 9
Bas. Nave 4, I
Dulling and pitting. Convex curving side; upper side of base concave, underside convex; slightly outsplayed base-ring with rounded edge. Cast, rotary-polished on both surfaces base-ring applied to bottom. D base-ring 3.7 cm; A Th base .4 cm.

158. C71.202
VI-D, Sounding 2, 0

Tiny fragment of base and base-ring (edge missing) of bowl or other vessel; opaque red. Green weathering and pitting. Bottom of vessel flat; slightly outsplayed base-ring. Technique of manufacturing probably identical to **No. 154**. D base-ring uncertain; A Th base .2 cm.

159. There is no description or illustration for **No. 159**.

160. Additional fragments of opaque colored fine wares.

- **a.** CG 449, Building L Exterior NW, 0. Tiny fragment of side of bowl; opaque light blue.
- **b.** CF 2439A, A.P. Room 26, I–III. Tiny fragment of vessel; opaque jade green.
- **c.** CF 2407, Temple B (forum), Exterior NE, I. Tiny fragment of vessel; opaque jade green.

*Fine Wares: Colorless with Broad, Overhanging Rims (**Nos. 161–201**)*

When this type was first described in 1936, few examples outside Egypt were known, and the type as a whole was considered to be of Alexandrian manufacture (Karanis 1936, 50). Subsequently, excavations throughout the Roman world, from Hadrian's wall in Britain to Dura-Europos in the Syrian desert, have shown that these cast colorless plates, dishes, and bowls were part of an "international" style of fine table wares that were immensely popular throughout the empire (and beyond) during the late first century A.D. and the first half of the succeeding century. Of all the glassware manufactured by the Roman industry, this type is perhaps the most elegant. The crisp, metallic profiles and skillfully cut and polished surfaces combine with a clear colorless glass to create a stunning effect. When facet-cut, the effect is greatly enhanced (here, **No. 193**). Nowhere is this more apparent than at the Cave of the Letters on the western shore of the Dead Sea in Israel, where three examples in pristine condition were discovered (Yadin 1963, 105–110, nos. 10–12, pls. 39–40). Also, of all the glass produced by the Romans, this type is perhaps the last to have been manufactured in quantity by casting rather than a blowing technique.

Aside from examples in Egypt and Israel, others (often in large numbers) are known from sites throughout the Roman Empire. Among them are Fishbourne, Caerwent, and Richborough in England; Cramond in Scotland; Maastricht and Nieuwenhagen in Holland; Tongres in Belgium; Montmaurin in France; Trier, Straubing-Sorviodurum, and Frixheim-Anstel in Germany; Iruña and Mulva in Spain; Boca do Rio, Conimbriga, Aljustrel, and Constancia in Portugal; Cosa, Luni, Rome, and Ostia in Italy; Carthage and Tipasa in North Africa; Dura-Europos in Syria; Novae in Bulgaria; Salona and Zadar in Croatia; and Aksum in the Sudan (for references, see list below).

In inspiration, these vessels fall into the same tradition as the cast and lathe-polished colored translucent and opaque fine wares discussed in the previous section. Similar manufacturing and finishing techniques seem to have been employed, and the same attention to quality is evident in their manufacture. Also, like their colored counterparts, the shapes (although decidedly different) are inspired by contemporary ceramics or metalwork, especially bronze and silver. At least this seems to be the case based on preserved examples of metal vessels, such as those illustrated in Mutz 1972, 87–89. Given the overwhelming predominance at this time of the rounded forms of blown table wares, the abrupt angularity of these cast vessels must have accorded them special distinction.

At Cosa, as elsewhere, three shapes are dominant: a large shallow bowl (**Nos. 161–168**), a shallow medium-sized plate (**Nos. 169–174**), and a small hemispherical bowl (**Nos. 188–192**). The large bowls vary from 22 to 30 cm in diameter and have high base-rings and pronounced rims with overhangs. On most pieces the upper surface of the rim is decorated with a wide rotary-polished depression flanked by rounded ridges, and with a narrower cut groove or depression on the underside of the rim at the junction with the side. Not only do **Nos. 169–164** at Cosa represent fragmentary examples of these bowls, but so do **Nos. 175–186**, portions of base-rings from comparable vessels.

The plates vary in diameter from 17 to 18 cm. They too rest on high base-rings and have broad rims, but these lack the characteristic overhang. On three examples (**Nos. 170–172**) cut grooves occur on the upper side of the rims. The best-preserved examples of these plates are those found in the above-mentioned Cave of the Letters, dating to the years before A.D. 132–134.

No. 188 is an outstanding example of the small bowls that also are typical of these colorless fine wares. It, too, exhibits a high base-ring and broad rim with an overhanging edge. Polished depressions occur on the upper surface of the rim and on the underside, at the junction of the rim and body. (It should be noted, however, that some examples like **Nos. 194–197** lack both the overhang and the depressions.) **Nos. 198–201** are fragmentary base-rings belonging to comparable bowls. **No. 193** is also a bowl of this sort, but like specimens from other sites it is facet-cut on both surfaces of the rim as well as on the exterior of the body. Even the overhang is decorated with vertical cuts of facets. On other faceted pieces, the overhang is cut with an "egg and dart" motif. For examples of such facet-cut bowls, both large and small, see Conimbriga 1976, 176–177, no. 148, pl. 40, found in a Trajanic level; Tipasa 1967, 94, no. 198, pl. 9.3; also the large shallow bowl from the Cave of the Letters and another from Luni (Luni II 1977, CM 12190, pl. 155.15).

Yet another shape characteristic of this type is that of an oval dish, often with elaborate "ear-handles" and facet cutting on the rim and body. **No. 175** may have been part of such a vessel although no cutting is preserved. The same holds true of **No. 247**, a flat facet-cut section of floor, which is reminiscent of the superior cutting on a previously unpublished dish of this class. Another well-preserved example of these oval dishes is now housed in the Museo Vetrario in Murano but originally was found at Zadar in present-day Croatia (Mariacher 1966, 6, fig. 9).[27]

Regardless of shape, the walls of these vessels usually taper in thickness from the rim to the base. Floors are either flat (**Nos. 169** and **188**) or slightly convex (**Nos. 176–181**); occasionally they are decorated with grooves (**No. 181**). Alternatively, they display a small circular groove (**No. 200**) and/or raised dot at the center of the bottom, the result of the lathe attachment.

Aside from their distinctive shapes and careful manufacturing methods, the most obvious trait exhibited by these fine wares is the colorless glass employed to make them. The vogue for decolorized fabrics swept the marketplace in the mid-first century A.D., probably beginning as early as the forties. Pliny the Elder, writing in the sixties, actually states that "the most highly prized glass is colorless and transparent, as closely resembling rock crystal as possible" (*HN* 36.192). Coincident with the introduction of colorless wares was the rapid abandonment of intentionally colored vessels, both polychrome and monochrome, which had proven so popular during the first century B.C. and early decades of the first century A.D. Indeed, it can be argued that after ca. A.D. 50 the finest table wares invariably were colorless, whereas more serviceable or cheaper vessels employed naturally colored fabrics.

Parallels and Dated Examples of colorless fine wares

Cosa: **No. 161**, Deposit V, ca. 25/15 B.C. to ca. A.D. 40/45. **Nos. 165**, **167**, **169**, **171**, **173**, **191**, **198**, Deposit IX, late first to early third century A.D.

[27] For additional illustrations of colorless oval bowls with ear-handles and/or facet cutting, see Fremersdorf 1967, 63, pl. 22, with another piece from Leiden shown on pl. 23; also Dusenbery 1971, 11–12, no. 7, fig. 4.

Luni I 1973, 774, KA 360, fig. 215.13.

Luni II 1977, Group 32, 280–281, CM 4379, fig. 155.11 and CM 12190, fig. 155.15 (large bowls); CM 3286/1 and CM 11466/1, fig. 155.8 (small bowls); also four other fragments of rims and two fragments of base-rings (p. 285). Found in strata of the second century A.D. Also see p. 397, no. CS 1361, fig. 198.2, and 552–533, nos. K 2658 and K 2949/4, fig. 285.10, for additional examples.

Villa Potenza 1971, 410, no. 48, fig. 11.1. Rim of large bowl.

Mariacher 1966, 6; oval facet-cut bowl; also p. 15, fig. 10 for a circular cut dish.

Adria: circular facet-cut bowl, cited in Fishbourne 1971, 334. Unpublished.

Rome: Antiquarium Comunale. Uncatalogued rim fragments of large bowl and small hemispherical bowl. Unpublished.

Vatican Coll. 1975, 62–63, nos. 570–571, pls. 27–28. Provenance unknown but presumably Rome or Italy.

Ostia II 1970, 250–251, pl. 43, figs. 617–618.

Ostia IV 1977, 25, pl. 8.30; 26, pl. 9.36; 322, pl. 46.322–323. Finds mostly belonging to levels of the third to fifth century A.D.

Herculaneum: Antiquarium. Inv. no. 918, two large joining fragments of a broad shallow bowl (Est. D 25 cm) with an elaborate, pierced ear-handle. Found in the Casa dei Due Atri, no. 6. Antedates A.D. 79.

Paestum: Museo Archeologico. Three fragments of plates from the Sanctuary of Bona Dea, University of Michigan–University of Perugia excavations. Unpublished.

Salona 1975, 155–156, no. 39, pl. 29. Facet-cut rim with floral motif cut on the underside.

Limburg 1971, 22, no. 62, fig. 4. Bowl probably from Maastricht. Numerous parallels cited on p. 22, including many from sites mentioned in the opening paragraphs of this section.

Charlesworth 1959, 38–40, fig. 3. Cites examples from Britain and calls them predominantly Flavian–Trajanic in date, noting that no fragments appear at Camulodunum (ca. A.D. 43–65).

Caerhun: Bowl with broad rim, probably of this type, from a lave dated ca. A.D. 90–100; cited in Charlesworth 1959, 40.

Verulamium 1972, 331–332, 336, nos. 26, 28, 33–34, fig. 138. Fragments from Period 2 (A.D. 75–100) and Period 3 (A.D. 100–270).

Caerleon: Unpublished fragments, cited in Limburg 1971, 22, no. 62, and dating before A.D. 140; also a fragment in a Hadrianic context.

Conimbriga 1976, 172–174, nos. 96–116, pls. 37–38. Fragments found in Flavian or Trajanic levels but considered no earlier than A.D. 50/60, based on the premise that colorless glass was not in general use until then.

Yadin 1963, 105–110, nos. 10–12, pls. 39–40. Three examples securely dated before A.D. 132–134.

Karanis 1936, 50, 60–62, nos. 73–77, pl. 11. Mostly medium-sized plates.

Carthage: University of Michigan excavations. Two unpublished rim fragments.

Tipasa 1967, 94, no. 198, pl. 9.3.

Aksum, Sudan: Fragment of rim of large bowl. Unpublished.

Dura-Europos 1963, 22, nos. 76–86, pls. 2–3, plus other fragments.

Based on the evidence cited above, these table wares occur principally in Flavian, Trajanic, and Hadrianic contexts, with some pieces found in strata of the later second or early third century A.D. The finds from Cosa, mostly from Deposit IX, lend support to such a chronology but add little precision to the dating of the type. The fragment (**No. 161**) recovered from Deposit V (ca. 25/15 B.C. to ca. A.D. 40/45) is in all probability intrusive in so early a locus. Magdalensberg, abandoned ca.

A.D. 45, and those levels at Camulodunum predating A.D. 43 likewise lacked examples of the type. On the other hand, the presence of a broad shallow bowl from the Casa dei Due Atri at Herculaneum conclusively proves that these vessels were already in production during the seventies, a date somewhat earlier than previously thought (see list below). Although the greatest popularity of the type seems to be in the following four decades, the latest closely dated pieces, such as the vessels from the Cave of the Letters, reveal that examples were in use until the mid-second century A.D.

The location of the factories producing these vessels has traditionally been assigned to Alexandria in Egypt (Karanis 1936, 50; Yadin 1963, 110; Limburg 1971, 22) or to the Near East generally (Dura-Europos 1963, 57). The large number of examples now known from Italy, North Africa, and the provinces of central and northwestern Europe, however, undermine the certainty of such claims. Indeed, because the type is so plentiful and widespread, it seems unlikely that a single center of production could have been responsible for all examples. On the other hand, if there were but one source, the organization of trade responsible for the dispersal of the type throughout the length and breadth of the Roman world would be all the more remarkable.

Large Shallow Bowls, Medium Plates, and Oval Bowls

161. CE 2182 — Pl. 10
A.P. Room 16, I — Deposit V
ca. 25/15 B.C. to ca. A.D. 40/45

Small fragment of rim (incomplete) of large shallow bowl; colorless. Milky weathering and pitting. Broad, outsplayed rim with rounded overhanging edge. On upper side of rim, a broad cut depression with ridges at edge and at junction of rim and side. Cast, rotary-polished and cut on both surfaces. D ca. 28–30 cm; A Th .35 cm.

162. C71.197 — Pl. 10
VI-D, E, SW Perimeter Trench, 0

Fragment of rim (incomplete) of large shallow bowl; colorless. Iridescence and pitting. Shape and technique of manufacturing as in **No. 161**. D ca. 28–30 cm; A Th .35 cm.

163. CD 1144 — Pl. 10
A.P. Room 16 Exterior, II

Fragment of rim and indications of side of large shallow bowl; colorless. Milky weathering and pitting. Shape and technique of manufacturing as in **No. 161**. D ca. 23 cm; Max Th .4 cm.

164. CD 1105 — Pl. 10
A.P. Room 16 Exterior, III

Fragment of rim and side of large shallow bowl; colorless. Milky weathering and pitting. Broad, outsplayed rim with rounded overhanging edge. On upper side of rim, a broad cut depression with ridges at the edge and junction of rim and side. On underside, at junction of rim and side, a rounded cut depression. Cast, rotary-polished and cut on both surfaces. D ca. 22 cm; Max Th rim .4 cm; Min Th side .2 cm.

165. CF 2454 — Pl. 10
A.P. Room 25D, II — Deposit IX
Late first to early third century A.D.

Fragment of rim and indications of side of large shallow bowl; colorless. Milky weathering and pitting. Shape and technique of manufacturing as in **No. 164**. D ca. 22 cm; Max Th .4 cm.

166. C73.74
A.P. III SW, Street 5, Sounding, 0
Small fragment of rim (incomplete) of large shallow bowl; colorless. Iridescence and pitting. Shape and technique of manufacturing as in **No. 161**. D ca. 22–24 cm; Max Th .4 cm.

167. CF 2460
A.P. Room 25D, II — Deposit IX
Late first to early third century A.D.
Small fragment of rim (incomplete) of large shallow bowl; colorless. Milky weathering and pitting. Shape and technique of manufacturing as in **No. 161**. D uncertain. Max Th .3 cm.

168. Additional fragments of similar large shallow bowls; diameters uncertain.
a. CE 2124, A.P. Room 6, I. Tiny fragment of rim; colorless.
b. CE 2127, A.P. II, Room 50, I. Tiny fragment of rim; colorless.
c. CF 2461, Forum ENE, 0. Tiny fragment of rim, colorless.
d. CF 2453, Forum ENE, 0. Tiny fragment of rim; colorless.
e. CG 579, Forum East Corner, NE Passage, 0-I. Tiny fragment of rim; colorless.

169. CF 2311 — Pl. 10
A.P. Room 25C, II — Deposit IX
Late first to early third century A.D.
Two joining fragments of rim, side, floor, and base-ring of shallow plate (profile restorable); colorless. Considerable milky weathering and pitting on one fragment, iridescence and pitting on the other. Fairly broad, nearly horizontal rim with rounded edge (no overhang); short convex curving side; flat floor; high, slightly outsplayed base-ring. On upper side of rim, a broad cut depression. Cast, rotary-polished and cut on both surfaces; base-ring cast as part of vessel. H 2 cm; D ca. 18.5 cm; A Th .2 cm.

170. CF 2394 — Pl. 10
A.P. Room 16 Exterior, III
Fragment of rim, side, and floor of shallow plate; colorless. Slight iridescence and pitting. Shape as in **No. 169**. On upper side of rim, two shallow cut grooves .3 and 1 cm from edge of rim. Technique of manufacturing as in **No. 169**. D ca. 17 cm; Max Th .3 cm.

171. CF 2459
A.P. Room 25C, III — Deposit IX
Late first to early third century A.D.
Two small joining and two non-joining fragments of rim, side, and floor of shallow plate; colorless. Milky weathering, pitting, and calcium encrustation. Shape identical to **No. 169**. On upper side of rim, two shallow grooves .3 and 1 cm from edge of rim. Technique of manufacturing as in **No. 169**. D uncertain; Max Th .3 cm.

172. CF 2393 Pl. 10
A.P. Room 16 Exterior, III
Fragment of rim, side, and floor of shallow plate; colorless. Iridescence, pitting, and discoloration. Shape identical to **No. 169**, except that the floor slants downward toward the center. On upper side of rim, shallow grooves as in **No. 170**. Technique of manufacturing as in **No. 169**. D ca. 17 cm; Max Th .25 cm.

173. CF 2455
A.P. Room 25C, II Deposit IX
Late first to early third century A.D.
Fragment of rim, side, and floor of shallow plate; colorless. Milky weathering and pitting. Shape identical to **No. 169**. On upper side of rim, a broad cut depression. Technique of manufacturing as in **No. 169**. D uncertain; Max Th .3 cm.

174. C74.4
VIII-F, Sounding, 0
Fragment of rim (edge missing), side, and floor of shallow plate; colorless. Iridescence and pitting. Shape identical to **No. 169**. On upper side of rim, a broad cut depression. Technique of manufacturing as in **No. 169**. D ca. 17–18 cm; Max Th .2 cm.

175. CE 2129 Pl. 10
A.P. II, Room 62, I
Small fragment of rim, side, and handle (incomplete) of large shallow bowl, probably oval in shape; colorless. Iridescence and pitting. Broad, nearly horizontal rim with rounded, thickened edge; convex curving side; projecting outward from rim, part of a cast and cut "ear-handle." On upper side of rim, a broad cut depression; on underside at junction of rim and side, a shallow rounded depression. Technique of manufacturing as in **No. 169**. Max D 20–25 cm; Max Th .3 cm.

176. C74.3 Pl. 10
VIII-F, Sounding, 0
Fragment of floor and base-ring of large shallow bowl or other vessel; colorless. Iridescence and pitting. Upper side of floor slightly concave; short rounded base-ring. Cast, rotary-polished on both surfaces; base-ring cast as part of vessel. H floor .8 cm; D base-ring ca. 18 cm; Max Th floor .3 cm.

177. CF 2440 Pl. 10
A.P. Room 25 Exterior, III
Fragment of floor and base-ring of large shallow bowl or other vessel; colorless. Iridescence and pitting. Upper side of floor slightly concave; outsplayed base-ring with rounded edge. Cast, rotary-polished on both surfaces; base-ring cast as part of vessel. H floor 1.1 cm; D base-ring ca. 16–17 cm; Max Th floor .2 cm.

178. CE 2125
A.P. Room 22, I
Small fragment of floor and base-ring of large bowl or other vessel; colorless. Iridescence and pitting.

Shape and technique of manufacturing as in **No. 177**. H floor 1.2 cm; D base-ring ca. 16–17 cm; Max Th floor .2 cm.

179. CF 2418 Pl. 10
A.P. Room 25 Exterior, I
Fragment of floor and base-ring of large bowl or other vessel; colorless. Iridescence, pitting, and discoloration. Shape and technique of manufacturing as in **No. 177**, except the floor is polished very thin at center. H floor 1 cm; D base-ring ca. 13 cm; Max Th floor .1 cm.

180. CD 1103
Bas. ENW 2, I
Two joining fragments of floor and base-ring of large bowl or other vessel; colorless. Iridescence, pitting, and discoloration. Shape and technique of manufacturing as in **No. 177**. H floor 1.2 cm; D base-ring ca. 12 cm.; Max Th floor .2 cm.

181. CF 2416 Pl. 10
A.P. Room 16 Exterior, II
Fragment of side, floor, and base-ring of bowl or other vessel; colorless. Iridescence and pitting. Convex curving side; flat floor slants downward toward center. On upper side of floor, a thin horizontal groove at junction of side and base. Technique of manufacturing as in **No. 177**. H floor 1.1 cm; D base-ring ca. 12 cm; Max Th floor .2 cm.

182. CF 2310 Pl. 11
A.P. Room 25C, II Deposit IX
Late first to early third century A.D.
Five joining fragments of floor and base-ring of bowl or other vessel; colorless. Iridescence and pitting. Shape and technique of manufacturing as in **No. 177**. H floor 1.4 cm; D base-ring ca. 10 cm; Max Th floor .2 cm.

183. CF 2324 Pl. 11
A.P. Room 16 Exterior, V
Ten joining fragments of floor and base-ring of large bowl or other vessel; colorless. Milky weathering and pitting. Flat floor slants downward toward center, with indication of convex curving side; high, slightly outsplayed base-ring with flat edge. On upper side of floor, a broad cut circular ridge (and vestiges of a second ridge at center of base). Technique of manufacturing as in **No. 177**. H floor 2.1 cm; D base-ring ca. 11 cm; Max Th floor .3 cm.

184. CF 2462
VI-D, E, 0
Small fragment of floor with high base-ring of large bowl or other vessel; colorless. Milky weathering and pitting. Shape and technique of manufacturing as in **No. 183**. H floor 1.8 cm; D base-ring uncertain; Max Th floor .3 cm.

185. CE 2132
A.P. II, Room 50, II

Small fragment of floor with high base-ring (incomplete) of large bowl or other vessel; colorless. Iridescence and pitting. Shape and technique of manufacturing as in **No. 183**. Restored H floor 2 cm; D base-ring uncertain; Max Th floor .25 cm.

186. CD 2441
A.P. Room 26, I–III
Small fragment of floor and base-ring of large bowl or other vessel; colorless. Milky weathering and pitting. Shape and technique of manufacturing as in **No. 183**. Restored H floor 2.7 cm; D base-ring uncertain; A Th floor .3 cm.

187. CD 1104 Pl. 11
Bas. ENE 1, I
Fragment of rim, side, and base of dish (profile restorable); colorless with distinct greenish tinge. Iridescence and pitting. Outsplayed rim with rounded edge; convex curving side; flat or slightly concave floor; rounded base-ring protrudes outward from junction of side and floor to form resting surface. Probably cast; rotary-polished on both surfaces. H 2.6 cm; D ca. 18 cm ; A Th .22 cm.

Small Bowls

188. C69.167 Pl. 11
VIII-E, F, 0
Five joining and three non-joining fragments of rim, body, and base of small bowl (profile restorable); colorless. Iridescence and pitting, slight discoloration. Excellent craftsmanship. Broad, outsplayed rim with rounded overhanging edge; convex curving side; slightly concave floor; high, slightly outsplayed base-ring with rounded edge. On upper side of rim, a broad cut depression. On exterior, a horizontal cut depression at junction of rim and side; also a small circular groove on underside of floor at the center. Cast, rotary-polished and cut on both surfaces; D base-ring 6.2 cm; Max Th rim .3 cm; A Th side .12 cm.
Bibliography: Grose 1973, 179, pl. 189, no. 7.

189. C69.429 Pl. 11
S.H. Room 18, 0-I
Fragment of rim and side of small bowl; colorless. Milky weathering and pitting. Strain cracking. Shape of rim and side as in **No. 188**; cut depressions similarly placed. Technique of manufacturing also identical. D ca. 12 cm; Max Th rim .3 cm; Min Th side .15 cm.

190. C70.618 Pl. 11
S.H. Street 5, 0-I
Fragment of rim and side of small bowl; colorless. Iridescence and pitting. Shape of rim and side as in **No. 188**; depressions similarly placed. Technique of manufacturing also identical. D ca. 12 cm; Max Th rim .3 cm; Min Th side .15 cm.

191. CF 2456
A.P. Room 25C, III Deposit IX
Late first to early third century A.D.
Small fragment of rim (incomplete) and side of small bowl; colorless. Iridescence and pitting. Shape

of rim and side as in **No. 188**; depressions similarly placed. Technique of manufacturing also identical. D uncertain; Max Th rim .3 cm.

192. CF 2457
A.P. Room 25 Exterior, I
Fragment of rim and side of small bowl; colorless. Iridescence and pitting. Shape of rim and side as in **No. 188**, but cut depressions less pronounced. Technique of manufacturing also identical. D ca. 12 cm; Max Th rim .35 cm.

193. C74.2 Pl. 11
VIII-F, Sounding, 0
Two joining fragments of rim and side of small bowl; colorless. Silvery iridescence, pitting, and some corrosion. Shape of rim and side as in **No. 188**. On overhanging edge of rim, numerous closely spaced, wedge-shaped cuts forming a scalloped pattern. On underside of rim, a row of oval facets set side by side. Cast, rotary-polished on interior and on rim; facet-cut on edge, underside of rim, and on exterior of side. D ca. 12 cm; Max Th rim .3 cm; A Th side .2 cm.

194. CE 2126 Pl. 11
A.P. II, Room 60, I
Fragment of rim and side of small bowl; colorless. Milky weathering and pitting. Broad, nearly horizontal rim with rounded edge (no overhang); nearly straight side, tapering diagonally downward. No cut depressions on or under rim. Technique of manufacturing as in **No. 188**. D ca. 12 cm; Max Th rim .3 cm; A Th side .2 cm.

195. C68.694 Pl. 11
Forum ENW, 0
Fragment of rim and side of small bowl; colorless with distinct greenish tinge. Iridescence and pitting. Outsplayed rim with rounded edge; nearly straight side, tapering downward. On exterior, a shallow cut depression at junction of rim and side. Technique of manufacturing as in **No. 188**. D ca. 12–14 cm; Max Th rim .3 cm.

196. CF 2445 Pl. 11
A.P. Room 25 Exterior, I–II
Small fragment of rim and vestiges of side of small bowl or other vessel; colorless. Iridescence and pitting. Broad, horizontal rim with cut depression on upper side. Technique of manufacturing as in **No. 188**. D ca. 12 cm; A Th rim and side .2 cm.

197. C70.628
S.H. Street 5, Sewer, I
Small fragment of rim and vestiges of side of small bowl; colorless. Iridescence and pitting. Shape of rim and side as in **No. 196**. Technique of manufacturing as in **No. 188**. D ca. 12 cm; A Th .2 cm.

198. CF 2420 Pl. 11
A.P. Room 25A, II–III
Late first to early third century A.D.

Three non-joining fragments of floor and base-ring of small bowl; colorless. Iridescence and pitting. Slightly convex floor; high, slightly outsplayed base-ring. On upper surface at center of floor, numerous circular rotary-polishing marks. Technique of manufacturing as in **No. 188**. H floor 1.4 cm; D base-ring ca. 6 cm; A Th floor .2 cm.

199. CE 2205
A.P. Room 12A, Cesspool, I
Small fragment of floor and base-ring of small bowl; colorless. Iridescence and pitting. Shape as in **No. 198**. Technique of manufacturing as in **No. 188**. H floor 1.3 cm; D base-ring uncertain; A Th floor .2 cm.

200. CD 1163 Pl. 11
Bas. ENE 1, I
Fragment of side, floor, and base-ring of small bowl; colorless. Iridescence and pitting. Convex curving side; almost flat floor; high, slightly outsplayed base-ring with rounded edge. On underside of floor at center, two concentric, circular grooves. Technique of manufacturing as in **No. 188**. H floor 1.1 cm; D base-ring ca. 4 cm; A Th floor .2 cm.

201. C74.5
VIII-F, Sounding, 0
Tiny fragment of floor and base-ring (incomplete) of bowl or other vessel; colorless. Iridescence and pitting. Convex curving side joins slightly concave floor; high, slightly outsplayed base-ring. Shape similar to **No. 200**. Technique of manufacturing as in **No. 188**. D base-ring uncertain; A Th floor .25 cm.

*Colorless Vessels: Miscellaneous (**Nos. 202–204**)*

In addition to fragments belonging to the above-mentioned type, there exist others at Cosa that are cast and colorless and finished by rotary-polishing but that are sufficiently different in shape so as to suggest other traditions. **No. 187**, for example, is a dish with a form for which there is no known parallel. **Nos. 202** and **204** are bases, one from an exceptionally large vessel, the other from a small bowl or beaker, but neither closely resembles the small bowls with overhanging rims described above. **No. 203**, an unusual rim fragment, is of special interest because of its context. Located in Deposit V, as was **No. 161**, it can be assigned to the years ca. 25/15 B.C. to ca. A.D. 40/45. As such, it is one of the earliest examples of fine colorless glass at Cosa and suggests that the transition to colorless fabrics was probably under way by the forties A.D. A possible parallel to this vessel was found at Villa Potenza 1971, 410, no. 48, fig. 11.9.

COLORLESS VESSELS: MISCELLANEOUS

202. CF 2458
A.P. Room 25C, II Deposit IX
Late first to early third century A.D.
Two small joining fragments of floor of vessel; colorless. Milky weathering and pitting. Slightly convex floor; no indications of base-ring preserved on fragments. On upper surface at center of

floor, two concentric, circular grooves. Technique of manufacturing as in **No. 188**. A Th floor .25 cm.

203. CE 2130 Pl. 11
A.P. Room 4, I

Fragment of rim and side of small bowl; colorless with greenish tinge. Iridescence and pitting. Upright rim with thickened, rounded edge almost forming a lip; convex curving side. On exterior, a horizontal groove below edge of rim. Cast, rotary-polished and cut on both surfaces. D ca. 8.5 cm; A Th side .3 cm.

204. CE 2183 Pl. 11
A.P. Room 16, I Deposit V
ca. 25/15 B.C. to ca. A.D. 40/45

Fragment of side, floor, and base-ring of small bowl; colorless with distinct greenish tinge. Iridescence and pitting. Convex curving side; slightly concave floor. Cast, rotary-polished on both surfaces; base-ring cast as part of vessel. D base-ring ca. 5.5 cm; A Th floor .25 cm.

EARLY IMPERIAL ROMAN MOLD-BLOWN DECORATED GLASS

Among glass finds at Cosa comparatively few were manufactured by any sort of mold-blowing process. Although this technique has the advantage of imparting decoration, potentially very intricate, simultaneously with shape, only twenty-eight fragments of figural or decorated mold-blown vessels were found at the site.[28] Perhaps this paucity of mold-blown glass is best explained by the fact that for much of its history the Roman industry in general made surprisingly little use of the technique, despite its versatility.[29] Aside from the actual types discovered at Cosa, mostly from the early first century A.D., few additional varieties of mold-blown, expanded mold-blown, or optic blown wares were made before the later third and the fourth centuries. Exceptions are head flasks (not represented at Cosa) and such serviceable storage wares as prismatic bottles (**Nos. 525–548**) or Mercury flasks (**No. 536**), here catalogued with free-blown containers. In marked contrast, glassmakers of the late Empire and early Byzantine periods made prominent use of closed and especially open molds and other forming devices in manufacturing their products.

The origin of the mold-blowing technique is a matter of debate. D. B. Harden long held that mold-blowing developed before free-blowing and, in fact, arose as the result of experiments by the late Hellenistic casting industry to find more efficient ways to fill molds (Harden 1933, 421; more recently, Harden 1969, 46–47). This theory, however, has not been universally accepted. F. Schuler disagreed on technical grounds, arguing that mold-blowing requires greater expertise on the part of the maker because the pipe must be held and manipulated vertically over the mold. By contrast,

[28] This figure does not include prismatic bottles or Mercury flasks, which are decorated in the sense that they often carry figural and geometric designs or inscriptions, usually on their bottoms but occasionally on their sides.

[29] The absence of glassblowing molds from Roman sites is less perplexing when one considers that a large percentage of figural mold-blown types was made only for a short period of time in the first century A.D., and probably at only a few major centers that have yet to be detected. Molds that do exist are ceramic and have been found at Macquenoise in Belgium (mold for a grape flask), at Borgo on Corsica (possible molds for sports cups), and at Cologne (bottom-mold for a prismatic bottle); see Price 1976, 199, for discussion and references.

free-blown vessels are manufactured by a seated glassblower, using wooden props and manipulating the pipe horizontally, as can be seen on the discus of a Roman lamp of the early first century A.D. It is therefore a simpler method, and in Schuler's opinion simpler techniques usually precede more complex ones (Schuler 1959).

Another objection, potentially more serious, arises from the significant discrepancy in date between the earliest known examples of mold-blown vessels, on the one hand, and the evidence for the introduction of glassblowing, on the other. On both eastern and western Mediterranean sites, free-blown glass first appears in strata assigned to the last decades of the first century B.C., then becomes increasingly ubiquitous in the first two decades of the next century (information summarized in Grose 1977, 9–29). Nonetheless, no excavation has yet yielded a mold-blown vessel from a dated context earlier than the second quarter of the first century A.D., one to two full generations after the invention of the blow pipe itself.[30] In fact, **No. 207** at Cosa is at present the *earliest* recorded mold-blown vessel from a securely dated deposit, and it belongs to the decade before ca. A.D. 40/45.[31] This would suggest that the requisite technical expertise emerged several decades (possibly even more) after the advent of free-blowing.

At Cosa, all figural mold-blown vessels fall into one of a half dozen well-known types or groups. In fact, the site yielded at least one example of almost every prominent category of mold-blown glass ascribed to the first century A.D. Most fragments are associated with types usually referred to as "Romano-Syrian" or "Sidonian," on the premise that they were first invented and produced by glassmakers of that region. This hypothesis was first put forward by Harden in 1935 for reasons that are discussed in greater detail in subsequent sections. Types include Romano-Syrian beakers with (and without) Greek inscriptions (**Nos. 206–209**), small "Sidonian" bottles (**No. 210**), and a series of mold-blown fluted bowls (**Nos. 211–219d**). Of comparable date and probably influenced by the Romano-Syrian production are other types, namely gladiatorial beakers such as **No. 205**, tall conical beakers with mythological scenes (**No. 220**), and possibly beakers with bossed and/or figural decoration (**Nos. 221–226**).

The close stylistic and chronological affinities shared by these types are noteworthy but puzzling. They suggest that in most instances allied families or groups of glassmakers are involved. Not only are most types roughly contemporary with one another, but they exhibit a common repertoire of shapes, colors, sizes, decorative devices, and methods of manufacture and finishing. Even patterns of distribution are, for the most part, similar. In function, virtually all examples are drinking vessels, usually beakers or small bowls, with the occasional handled cup. The sole storage wares are Sidonian bottles, Ennion-type jugs, and several rare bossed flasks (see "Bossed Beakers"), and most of these probably antedate the principal groups of drinking vessels (see pp. 103--106 and note 37 below). Indeed, the normal range of forms is limited to no more than five or six main shapes, all cups or beakers, and most of these make use of upright or off-set rims with unfinished or slightly polished edges. With few exceptions bases are absent, and the undersides of the vessels are generally adorned

[30] For mold-blown vessels from Italy attributed to the second and third quarters of the first century, see discussion under "Romano-Syrian Vessels," pp. 103–106 below. Also said to belong to the first half of the first century is a mold-blown flask in the shape of a barrel, recovered from a cemetery at Akko in Israel (Fortuna 1965, 19, figs. 6–7). Although the excavators believe that the grave in which it was found dates to the first half of the first century, based largely on the presence of small free-blown unguent bottles, the grave and its contents are equally attributable to the second half of that century, possibly even later. Finally, from Magdalensberg 1979, 144–145, no. 1055, pls. 9 and 22, comes a fragmentary mold-blown pyxis, beaker, or other mold-blown container (D = 5 cm) that must date before the abandonment of the site, ca. A.D. 45.

[31] Possibly as early is a fragment of an Ennion beaker found at Corinth in a context said to date to about A.D. 40 (McClellan 1983, 74).

with two or three concentric circular ridges. Most examples are naturally colored, in pale shades of green, bluish-green, olive, yellow, or purple—all colors customarily associated with the utilitarian glass of the late Empire but here dating to the first century. The brightly colored translucent and opaque fabrics favored by the Roman casting and free-blowing industries of the late first century B.C./early first century A.D. are seldom found, except among Sidonian bottles, fluted bowls, and the earliest production of Ennion.[32] Decolorized fabrics are equally rare, except among the latest mold-blown types, here mythological and bossed beakers.

Although such similarities strongly suggest that most types are related to one another in some fashion, or at least influenced by one another, it is not now possible to explain their interrelationships with any precision. Such affinities could represent the production of one or two major centers, such as Sidon and/or Italy (perhaps Campania), where several different types may have been manufactured simultaneously or in overlapping sequences, with individual types retaining or rejecting elements from their predecessor(s) while passing along certain features to the next generation of glasses. Alternatively, these common denominators could represent intentional copying by a large number of contemporary but rival glassworks, each producing different types and operating in several regions both in the East and in Italy. On the other hand, such similarities may represent little more than casual imitation by several factories of shapes and decoration popular at the time, not only in glass but also in other media, such as pottery or metals.

Despite such uncertainties, however, several conclusions can still be reached concerning the early Roman mold-blowing industry. First, by the second quarter of the first century, and continuing for several decades, an active mold-blowing industry flourished in the Roman world. Second, the types produced are as innovatively decorated as they are few in number, and most fall into related groups of drinking vessels. Third, the total number of vessels produced in this short period was sufficiently large that examples now survive in appreciable quantity.[33] Fourth, the distribution of these glasses, especially the Syro-Palestinian types, is considerably more extensive than previously has been appreciated. New finds from the Aegean, North Africa, and the Iberian Peninsula now show that these mold-blown glasses circulated widely in much the same way as certain of their contemporary cast and free-blown counterparts. Nonetheless, there still exists, as Harden observed in 1935, a telling concentration of many types (sports and myth beakers excepted) in Italy and the Syro-Palestinian region. This fact further emphasizes the special relationship between glassmakers of the two areas and suggests that one or both regions may have been responsible for the production of most of these wares. Unfortunately, no mold-blowing factories in either region have yet been identified, and the epigraphical evidence from the vessels themselves (see pp. 104–105 below) only establishes that eastern, Greek-speaking artisans of Sidonian and Cypriot background were involved in making these vessels. The inscriptions do not necessarily prove that the workshops were in those locations.

[32] In discussing the Ennion workshop, Harden 1935, 165, cleverly observed that such unusual shapes as footed jugs, amphoriskoi, and two-handled cups may well represent early experimental pieces that disappeared as the more standard shapes were adopted. The same can be argued for the relatively few molded vessels in bright colors. Such colors are principally associated with mold-blown types considered popular in the thirties and forties but not in the second half of the first century A.D. As the fashion for color waned at midcentury (pp. 81–82, 88–91 above), mold-blowing workshops, along with casting and free-blowing factories, quickly abandoned intentionally colored fabrics in preference for naturally colored and colorless glass.

[33] One hundred fifteen Romano-Syrian beakers are listed in Harden 1935 and 1945; elsewhere he mentions the existence of over one hundred sports cups (Fishbourne 1971, 337). McClellan 1983, 76–78, adds thirty-eight additional Romano-Syrian beakers. Although not yet counted, Sidonian bottles, mold-blown fluted bowls, and bossed beakers are equally ubiquitous on archaeological sites and in museum collections.

Finally, all types (with the possible exception of the bossed beakers) appear, almost inexplicably, to have come to a fairly abrupt end sometime in the Flavian period. Even more surprising, they are not succeeded in the late first/early second centuries by other comparable, decorated mold-blown wares. Although one hesitates to draw correlations between historical events and the cessation of particular industries, it is noteworthy that both areas of alleged production suffered major calamities in the late sixties and seventies. The First Jewish Revolt, culminating with the destruction of Jerusalem in A.D. 70, undoubtedly disrupted or destroyed many Syro-Palestinian craft industries. Likewise, in Campania the eruption of Mount Vesuvius in A.D. 79 annihilated the agricultural and industrial production of much of the region, a district that Pliny (*HN* 36.194), writing a few years earlier, implied was an important center of glass manufacture.

Sports Cups (**No. 205**)

Sports beakers (*Zirkusbecher*) constitute one of the few categories of decorated Roman glassware with analogues to Roman social history. These beakers form a homogeneous group of figural mold-blown drinking vessels decorated with scenes of chariot races, gladiatorial combats, athletic contests, or combinations of the above. Over one hundred examples survive (Fishbourne 1971, 337), and, of these, beakers depicting racing quadrigae, often with famous charioteers named, are the most frequently found. A somewhat smaller series shows groups of professional gladiators (usually four pairs) locked in combat, and, once again, names in Latin are often ascribed in molded relief in a band near the rim or at the bottom of the cup (or occasionally within the figural frieze). Typical of these is a beaker from Chavagnes-en-Paillers in the Vendée that lists eight celebrated fighters: Spiculus, Columbus, Calamus, Holes, Tetraites (spelled Petraites on other cups), Prudes, Proculus, and Cocumbus. Among these, Spiculus and Columbus achieved popularity under Caligula and Nero (Suet. *Gaius* 55; *Ner.* 30.47). The same is probably true for Tetraites and Prudes (Petron. *Sat.* 52.3, 71.6; cf. Rowell 1958).

Sports beakers in general, and gladiatorial beakers in particular, occur in three common shapes: short conical beakers, bulbous beakers (akin in form to **No. 206**), or cylindrical beakers, such as the example published here (**No. 205**). These forms occur with little variety and closely reflect the most common shapes of Romano-Syrian beakers and other mold-blown wares described in other sections of this catalogue. Colors also match those of other first-century mold-blown types. Sports beakers almost always are naturally colored, usually in pale yellow, yellowish-green, bluish-green, light green, olive, or pale purple fabrics. A few are colorless, but none appear in brightly colored glass. Rims are usually upright or off-set, with unfinished or polished edges. Bottoms are flat and decorated with concentric circular ridges.

No. 205 at Cosa falls into a standard subclass of such gladiator cups. The fragment preserves the lower side of a yellowish-green cylindrical beaker with three men dressed as *murmillones* depicted in the bottom register of the vessel. A close intact parallel is the example from Chavagnes, mentioned above, and illustrated in Vindonissa 1960, 65, no. 24, pl. 10. The two beakers, in fact, may derive from the same mold, although no portion of an inscription survives on the Cosan fragment. Other examples, presumably with similar inscriptions and fighting scenes, mostly in cylindrical shapes, are known from Vertillum (near Chatillon-sur Seine), Lillebonne, Vetera, Heimersheim, Rheinhessen, Vienna (without find-spot), and possibly Trier, Leicester, and Camulodunum (see extensive list cited in Vindonissa 1960 and Fishbourne 1971). Additional Italian examples, closely dated pieces, and newly published sports cups are cited below.

Parallels and Dated Examples of sports Cups

Luni I 1973, 479, no. C132, pl. 82.19.
Luni II 1977, 287; CM 5820/2, CM 5820, CM 5820/1; pls. 158.20, 158.18, 158.19. Three fragments apparently from different vessels.
Gubbio: *NSc* 8, II, 1948, pp. 53–55, fig. 9. Conical cup with the name of Calamus preserved. From a first-century A.D. context.
Palatine (Rome): *NSc* 8, III, 1949, p. 52, fig. 4. Fragment of uncertain shape preserving the letters [COL]UMBUS. Dated before A.D. 80.
Carmona 1931, 146–147, no. 137, pl. 75.
Conimbriga 1965, 30–32, nos. 18–20, pl. 1. Two of the three fragments are probably from cylindrical cups similar to **No. 205**.
Sabodell (near Barcelona): J. Guidol-Richard, *Els vidres Catalans* (Monumenta Cataloniae, III), (Barcelona 1936), 23, cited in Harden 1969. 51, n. 38.
Vindonissa 1960, 56–67, esp. 62; nos. 156–159, pl. 10.21. Diverse examples, with one cup dating before A.D. 60/75 and another from a Flavian context. Full discussion, with additional parallels cited, pp. 65–57.
Berger 1977, 63–68. Additional examples from Vindonissa with full discussion of new material from other sites.
Vetera 1917, 166, fig. 2. Found in a Claudian/Neronian context.
Camulodunum 1947, 229–330, nos. 50–52, pl. 86. From Neronian levels, ca. A.D. 49–61.
Fishbourne 1971, 338–339, no. 36. Fragment assigned to Period I, ca. A.D. 43–75. Discussion with many other pieces cited.
Ville 1964, 722–733. He lists many examples and discusses the names and identities of the gladiators.

Dating evidence for sports beakers, including the gladiatorial group, places them firmly in the mid-first century A.D., possibly commencing sometime in the second quarter of the century. Although it is tempting to equate the names from the inscriptions with gladiators known to have achieved fame under Caligula and Nero, thus dating the beakers accordingly, close inspection of the inscriptions reveals sufficient inconsistencies and errors to warrant a more cautious assessment of their chronological value. Not only are names conflated or misspelled (e.g., Petraites for Tetraites), but on several beakers the names are so paired that a certain Columbus, known from literary sources to have died under Caligula, has as his opponent Spiculus, a *murmillo* who made his debut over a decade *later* under Nero (Ville 1964, esp. 731–732). Such discrepancies suggest that the mold-makers were little concerned by absolute accuracy and that new molds were carelessly fashioned, probably using ones already worn out as models. These errors suggest that the cups may have been made over a number of years or decades, even long after the deaths of the gladiators they celebrated. Nevertheless, the stratigraphical and epigraphic evidence still demonstrates that the beakers belong to the middle decades of the first century and that they are contemporary with most other Romano-Syrian types (even if they perhaps begin slightly later). This in turn dictates that they too evolved shortly after the invention of mold-blowing itself.

Distribution, once thought to be restricted primarily to the provinces of central and northwestern Europe (also North Africa; Vindonissa 1960, 67, no. 47), must now take into account Italy. The evidence from Gubbio, Rome, and especially Luni and Cosa, demonstrates that these drinking wares were as popular among the residents of small Italian communities as they were among legionaries at frontier stations. No longer need the type be regarded as exclusively British or Gaulish and attributed to factories in those provinces (Kisa 1908, 726–750; Harden 1940; and Ville 1964, 722). Manufacture could as easily have occurred in Italy. In fact, ceramic molds with gladiators have been found on Corsica according to Price 1976, 119, although it is uncertain whether they were intended for glass or for ceramic vessels. More compelling is a recent study of the race track paraphernalia depicted on circus beakers. According to J. Humphrey the structures represented can only reflect the actual architecture of the circuses at Rome because comparable race tracks had yet to be built in the provinces of Europe.[34] Also, it would be surprising (but not impossible) to find British or Gaulish workshops fabricating mold-blown wares so shortly after the initial invention of the technique itself. The high incidence of examples on military sites in central and northwestern Europe is best explained by the fact that virtually all fine or novel utensils used by the soldiers of this era were requisitioned from Italy and the Mediterranean world.[35]

Sports Cup

205. C70.342 Pl. 12
S.H. Street M, 0-I

Fragment of lower side and bottom of cylindrical beaker depicting a scene of gladiatorial combat; yellowish-green. Iridescence and pitting. Vertical side terminates in a rounded molding that projects outward, then turns in toward bottom. Extant fragment shows portion of a lower horizontal frieze bordered on the bottom by a horizontal curved molding, with indications of a similar molding at the top of the frieze. On the fragment one gladiator (a *murmillo*) thrusts his sword at another gladiator (partly preserved); the leg and shield of a third man are at the far left. A foreshortened shield lies between the two figures. Blown into a patterned mold. H fragment 4.4 cm; D body ca. 8 cm; A Th .3 cm.
Bibliography: Grose 1977, 179, pl. 189, no. 8.

Romano-Syrian Vessels with Mold-Blown Inscriptions (***Nos. 206–209***)

The term "Romano-Syrian" glass with mold-blown inscriptions was devised in 1935 by D. B. Harden to describe a family of closely allied vessels with Greek inscriptions and other emblemata raised in relief (Harden 1935, 163–186; supplemented by Harden 1945, 81–95). This series of vessels carries complimentary inscriptions that may be freely translated as "success to you" or "seize the victory." Sometimes the names of the makers (or possibly the factory owners) are also found; examples include such men as Ennion, Jason, Meges, Neikaios, and Aristeas, often as part of the formula "Ennion made me." Harden's first article classified seventy-seven examples divided into eleven main groups according to variant inscriptions, as well as numerous subgroups based upon differences of shape, decoration, or nuance of epigraphy.[36] Harden's second article (in 1945) added sixteen more examples and drew up a list, equally long, of other vessels comparable in shape and ornamentation but lacking inscriptions.

[34] Information furnished by John Humphrey, author of *Roman Circuses: Arenas for Chariot Racing* (London 1986).

[35] The single example of a sports cup allegedly acquired in Egypt need not dictate an eastern origin or inspiration for the group as a whole, as has been suggested by Harden 1982, 31–34, figs. 1–2. The presence of artifacts of western manufacture in Imperial Egypt ought to be expected. See, for example, the fragmentary gold-band bottle found at Karanis and now in the Kelsey Museum of Archaeology, University of Michigan (see p. 40 above).

[36] Subsequently, two additional inscribed groups have been identified, raising the total to thirteen; McClellan 1983, 72.

Considering the distribution of pieces with known provenance, the long-recognized role of the Syro-Palestinian area as a center of glassmaking, and the presence of Greek inscriptions, Harden argued that most of the workshops were located along the Syro-Palestinian litoral. He further argued, based on the large number of Ennion's cups known from Italian sites, that Ennion at some later date either migrated to or established a branch factory in Italy, perhaps in the northern part of the peninsula, where many identical pieces have been discovered (Harden 1935, 164–165; most recently, Harden 1969, 46–48).[37]

Subsequently, this argument was accepted by many specialists, but the publication of new finds from around the Mediterranean probably complicates rather than elucidates the question of factory location.[38] In fact, recent studies of inscribed beakers from the Iberian Peninsula (Price 1974, 69, fig. 1:1a, b) and Greece (McClellan 1983, esp. 76–78) do just that. By demonstrating that Romano-Syrian mold-blown beakers are to be found as far west as Spain and in relatively large numbers in the Aegean, they appear somewhat to weaken Harden's original proposition. On the other hand, it can just as easily be argued that the new discoveries are not necessarily inconsistent with the past information. If the glass was indeed manufactured in the Levant, finds in the Aegean are to be expected, and the examples from Spain (or North Africa) are entirely compatible with contemporary trade patterns in other types of glass (for examples, see "Ribbed Bowls: Colored and Naturally Colored" and "Linear-Cut Bowls: Colored and Colorless," pp. 54–58 and 68–70 above). Furthermore, the fragments from Cosa published here, and those from Luni (see list below), amply demonstrate how common these wares are on Italian soil and thereby affirm the possibility of an Italian production. Thus, while the new finds do not substantiate Harden's theories, they do not repudiate them either.[39] If anything, they continue to corroborate the close links between glassmakers of the Syro-Palestinian area and Italy in the first century A.D., a tie already discussed in previous sections (cf. Grose 1983, 45).

At Cosa four fragmentary vessels may be classified as "Romano-Syrian." Two bear complimentary inscriptions in Greek (**Nos. 206–207**). The other two (**Nos. 208–209**) may have carried inscriptions, but no lettering is preserved on the surviving fragments. Of the inscribed vessels, **No. 206** readily falls into Harden's class G and probably belongs to subclass G i or G ii (Harden 1935, 174–175). It is a bulbous beaker in pale yellow that carries the inscription "your very good health," of which only three letters are preserved: ΦPA. The letters are in a continuous horizontal zone framed above and below by horizontal ridges. On the lower side of the fragment is a band of vertical hairpin fluting. The inscription is of interest for having given rise to a most peculiar thesis: that similar cups with Greek drinking slogans inspired the words of Jesus Christ as he spoke to his disciples during the Last Supper (Harris 1927, 289–291). On several obvious grounds, this is untenable.

[37] Most published examples of Romano-Syrian beakers from Italy are concentrated in the valley of the Po or adjacent regions. Of eleven vessels signed by Ennion, six are from the Veneto and another five from the Piemont, Emilia, or Lombardy. A single example is from Solunto in Sicily (Laviosa 1958, 172). Examples of related Romano-Syrian beakers, such as **Nos. 206–209** at Cosa, are also known from Tarquinia and Cornus on Sardinia (Laviosa 1958, 172; Canivet 1969, 23). Also see Aquileia 1968, 97–106, for discussion of Italian examples. For Syro-Palestinian pieces, see most recently Israeli 1964, 34–41, and Israeli 1983, 65–69.

[38] Hayes challenged this viewpoint, suggesting the possibility of a reverse migration of glassmakers from Italy to Syria in the mid-first century. According to this hypothesis these might actually be "the sons or grandsons of migrants to Rome," returning to Syria and establishing new factories there (ROM Coll. 1975, 29–30 and 33, n. 2). In part, Hayes supports this contention on the fact that several vessels naming certain glassmakers, in particular Aristeas and possibly Artas as well, are known *only* from Italian sites. On the other hand, a beaker not known to Hayes actually states that Aristeas was a Cypriot (McClellan 1983, 78, no. M2.)

[39] Only the discovery of actual production centers or confirmation from firm literary or epigraphical evidence will finally resolve the issue of provenance.

Among Italian or dated parallels are beakers from Pizzighettone near Cremona, included in Harden's class G ii (Sangiorgi Coll. 1914, 22, no. 133, pl. 19; Harden 1935, 174) and another said to be from Italy (or possibly North Africa), now in the Leyden Museum and belonging to Harden's class G i (Harden 1935, 174, and illustrated in Harris 1927, 291, no. 3, fig. 3). From Siphnos in the Aegean comes a bulbous beaker with an identical inscription but an unusual shape. It was found in a burial dated (on tenuous grounds) to the last quarter of the first century A.D. (Harden 1945, 87–88; Brock and Mackworth-Young 1949, 86, pl. 28.3–5).

No. 207 at Cosa is a cylindrical, naturally colored light green beaker. The fragment is decorated at the rim by three raised horizontal ridges, below which is a horizontal palm branch, and then an inscription of which only the partial letters NH survive. The shape and preserved ornamentation are identical to Harden's class F (Harden 1935, 171–183), one of his largest classes, but the surviving letters do not fit the customary inscription for this class: "your very good health."[40] In fact, the combination of letters NH suits none of the common inscriptions published by Harden. It is probable, however, that the letters are part of the word ΕΥΦΡΟCΥΝΗ, which occurs on at least two beakers, one from Cyprus now in the British Museum (Harden 1935, 176, group J), and another in the Corning Museum of Glass (*JGS* 1965, 121, fig. 2).[41] Alternatively, the inscription may have read "let the buyer be remembered" (ΜΝΗCΘΗ Ο ΑΓΟΡΑΖΩΝ),[42] as a beaker in the Boston Museum of Fine Arts that is said to have come from Asia Minor (Harden 1935, 170–171, group E). This same inscription is also found on a group of two-handled cups but always in a *tabula ansata* with a second inscription on the opposite side (also in a *tabula ansata*) naming a glassmaker such as Ennion (see Harden 1935, 166–167 and, more recently, Pavia Coll. 1983, 33–35, no. 9, color pl. 1, with references). Among these, the beaker at the Corning Museum appears to be the closest parallel to the fragment at Cosa, but there still are sufficient differences to suggest that the letters on the Cosan piece may possibly belong to an inscription that has not yet been recorded.

Despite the unusual character of the inscription, **No. 207** is in all other respects a standard Romano-Syrian beaker. This is of special significance because it was retrieved from a storeroom of a shop (here Deposit VII) that contained a great quantity of glass awaiting sale just prior to the destruction of the building, ca. A.D. 40/45. Whether the beaker was new merchandise or already broken in pieces on the floor (see chapter 2), it represents one of the few closely dated examples of Romano-Syrian beakers (or of mold-blown vessels in general) and places the cylindrical shape and decorative motifs of Harden's class F as early as the thirties or forties A.D. Indeed, among published examples, **No. 207** is the earliest securely dated beaker not only of its class but of all Romano-Syrian mold-blown beakers.[43]

Among other dated beakers from Italy is a two-handled cup signed by Ennion, now in Turin. It was discovered in 1873 at Capella S. Giorgio, Caresana, near Vercelli, and is said to have been found with a coin of Claudius struck in either A.D. 41 or 46 (Calvi 1965, 12, fig. 4; Laviosa 1958, 173,

[40] Two other subgroups of this class, G iii and G iv, have floral sprays in horizontal bands above and below the inscription. Vessels of these subclasses are also taller than those of the first two subgroups. Since the Cosa fragment preserves no floral zone and appears to be part of a short bulbous beaker, it probably is identical either to G i or G ii. It should be noted that the preserved alpha on **No. 206** has a V-shaped crossbar.

[41] For identical cylindrical beakers from Italy, but with the inscription *Ennion*, see Mariacher 1966, cover; and Laviosa 1958, 171–173. The first publishes an example from Zadar; the second a beaker now in the Museo Archeologico in Florence.

[42] I wish to thank D. B. Harden and G. D. Weinberg for independently suggesting this restoration. McClellan has also suggested this restoration for the Cosa piece: McClellan 1983, 72 and 76.

[43] A fragment of an Ennion cup is also attributed to a level of "fill" at Corinth dated around A.D. 40; McClellan 1983, 75.

no. 11). Somewhat later is a fragment found at Luni, considered from a stratum dated to ca. A.D. 50–65/70 (Luni II 1977, 287, CM 5109 and 5269, pls. 157.4 and 158.17). Other examples of dated Romano-Syrian vessels from various Mediterranean sites are uniformly from Flavian contexts, such as the Siphnos beaker cited above. Based on this meager evidence, it appears that Romano-Syrian beakers were first devised sometime in the second quarter of the first century A.D. and continued to circulate into the Flavian period. It also seems probable that their production coincided with the initial invention and exploitation of the mold-blowing technique itself.

Romano-Syrian Vessels with Mold-Blown Inscriptions

206. C67.554 Pl. 12
Arx W. Slope, Trial Trench I, IV

Three joining fragments of side of beaker of Romano-Syrian type; pale yellow. Slight iridescence and pitting. Blowing spirals. Convex curving side. Shape and decoration probably identical to type G ii in Harden 1935. Fragment divided into two horizontal friezes by two parallel rounded ridges. The upper frieze preserves the Greek letters ΦPA, presumably part of the inscription "your very good health." The lower frieze is composed of multiple vertical ribs (shaped like hairpins), raised in molded relief. At the top of the upper frieze is a partially preserved band also consisting of two horizontal, rounded ridges. Blown into a patterned mold. Max D body ca. 10 cm; A Th .07–.1 cm.
Bibliography: Grose 1973, 179.

207. CE 2035 Pl. 12
A.P. Room 22, II Deposit VII
Decades prior to ca. A.D. 40/45

Fragment of rim and side of cylindrical beaker of Romano-Syrian type; natural light green. Slight iridescence and pitting. Upright rim with short out-turned lip, with knocked-off edge; vertical side, progressively becoming thicker as it moves toward bottom. Shape probably identical to type F in Harden 1935. On fragment, portions of three zones preserved. The uppermost consists of a band of three superimposed, horizontal, rounded but uneven ridges; the next zone consists of horizontal palm branches paired in reverse directions; the third zone preserves the Greek letters NH, which are probably part of the inscription EYΦPOCYNH. A vertical palm branch masks the seam of the body-molds. Blown into patterned mold; edge of rim probably polished smooth. H fragment 4.5 cm; D ca. 6.2 cm; A Th .2 cm.
Bibliography: Grose 1975, 37–38, no. 8, fig. 2.

208. C68.664 Pl. 12
Arx W. Slope, 0

Fragment of lower side of cylindrical beaker of Romano-Syrian type; natural light green. Iridescence and slight pitting. Strain cracking. Nearly vertical side turns inward toward bottom. Shape and decoration probably similar to type F in Harden 1935. Extant fragment is divided into two partially preserved squarish panels, each containing two concentric circles raised in molded relief around a central raised dot. The panels are separated by a vertical palm branch or leafy garland. Above the panels is a band of two (?) horizontal, rounded ridges; below the panels is a single horizontal ridge and a portion of what seems to be a horizontal palm branch. This branch appears to be intersected by yet another branch in the lower right corner of the fragment. Blown into a patterned mold. D body ca. 7–8 cm; A Th .1 cm.

209. CD 1084 Pl. 12
Bas. ENE 3, I

Small fragment of side near rim of cylindrical beaker of Romano-Syrian type; natural light yellowish-green. Iridescence and slight pitting. Nearly vertical side turns slightly outward near rim, as also is true of **No. 207**. Shape probably similar to type F in Harden 1935. At the rim, a band of three horizontal ridges raised in molded relief; below the ridges is a partially preserved horizontal palm branch. Blown into a patterned mold. D body ca. 6.5 cm; A Th .13 cm.

*"Sidonian" Bottles (**No. 210**)*

No. 210 is a part of a small, dark blue bottle that belongs to a varied and numerous series of similar small bottles, jars, and jugs, as well as small lidded pyxides, all related in some manner to the Romano-Syrian beakers and bowls dealt with in the previous section. These too are believed to have been manufactured in the Syro-Palestinian area, hence the name "Sidonian," and possibly in Italy as well (for discussion, see Harden 1969, 47; ROM Coll. 1975, 32–33; Calvi 1965, esp. 14–15; Newark Coll. 1976, 66–71). However, despite similarities of provenance, distribution, and in some instances decorative motifs, they differ from the Romano-Syrian beakers in form, function, and, above all, color. In fact they make use of the widest possible range of translucent and opaque fabrics. In some instances they even employ polychrome splashed decoration. Seldom are they naturally colored or colorless. As such, they are possibly closer in date to the free-blown, colored wares of the first half of the first century A.D. than they are to the main series of Romano-Syrian mold-blown wares. If this is true, then it would appear that the majority of these bottles were manufactured before the middle of the first century, at a time when bright colors were still in favor.

No. 210 appears to have been hexagonal in shape, as also is true of most Sidonian bottles. Some are miniature handleless flasks, others jugs or two-handled jars. On most examples, each of the six sides are decorated with emblems thought to be thematically related. For example, Auth in Newark Coll. 1976, 69, no. 64, describes a hexagonal flask with objects connected with athletic activities or victories. These include a Panathenaic amphora, discus, victor's wreath, strigil, jug, and a shield. The Cosan piece once exhibited a series of six different vessels, one to a side, including a patera (seen on the fragment), a footed oinochoe (partly preserved), followed by a krater, ladle, kantharos, and a second krater, assuming, of course, that the bottle resembled the decorative program found on comparable intact pieces. The actual theme or intention expressed by these six objects, unfortunately, is unclear. Clairmont in Dura Europos 1963, 39, suggested that the vessels symbolized cult practices of an uncertain nature.[44]

Parallels and Dated Examples of "Sidonian" Bottles

Pavia Coll. 1983, 34, no. 8, pl. 2. Opaque light blue flask, presumably found in Italy.

Luni II 1977, 286–287, CM 5109 and CM 5269, pls. 157.4 and 158.17. Found in strata dated to A.D. 50–65/70.

[44] Eisen 1927, I, 250–252, also identifies these vessels as ritual objects that belong to his "processional series" of Sidonian bottles. Hayes 1975, 48, cites additional bottles of comparable design.

Cumae 1913, col. 740, pl. 118.9. Blue jar with opaque white handles; motifs differ somewhat from those on the Cosan fragment.

Pompeii: Museo Nazionale, Naples. Jug illustrated in Spinazzola 1928, 222, middle shelf. Predates A.D. 79.

Magdalensberg 1979, 144–145, no. 1055, pls. 9 and 22. Fragment of a bottle or jug, less likely a pyxis, with an ivy motif. Presumably antedates the abandonment of the site in ca. A.D. 45. If a jug, it probably is identical to a comparable vessel from Cyprus (Vessberg 1956, 151, no. 19, fig. 47); if a pyxis, a possible intact parallel is illustrated in Yale Coll. 1980, 47–48, no. 124. Additional parallels for jugs with ivy sprays are cited in Harden 1945, 84–86.

Harden 1945, 84, pl. 7. Jug found with an inscribed Romano-Syrian beaker similar to **No. 207** at Cosa. From a tomb at Yahmour (Syria) dated to the first century A.D.

Dura Europos 1963, 39, nos. 146–147, pl. 21. Flask with motifs similar to those of the Cosan piece.

Constable-Maxwell Coll. 1979, 95, no. 151, ill. opposite p. 94. Handleless blue flask with molded motifs comparable to those preserved on the fragment at Cosa. Provenance unknown. Additional Sidonian bottles exhibiting other motifs are also illustrated.

ROM Coll. 1975, 48, no. 84. Purple handleless flask with comparable motifs. Provenance unknown.

Bomford Coll. 1976, 19, no. 49. Handleless blue flask with comparable motifs.

BMFA Coll. 1968, no. 31. Greenish jug with a slightly different group of vessels depicted on each of its six sides.

Oppenländer Coll. 1974, 142–159, nos. 401–446. Numerous Sidonian bottles and pyxides illustrating a great many decorative motifs.

Eisen 1927, I, 233–237, pls. 50–58. Additional examples of this group with an extensive discussion, albeit exceedingly arcane and impressionistic.

Few Sidonian bottles have been recovered from closely dated deposits, but customarily the group is ascribed to the first century A.D., usually to the mid- or later decades of that century (see, for examples, ROM Coll. 1975, 33). A fragment from Magdalensberg supports such a chronology (see list above) but also indicates that these vessels were in production at least as early as the forties and are contemporary with other types of Romano-Syrian mold-blown wares. In fact, their preference for deliberately colored fabrics best suits the first half of the century, as noted above. The Magdalensberg fragment also demonstrates that such bottles were being circulated in Italy and its adjoining provinces soon after the type was invented, a fact that might suggest manufacture in Italy as well as in the eastern Mediterranean.

"Sidonian" Bottle

210. C66.810 Pl. 12

S.H. Street M, 0-I

Small fragment of lower side of cylindrical or hexagonal bottle of Romano-Syrian or Sidonian type; dark blue. Iridescence. Blowing spirals. Vertical side with projecting, rounded molding near bottom. Extant fragment preserves portions of two squarish panels. One panel contains a partially preserved pattern of two concentric circles around a central dot (?), all raised in molded relief. This motif has been identified by some specialists as representing a patera or a shield. The other panel contains a partially preserved footed vessel, possibly an oinochoe, raised in molded relief. The panels are divided by a vertical ridge. Blown into a patterned mold. D body ca. 4–5 cm; A Th .08 cm.

Fluted Beakers and Bowls (***Nos. 211–219***)

An extensive series of beakers and bowls decorated solely with vertical hairpin ribs or flutes comprises another homogenous group of mold-blown glasses. These too are thought to be related to the types of Romano-Syrian vessels already described (see Camulodunum 1947, 299–300, no. 48; Vindonissa 1960, 55; ROM Coll. 1975, 33, no. 82). In shape, most examples closely resemble bulbous beakers such as **No. 206**, an especially common form among Romano-Syrian inscribed vessels. This shape is also reminiscent of certain contemporary free-blown bowls with pinched ribs (**Nos. 249–258**), as well as certain kinds of cast ribbed bowl (here **No. 34**). The fluting on the mold-blown examples, however, is comparable to that found on metalwork and differs markedly from the pronounced ribs on cast and free-blown wares. Additional shapes include taller bowls, of which **No. 211** may be an example, or broad shallow bowls such as **No. 203** (for intact pieces, see Aquileia 1968, 101–102, 105–106, and the other references in the list below). Less common are jars (Oppenländer Coll. 1974, 160, no. 449), which in all probability also belong to this type. Unlike their Romano-Syrian counterparts, however, the fluted bowls tend to be larger, have thicker walls, and aside from hairpin flutes, are otherwise undecorated. In all instances rims are either upright or off-set, with an unfinished or slightly polished edge. The bottoms, often quite thick, are slightly concave and usually exhibit several concentric, circular ridges.

Although most mold-blown fluted bowls are naturally colored (often light green [**No. 215**], bluish-green [**No. 211**], pale yellowish-green [**No. 213**], light purple, or olive), decolorized pieces such as **No. 217** exist, as do deliberately colored examples. Dark blue (**No. 214**) is the most abundant. As with Sidonian bottles, it appears that fluted bowls also made the transition at midcentury from colored to naturally colored fabrics.

Intact examples are found in many museum collections, but few have been published from excavated sites, making mold-blown fluted bowls seem less common than is actually the case.

Parallels and Dated Examples of Fluted Beakers and Bowls

Aquileia 1968, 101–102, 105–106, nos. 249–250, pl. 16.1–2, with a list of similar pieces on p. 101, n. 188.

Rome: Antiquarium Comunale. Several fragments of fluted bowls. Unpublished.

Pompeii: Museo Nazionale, Naples. Illustrated in Spinazzola 1928, 222. Pre-A.D. 79.

Murano: Museo Vetrario, inv. no. 5074. Broad, shallow bowl from Zara. Unpublished.

Histria II 1966, 219, pls. 81 and 101. Tall beaker found in a burial attributed to the Neronian–Vespasianic period.

Tarentum: Museo Nazionale, inv. no. 22722. Broad shallow bowl from a burial in Contrada Corti Vecchie, Tomb 20, cited in Newark Coll. 1976, 48, no. 39.

Fremersdorf 1958, 32, no. 29.157, pl. 34. Tall bulbous beaker found with a coin of Nero.

Camulodunum 1947, 300, no. 48, pls. 136–137. Fragment from Period IV, A.D. 10–65.

Verulamium 1972, 196, nos. 1–2, fig. 74. Two fragmentary examples.

Magdalensberg 1979, 34, no. 318, pl. 2. Presumably a mold-blown fluted bowl. It must antedate the abandonment of the site ca. A.D. 45.

Vindonissa 1960, 55–56, nos. 139–143, pls. 9, 18. Nineteen fragmentary examples from contexts of the first century A.D., including several from Tiberian/early Claudian levels.

ROM Coll. 1975, 33, 47–48, no. 82, fig. 2, pl. 4. With additional parallels listed from western and eastern Mediterranean sites.

JGS 4 1962, 140, no. 8. Broad, shallow bowl said to come from western Anatolia.
Conimbriga 1976, 167, no. 43, pl. 35. With other examples cited, both from western and eastern Mediterranean sites.
Victoria and Albert Museum: Bowl said to come from Jerusalem, acc. no. 899-1902; another from Cyprus, acc. no. C35-1961. Unpublished.

Although mold-blown fluted bowls have generally been assigned to the second half of the first century A.D. (ROM Coll. 1975, 33), sufficient examples are now recorded from earlier contexts to indicate that they, like most other types of early mold-blown glass, were being manufactured by the second quarter of that century. Their pattern of distribution, not surprisingly, resembles that of other Romano-Syrian wares as well, encompassing the Aegean and Syro-Palestinian regions, as well as Italy and its adjoining provinces. This in turn may suggest that production occurred in both Syro-Palestinian and Italian workshops.

Fluted Beakers and Bowls

211. CF 2431 Pl. 12
A.P. Room 16 Exterior, V
Fragment of rim (edge missing) and side of deep bowl; natural bluish-green. Slight iridescence and pitting. Off-set rim, turns outward to join side; straight side, tapering downward. Uneven interior surface. On exterior of side, multiple vertical flutes raised in molded relief. Blown into a patterned mold. D uncertain; A Th .2 cm.

212. CF 2400 Pl. 12
A.P. Room 16 Exterior, V
Two non-joining fragments of rim (edge missing) and side of broad shallow bowl; natural bluish-green. Slight iridescence and pitting. Off-set rim, turns outward to join side; almost straight side, tapering downward. Interior surface uneven. On exterior of side, multiple vertical flutes raised in molded relief. Blown into a patterned mold. D near rim ca. 9 cm; A Th .15 cm.

213. CF 2334 Pl. 12
A.P. Room 25, I
Fragment of rim (edge missing) of broad shallow bowl; pale yellowish-green. Slight iridescence and pitting. Off-set rim slopes diagonally downward, then turns outward to side; straight side set obliquely to rim. Interior surface uneven. On exterior of side, multiple vertical flutes raised in molded relief. Blown into a patterned mold. D uncertain; A Th .3 cm.

214. C66.803 Pl. 12
V-D, North, 0
Two joining fragments of lower side and bottom of bowl; dark blue. Iridescence and pitting. Convex curving side with smooth interior surface; concave bottom. On exterior of side, multiple vertical flutes raised in molded relief. Fluting extends to the first of three rounded, concentric ridges that encircle a raised central dot at the center of the bottom. Blown into a patterned mold. Max D bottom 6.5 cm; A Th bottom .3 cm.

215. C70.572 Pl. 12
S.H. Street M, 0-I

Fragment of side near bottom of bowl; natural light green. Slight iridescence and pitting. Convex curving side with smooth interior surface. On exterior of side, multiple vertical flutes raised in molded relief. Fluting terminates 1 cm above a band of two concentric ridges that encircle the bottom. Blown into a patterned mold. D lower side ca. 10 cm; A Th lower side .4 cm.

216. CE 2027
A.P. Room 22, I
Fragment of side near bottom of bowl; greenish-yellow. Slight iridescence and pitting. Convex curving side with uneven interior surface. Shape and decoration as in **No. 215**. Blown into a patterned mold. A Th .3 cm.

217. CD 1083 Pl. 12
Bas. ENW 2, II
Fragment of side near bottom of bowl; colorless with grayish tinge. Slight iridescence and pitting. Convex curving side with smooth interior surface. Shape and decoration as in **No. 215**. Blown into a patterned mold. A Th .4 cm.

218. CF 2430
A.P. Room 25C, I
Fragment of side near bottom of bowl; natural greenish-blue. Slight iridescence and pitting. Convex curving side with smooth interior surface. On exterior side, multiple vertical flutes raised in molded relief, similar to those on **No. 217**. Blown into a patterned mold. A Th .33 cm.

219. Additional fragments of mold-blown fluted bowls.
a. C70.630. S.H. Room 21, 0. Tiny fragment of side; pale yellowish-green.
b. CE 2313. A.P. Exterior NW, I. Tiny fragment of side; natural bluish-green.
c. CE 1984. A.P. Room 4, I. Tiny fragment of side; greenish-yellow.
d. CE 2141. A.P. Room 15, II. Tiny fragment of side; natural bluish-green.

*Beakers with Mythological Scenes (**No. 220**)*

Aside from Vindonissa, Cosa is the only controlled excavation to have yielded an example of a most unusual group of decorated mold-blown beakers. These mythological beakers (*Götterbecher*) constitute a fairly uniform series of tall conical vessels with tapering sides, knocked-off rims, and flat bottoms, usually with several concentric ridges. On all examples the vessels are divided by columns into four rectangular panels. The columns are joined either by leafy garlands suspended from their capitals or by architectural pediments, thereby creating four architectonic niches, or *naiskoi*. Within each niche stands a figure, usually identified as a season (Horos) or a god (possibly a goddess). Based on size, two subclasses exist. One is quite large and thick-walled, as is true of a beaker in the Yale University Art Gallery (Yale Coll. 1980, 56–59, no. 137); the other shorter and possessing thinly blown sides, such as the example from Cosa published here (**No. 220**).

Over twenty such beakers are known, and most are collected and described by G. D. Weinberg (1972). In this article Weinberg identifies three main groups based on her identification of the figures, their attributes, and other stylistic criteria. The fragment from Cosa appears to belong to her class II (Weinberg 1972, 38). Beakers of this group employ four panels separated by Ionic columns crowned by garlands. Within each panel is a figure (all presumably male) with full front torso and profile head. On intact pieces, two of the figures are consistently identified as Poseidon and Dionysos.[45] The other two are thought to represent Summer and Autumn, though disagreement exists (Hayward 1962, 55–57, no. 9, fig. 13; Weinberg 1972, 42–42; Yale Coll. 1980, 58). The partially preserved figure on the Cosan fragment is that of Dionysos, identified by the back of his head and the top of a thyrsus that he is carrying. The fragment also preserves the upper part of an Ionic column with a garland suspended from its capital.

On complete examples, Dionysos is wearing a chiton, girdled at the waist, carrying a thyrsus in his left hand and in his right, a vessel, possibly a kantharos. A small animal, gazing upward, sits in front of his right leg. Usually the animal is considered a panther, a creature regularly associated with the god in classical mythology.[46] An intact beaker of this sort is illustrated in Yale Coll. 1980, 58, esp. fig. 137C, a detail of the god Dionysos. This vessel, however, is thick-walled and far larger in size in decided contrast to the fragment from Cosa. A closer parallel is a smaller beaker now in the Museum für Kunst und Gewerbe, Hamburg (Weinberg 1972, 40–41, no. 6), said to be from Asia Minor.

In addition to these two pieces, which are part of Weinberg's class II, four smaller intact and three fragmentary beakers have been published. Provenances, where known, range from Baden in Switzerland, Nijmegen in Holland, and Crnelo near Sticna (28 km east of Ljublana), to one purchased in Kerch and now housed in the Hermitage Museum, another said to be from Asia Minor, and yet another thought to be from Hama in Syria (see Weinberg 1972 for references). The beaker from Slovenia was discovered in a grave also containing a small blown ribbed bowl similar to **Nos. 249–258** at Cosa, and typical of comparable bowls dated to the early and mid-first century A.D. (pp. 122–124 below). Although these beakers have been assigned to the latter part of the first century, this blown ribbed bowl may suggest that the series began somewhat earlier in the century. Although the fragment from Cosa contributes no new chronological information about the type, it extends the geographic range of the mythological beakers to Italy.

Beaker with Mythological Scenes

220. CF 2450 Pl. 12
A.P. Room 25 Exterior, I

Fragment of upper side of tall conical beaker; colorless. Slight iridescence and discoloration. Almost vertical, straight side, tapering slightly downward. Interior surface uneven. On the right side of the fragment is preserved the upper portion of an Ionic column and capital, with a garland of leaves suspended from the capital. Below the garland is the upper part of a thyrsus. On the left side of the fragment is a lock of hair belonging to the figure carrying the thyrsus. This beaker is probably

[45] This particular representation of the god has been compared to similar images of Dionysos found on Arretine pottery (see discussion in Weinberg 1972 and Yale Coll. 1980, 58). A further comparison, not previously mentioned, can be drawn between the figures on glass beakers and on certain appliqués found on Gaulish sigillata, especially on the earlier southern Gaulish wares. Unlike the freer and more elegant figures on Arretine pottery, those on Gaulish sigillata are more awkward and given to rigid, stereotyped poses. The representation of Dionysos, as found in Weinberg's class II, in many respects resembles the Dionysos found on such sigillata vessels (Déchelette II 1904, 54, nos. 302–303). They exhibit similar poses and attributes but differ in that the god is clothed on glass vessels, nude on ceramic.

[46] The identification of the animal as a panther, despite its diminutive size, finds support in a comparable representation on a Roman wall painting from Pompeii (?) in which Dionysos appears with a small creature, here undoubtedly a panther. In the painting the animal's misshapen body faces away from the god, but its head turns back and gazes upward at Dionysos, who is pouring a libation from a patera above the animal's head. The god holds a thyrsus in his left hand. In fact, this is very close to the representation seen on glass vessels (Mount Holyoke College Art Museum, acc. no. C.P1. 2-5. 1958).

identical in shape and decoration to class II in Weinberg 1972. Blown into a patterned mold. D uncertain; A Th side .08 cm.
Bibliography: Grose 1973, 179, pl. 189, no. 9.

Bossed Beakers (*Nos. 221–226*)

Typical mold-blown products of the mid- to late first century A.D. are bossed or "lotus-bud" beakers. These beakers take the form of small (**No. 226**), medium (**Nos. 221** and **225**), and exceptionally large (**No. 222**) conical vessels with upright rims and straight sides that taper downward to a flat bottom. The undersides of the bottoms are usually decorated with concentric circular ridges but occasionally with other patterns. Although other shapes also employ bosses—for example, squat short-necked bottles (ROM Coll. 1975, 48–49, no. 86, pl. 7), long-necked bottles (Corning Coll. 1980, 52, no. 45), or small beakers with convex curving sides (Wheaton Coll. 1971, 12–13, fig. 5)—the tall conical beaker is predominant. It closely resembles myth beakers and shares common features with other beakers of Romano-Syrian type.

Decoration takes the form of convex, almond-shaped bosses raised in relief. These have been variously interpreted as representing lotus or palm buds (Newark Coll. 1976, 77–76, nos. 78–79) or merely as imitations of decoration found on contemporary metalware (Isings 1957, 45, Form 31). Bosses on a single vessel are generally of uniform size and shape (**No. 225**), but occasionally they are sized, often with the larger bosses near the rim but sometimes the reverse, as is true of **No. 226**. On some pieces the wider ends of the bosses are at the top (**No. 222**), while on other examples they are at the bottom (**Nos. 225–226**). Bosses may be formed from a single convex curve (**No. 226**), from two (a double boss, **No. 225**), or even three (a triple boss, **No. 222**). On the body of the vessel, the bosses may be set randomly or in a pattern, most often in a quincunx (**No. 225**) but occasionally in parallel rows. On at least one example, the bosses are laid on their sides and aligned horizontally.

Aside from simple bossed beakers there are more elaborate examples that combine bosses with other relief decoration. On occasion, bosses are situated within a diamond-shaped trellis (Vindonissa 1960, 52–54), or placed amidst grape vines and birds (Isings 1957, 45), or simply set randomly among vine tendrils, as is true of **No. 221**. Occasionally raised dots are interspersed among the bosses (again Vindonissa, cited above), or the bosses are pyramidal in shape (Fishbourne 1971, 338–339, no. 37, fig. 138, pl. 28). Berger, in discussing the many examples found at Vindonissa, distinguished five subclasses based on the nature and disposition of the decoration.

In color, bossed beakers reflect other classes of first-century mold-blown glass, especially those types that are considered predominantly Neronian or Flavian. Most examples are naturally colored, ranging from pale yellow and yellowish-green to light green, bluish-green, olive, and pale purple. Many others are decolorized, a fact which suggests that many bossed beakers were manufactured in the second half of the first century, possibly continuing into the second century. No specimen known to the author is deliberately colored and therefore reminiscent of the mold-blown types of the second quarter of the first century A.D.

Parallels and Dated Examples of Bossed Beakers

Luni I 1973, 479, no. C 284, not ill.
Luni II 1977, 288; CM 7465, CM 5822; pls. 157.5 and 157.6. One fragment exhibits bosses among vine tendrils, the second only tendrils.

Rome: Antiquarium Comunale. Several fragmentary beakers. Unpublished.
Vatican Coll. 1975, 69–70, no. 677, pl. 30. Provenance unknown but presumably Rome or Italy. On the underside of the bottom, raised spoke-like ridges.
Ostia III 1973, 200, pl. 40.308. Presumably a fragment of a bossed beaker.
Bologna Coll. 1979, 46–47, no. 33.
Adria: An excavated beaker of the first century A.D. in the Bocchi Collection, cited in Bologna Coll. 1979, 46.
Pompeii: Museo Nazionale, Naples. At least ten examples illustrated in Spinazzola 1928, or Eisen 1927, I, 289–290, fig. 131; others from Pompeii described in Isings 1957, 45–46, Form 31. Several examples exhibit vine tendrils as well as bosses. All antedate A.D. 79.
Zadar: Museo Vetrario, Murano. Beaker illustrated in Mariacher 1966, 11, pl. 12.
Bologna Coll. 1979, 46–47, no. 33, color pl. on p. 47. Conical beaker.
Biella (Piedmont) 1950, 29, fig. 5. Beaker with bosses and vine tendrils.
Crivelli 1979, unpaged ill. of Tomb 58.17, no. 143. Tall conical beaker.
Fishbourne 1971, 338–339, no. 37, fig. 138. Beaker with pyramidal bosses, with a general discussion of the type.
Verulamium 1972, 196–197, nos. 3–6, fig. 74. Three fragments representing one or more vessels with bosses set in a diamond grid; other parallels from Britain cited on p. 198 where all examples are thought to be no earlier than Flavian times.
Harden 1968, 54, no. 62. Beaker said to be from Syria.
Conimbriga 1976, 167–168, 177, nos. 44–45b, pl. 35. Three examples cited.
Price 1974, 66. Examples from Spain noted.
ROM. Coll. 1975, 48, no. 83. Beaker, with discussion and further parallels cited, from both western and eastern Mediterranean contexts.
Isings 1957, 45–46, Form 31. Many examples listed, together with chronological discussion.
Newark Coll. 1976, 76–77, nos. 78–79. Two beakers, with further references to other examples, including several in Asia Minor and the eastern Mediterranean.

Although few examples are from closely dated contexts, bossed beakers, as noted above, are regarded as belonging primarily to the second half of the first century A.D. (Isings 1957, 46). The absence of examples from contexts earlier than ca. A.D. 40/50, whether at Camulodunum, Magdalensberg, Cosa, or elsewhere, may, in fact, indicate that the type emerged only after midcentury. Their presence at Pompeii and Herculaneum, however, attests to their existence in the seventies.

The total number of examples and variants is exceptionally large, and their geographic range extends from Asia Minor and the Aegean south to Cyprus and the Syro-Palestinian area and west to North Africa, Italy, and the provinces of northwestern Europe. Like other mold-blown vessels, they are usually assigned to workshops operating both in Italy and in the eastern Mediterranean.

Bossed Beakers

221. CE 2076 Pl. 13
A.P. Room 25, I

Fragment of side of tall conical beaker; colorless with greenish tinge. Slight iridescence. Almost vertical, straight side, tapering downward. On exterior, portions of two double bosses, wider ends at bottom, set amidst a vine (?) tendril. Blown into a patterned mold. D uncertain; A Th .1 cm.

222. CD 1080 Pl. 13
Bas. ENE 3, I
Fragment of rim and side of large conical beaker; natural light green. Iridescence and pitting. Slightly inturned rim with knocked-off edge; almost vertical, straight side, tapering downward. On exterior of fragment, upper part of a large triple boss, wider end at top; also on the exterior, a number of horizontal polishing marks 1 cm below rim. Blown into a patterned mold and rotary-polished on and about the rim. D ca. 9.5 cm; A Th .17 cm.

223. C66.809
V-D, Street M, 0-I
Small fragment of side of tall conical beaker; natural light green. Iridescence and pitting. Almost vertical, straight side, tapering downward. On exterior, part of one large double boss, wider end at bottom. Blown into a patterned mold. A Th .2 cm.

224. CF 2401
Small fragment of side of tall conical beaker; colorless with greenish tinge. Iridescence and slight pitting. Almost vertical, straight side, tapering downward. On exterior, one large boss, wider end at bottom. Blown into a patterned mold. A Th .1 cm.

225. C65.674 Pl. 13
V-D, Casupola, 0
Fragment of side of tall conical beaker; colorless. Slight iridescence. Almost vertical, straight side, tapering downward. On exterior, portions of three double bosses, wider ends at bottom. Blown into a patterned mold. D ca. 8 cm; A Th .08 cm.

226. CD 1081 Pl. 13
Bas. ENE 4, II
Fragment of side of conical beaker; colorless. Slight iridescence. Almost vertical, straight side, tapering downward. On exterior, two small bosses, wider ends at bottom. Blown into a patterned mold. D ca. 5 cm; A Th .05 cm.

*Mold-Blown Vessels: Miscellaneous (**Nos. 227–229**)*

Nos. 227–229 are the bottoms of small vessels that were blown into molds, leaving concentric circular ridges on the underside. Because the pieces are too fragmentary to determine complete shapes, it is uncertain whether these pieces are related to the Romano-Syrian types of the first century A.D. or belong to other traditions of a later date.

Mold-Blown Vessels: Miscellaneous

227. CD 1118 Pl. 13
Bas. ENW 2, III
Fragment of side and bottom of small beaker or other vessel; natural bluish-green. Slight iridescence. Strain cracking. Nearly vertical side; concave bottom. On underside of bottom, two rounded, circular

ridges raised in molded relief around a raised central dot. Blown into a patterned mold. D bottom 4 cm; A Th side .1 cm; A Th bottom .2 cm.

228. CD 1085 Pl. 13
Bas. ENE 4 and 5, I
Fragment of side and bottom of small beaker or other vessel; pale greenish-yellow. Iridescence. Strain cracking. Vertical side; bottom slightly convex. On underside of bottom, a rounded, circular ridge raised in molded relief around a raised central dot. Blown into a patterned mold.

229. CD 1127
Bas. ENE 1, I
Fragment of bottom of small beaker or other vessel; pale greenish-yellow. Iridescence. Slightly concave bottom. On underside of bottom, a pattern as in **No. 228**. Blown into a patterned mold. D ca. 4 cm; A Th bottom .2 cm.

*Optic-Blown Vessels: Miscellaneous (**Nos. 230–231**)*

In the course of the second or third century A.D. several modified forms of mold-blowing were developed. One involved blowing a gather of glass into an open (or closed) mold in order to impart a design, usually a simple ribbed or geometric pattern, then further inflating the vessel once it had been removed from the mold. This caused the pattern to become less pronounced and, owing to the rotation of the pipe, often twisted in a uniform manner. These expanded mold-blown products were especially common in the late Empire. Their complete absence at Cosa only underscores how few inhabitants remained in the town after the third century (see chapter 1).

Another variation of the mold-blowing technique, as explained in chapter 3, required the glassmaker first to blow a gather into a patterned mold, then into a plain mold of approximately the same size. This second step caused the raised surface decoration to expand inwardly, leaving an impression on both surfaces. At Cosa, two examples (**Nos. 230–231**) of "optic blowing" were found. Neither fragment preserves an identifiable form, but both bear simple geometric designs. **No. 231**, possibly a beaker but lacking a parallel, was found in Deposit IX, belonging to the late first through early third century A.D. Had it not been found in this deposit, the fragment might have been assigned a late medieval date.[47]

OPTIC-BLOWN VESSELS: MISCELLANEOUS

230. CG 55 Pl. 13
Curia Exterior NW, Room 2, I
Fragment of side of vessel; colorless. Iridescence. Straight side, tapering downward. On exterior and interior, two partially preserved curved ridges, paired in reverse directions, raised in molded relief. Presumably blown into a patterned mold, then into a plain mold. D uncertain; A Th .05 cm.

[47] Information kindly furnished by Dr. Ludwig Berger. Given the considerable disturbance evident in the levels above Deposit IX, it is always possible that an intrusive piece of medieval date managed to work its way down into Roman strata.

231. CF 2307 Pl. 13
A.P. Room 25C, II Deposit IX
Late first to early third century A.D.

Fragment of side of vessel; colorless. Dulling and slight milky weathering. Straight side. Tapering slightly downward. On exterior and interior, a series of partially preserved, parallel ridges, disposed diagonally on body. Presumably blown into a patterned mold, then into a plain mold. D side ca. 7.2 cm; A Th .05 cm.

ROMAN BLOWN DECORATED GLASS

*Facet-Cut and Engraved Vessels (**Nos. 232–248**)*

Vessels at Cosa with cut decoration are not uncommon, but all are highly fragmentary, making it all but impossible to reconstruct the original patterns and designs, let alone to determine with any exactness the original forms of the glasses. The fragments, nonetheless, suggest a wide range of shapes, including plates, bowls, beakers, and jars (but no bottles or flasks), with dates spanning the late first through to mid-third century A.D. Some examples made use of cast blanks (**Nos. 235–242**, **244–247**), as is suggested by the rotary polished surfaces of the former and the slightly uneven, glossy inner surfaces of the latter. All pieces are colorless, and the quality of craftsmanship is at times exceptional, as is true of **Nos. 246** and **248**. One facet-cut vessel is catalogued with other examples of its type (**No. 193**, "Fine Wares: Colorless with Broad, Overhanging Rims"), whereas two fragmentary bases, here adorned with facets (**Nos. 238–239**), may in fact resemble **No. 276**, an almost intact bowl with coils but lacking cut decoration. Owing to the condition of the glass, few precise parallels can be located with certainty. Even those that are cited are often from excavations so distant from Cosa (Dura Europos, for example) that a common provenance is unlikely. Unfortunately, no examples of cut glass were retrieved from dated contexts at Cosa, but most pieces came from levels and areas of the town that also produced quantities of pottery, lamps, and coins of early to middle Imperial date.

The variety of cut glass present at Cosa is especially apparent when examining the fragments of plates and bowls from the site, **Nos. 232–242**. According to Clairmont (Dura Europos 1963, 65), it was not *until* the second century that the bowl (or plate) emerged as the preferred vehicle for the glasscutter's art. At the same time, such decorative elements as circular, oval, and grain-shaped facets or lattice-work cutting were also introduced to the cutter's repertoire. Based on these premises, it seems reasonable to conclude that most examples at Cosa date to the second century A.D., with some pieces as late as the third.

At Cosa, the best preserved bowl is **No. 232**, a vessel that appears to have been cast, then ornamented with off-set rows of lozenge-shaped or diamond-shaped facets. In form it may resemble the faceted bowl illustrated in Sangiorgi Coll. 1914, 45, no. 151, pl. 30, or even a saucepan found at Nijmegen (Isings 1964, 175, fig. 3). **Nos. 233** and **234** may also have first been cast as blanks and then cut. Unhappily, both are too fragmentary to suggest complete shapes. The pattern of **No. 233**, however, is echoed on the superb colorless bowl from the Cave of the Letters in Israel (Yadin 1963, 106–100, fig. 40, pl. 29). **No. 235**, a free-blown hemispherical bowl, is ornamented from the bottom to the rim with rows of oval facets, set either end to end or side by side in horizontal rows. Similar in concept but not identical in detail are several bowls described in Dura Europos 1963, 68–70, Group e, which also make use of thick-walled, hemispherical shapes and large circular or oval faceting. Such vessels are found not only in the Near East but also in the Rhineland, where many examples are thought to have been manufactured during the third or fourth century (Fremersdorf 1967, 69–71, nos. 33, 35, 37). **Nos. 236** and **237** make use of grain or rice-shaped facets superficially cut on the surface, while **No. 240** introduces a cut lattice. **Nos. 241** and **242** are insufficiently preserved for meaningful comment, although the cutting on the base of the latter is reminiscent of the underside of a faceted bowl in the Sangiorgi collection, noted above.

The only cut bowls with adequate comparanda are **Nos. 238–239**, thinly blown vessels that originally resembled **No. 267**, a deep hemispherical bowl, or possibly a shallower bowl comparable to one in the Ransom Collection at Cambridge University, originally from Curium on Cyprus (ill. in Karanis 1936, fig. 2a). In both instances the base makes use of two concentric, circular coils as base-rings. On the fragments at Cosa (as well as on the bowl from Curium), a pattern of grain-shaped facets is arranged symmetrically on the bottom around a central circular facet. Similar bases combining coils and facets are found at Luni II 1977, 556, K 3047, pl. 286.14; Dura Europos 1963, 80, 82–83, no. 309, pl. 30, also nos. 323 and 325, pls. 9 and 30; and Karanis 1936, 103 and 124, no. 337, pl. 15, with coils but lacking facets.

Among cut drinking vessels at Cosa are three examples of a well-known series of conical beakers (**Nos. 243–245**) of which some are tall, other shorter, but all possess upright rims and outsplayed rims and outsplayed base-rings carved from the vessel-blank. Exterior surfaces are adorned with contiguous hexagonal facets set either in a quincunx or in a honeycomb pattern. Most published examples are from the provinces of northwestern Europe, but a great many others are known from Italian sites and, to a lesser degree, from the eastern Mediterranean (see citations below). A surprisingly large number were exported outside the bounds of the empire and are now found in Germany and Scandinavia (Eggers 1951 and Ekholm 1936), or in a vast hoard of Roman glass found at Begram in Afganistan (Hamelin 1953, 128, pls. 7–8, and Hamelin 1954, 175–178).

Beakers of this sort first appear in the second half of the first century A.D., and vessels of comparable shape but different cutting (long, curving, often asymmetrical facets) appear at Pompeii (Ekholm 1936, 64, fig. 1; Spinazzola 1928, 228; and Norling-Christensen 1968, 410–427), and therefore antedate A.D. 79. Ekholm has argued that such vessels are prototypes of the standard series of facet-cut beakers, but in fact they are contemporary. A long-necked, handleless flask from Pompeii exhibits long, asymmetrical curving facets as well as hexagonal facets set in a honeycomb pattern (Norling-Christensen 1968, 420–421, pls. 10–11). This demonstrates that hexagonal faceting was known at Pompeii by A.D. 79.

The three faceted beakers at Cosa fall into the standard series of such vessels and probably belong to Berger's Form III (Vindonissa 1960, 67–74, pl. 11). In date, this group is principally Flavian or Trajanic (see evidence in Fishbourne 1971, 341). The provenance of such beakers is uncertain, but both Italy and Alexandria, as well as other eastern Mediterranean centers, have been suggested (Dura Europos 1963, 60–61; Fishbourne 1971, 341). Given their wide distribution within the empire and beyond, multiple places of production are quite possible.

Surprisingly, no examples of Berger's Form IV, a shorter beaker with oval faceting, has been recovered from Cosa (Vindonissa 1960, cited above).

Parallels and Dated Examples of Facet-Cut Vessels

Ostia III 1973, 198, no. 297, pl. 40. Faceted beaker, belonging to Berger's Form III; also 294, no. 689, pl. 73.

Luni I 1973, 146, CM 1493/1, pl. 83. Base of a faceted beaker.

Paestum: Museo Archeologico. Two fragmentary beakers from the Bona Dea Sanctuary. University of Michigan–University of Perugia excavations. Unpublished.
Pavia Coll. 1983, 86–87, no. 52. An undated beaker with long curving facets.
Conimbriga 1976, 176–177, and 183, nos. 149–152, pls. 40 and 58. No. 152 from a Flavian context. Parallels from the Iberian Peninsula also cited.
Ekholm 1936, 61–72. Examples from Scandinavia cited.
Eggers 1951, Type 185, pl. 14. Extensive list of examples from Scandinavian and German contexts. Eggers Type 185 is identical to Berger's Form III.
Werner 1953. Discussion of Eggers's beakers, Types 185–187.
Dura-Europos 1967, 60–63, nos. 236–247, pls. 7 and 25. Discussion, with Durene examples cited.
Vindonissa 1960, 67–74, nos. 165–179, pls. 10–11, 19. Some examples from Flavian contexts. Full discussion and many parallels listed.
Fishbourne 1971, 339–344, nos. 41-44, fig. 139, pl. 26. Discussion and parallels cited.
Nida-Heddernheim 1974, 55–62, nos. 146–149, pls. 9 and 21. Discussion with a few additional parallels provided.

Of the remaining pieces at Cosa, **No. 246** is a unique facet-cut jar (or other vessel), but presumably not a beaker, with almost vertical, thick walls and large oval facets beneath a rim that has been carefully carved away. No known parallels for the piece exist. The same is true of **No. 247**, a small fragment lacking any curvature. Either it was intended as inlay, or it is from a flat plate or tray. Finally, **No. 248** is a bowl made of a fine colorless glass, probably cast, and decorated with a band of abraded floral decoration lightly cut beneath two horizontal grooves. No precise parallel for this vessel is known either.

Facet-Cut Plates and Bowls

232. CD 1074 Pl. 13
Bas. ENW 2, III
Fragment of rim and wall of large bowl; colorless. Slight milky weathering and pitting. Upright rim with rounded edge; almost straight side, proceeding diagonally to bottom. On exterior, a wide horizontal groove below rim: below this, a row of lozenge-shaped facets; next, several off-set rows, partially preserved, of diamond-shaped facets. Possibly cast, rotary-polished on both surfaces; facet-cut on exterior. D ca. 20 cm; Max Th rim .4 cm; Min Th side .2 cm.

233. CD 1073 Pl. 13
Bas. ENE 1, I
Fragment of floor of plate or other vessel; colorless. Slight milky weathering and pitting. Upper side of floor is flat; underside is slightly convex. Shape uncertain. On underside, a circular band of round facets alternating with pairs of parallel grain-shaped cuts; the band is framed by two concentric circular grooves. Possibly cast, rotary-polished on both surfaces; exterior facet-cut. Max D circular grooves ca. 6 cm; Max Th .2 cm.

234. CD 1164 Pl. 13
Bas. ENW 2, III
Fragment of floor of plate or other vessel; colorless. Slight milky weathering and pitting. Nearly flat floor. Shape uncertain. On underside, a circular row of oval facets set end to end around center of floor. Within this row are indications of seven similar facets set side by side in an uncertain pattern. Possibly cast, rotary-polished on both surfaces; exterior facet-cut. Max Th .4 cm.

235. CF 2345 Pl. 13
A.P. Room 25 Exterior, III
Fragment of side and bottom of bowl; colorless. Iridescence and pitting. Bubbly material. Convex curving side; large circular facet at the center of bottom acts as a resting surface. On the exterior, a pattern of large oval facets arranged in superimposed rows: nearest the bottom are six oval facets set end to end; above this, a partially preserved row of oval facets set side by side around body. Blown, facet-cut on exterior. Max D fragment ca. 9 cm; Max Th .4 cm. Restored symmetrically.

236. CG 234 Pl. 14
IX-D, Building C, III
Fragment of rim of large bowl or other vessel; colorless. Iridescence and pitting. Horizontal rim with rounded edge. On upper side of rim, a row of grain-shaped facets set side by side. Blown, facet-cut on upper side of rim; underside of rim rotary-polished. D ca. 18 cm; Max Th .2 cm.

237. CF 2351 Pl. 13
A.P. Room 25 Exterior NW, I
Fragment of side and bottom of bowl, beaker, or other vessel; colorless. Iridescence and slight pitting. Straight side, tapering downward to a curved basal angle; bottom probably slightly concave. On exterior, a pattern of grain-shaped facets placed in off-set rows. Blown, facet-cut on exterior. Max D fragment ca. 8 cm; Max Th .2 cm.

238. CF 2434 Pl. 14
A.P. Room 25, I
Fragment of side and bottom of bowl or other vessel; colorless. Iridescence and pitting. Curving basal angle; flat bottom. Shape uncertain. On exterior, two concentric threads encircle bottom. Between the threads, a single elongated facet, presumably one of four. At center of bottom is a large circular facet with two elongated facets set at right angles to each other, each in its own quadrant. Blown, facet-cut on exterior, with applied colorless threads. Max D applied threads ca. 8 cm; Max Th bottom .3 cm. Restored symmetrically.

239. CF 2318 Pl. 14
A.P. Room 26A, II
Three joining fragments of side and bottom of bowl or other vessel; colorless. Iridescence and pitting. Curving basal angle; flat bottom. Shape uncertain. On exterior, two concentric threads encircle the bottom. Between the threads are two elongated facets. Pattern probably identical to **No. 238**. Blown, facet-cut on exterior, with applied colorless threads. Max D applied threads ca. 8 cm; Max Th bottom .3 cm.

240. CF 2452 Pl. 13
A.P. Room 25 Exterior, I
Small fragment of floor (?) of plate or other vessel; colorless. Iridescence and pitting. Crazing on surface. Nearly flat floor with slight curve on upper side. Shape uncertain. On exterior of fragment, a partially preserved, elongated facet set horizontally, probably one of several that encircle the bottom. Within this row of facets are two parallel cut grooves, and between the grooves, a latticework of cut lines and an arched groove. Original pattern uncertain. Blown, facet-cut on exterior. A Th .2 cm.

241. CF 2375 Pl. 14
A.P. Room 16 Exterior, V
Small fragment of side of bowl or other vessel; colorless. Milky weathering and pitting. Convex curving side. Shape uncertain. On exterior, two off-set rows of oval facets set side by side around the vessel. Blown, facet-cut on exterior. Max D fragment ca. 15 cm; A Th .2 cm.

242. CG 447 Pl. 14
X-E, Building L Exterior, 0
Small fragment of bottom of bowl or other vessel; colorless. Milky weathering and pitting. Upper side of bottom uneven. On exterior at center of bottom, a large round facet acts as a resting surface. On exterior several deep oval facets set side by side. Blown, facet-cut on exterior. Pattern restored symmetrically. Max D fragment ca. 6 cm; Max Th bottom .7 cm.

Facet-Cut Beakers and Jars

243. CD 833 Pl. 14
Bas. ENW 1, I
Fragment of side near rim of conical beaker; colorless. Iridescence and pitting. Almost vertical, straight side, tapering downward. On exterior at junction of rim and side, a horizontal cut ridge. Below the ridge are three partially preserved, off-set rows of hexagonal facets set in a quincunx. Possibly cast, interior rotary-polished; exterior rotary-polished and facet-cut. Max D fragment ca. 7 cm; Max Th .3 cm.

244. CE 1994 Pl. 14
A.P. Room 12-a-5, Cesspool, I
Fragment of side of conical beaker; colorless. Milky weathering and pitting. Crazing on both surfaces. Almost vertical, straight side, tapering downward. On exterior, four rows of partially preserved hexagonal facets set in a honeycomb pattern. Blown, facet-cut on exterior. Max D fragment ca. 7.5 cm; Max Th .25 cm.

245. C68.661 Pl. 14
Forum ENW, 0
Fragment of side of conical beaker; colorless. Iridescence and pitting. Almost vertical, straight side, tapering downward. On exterior, five rows of partially preserved hexagonal facets set in a quincunx. Blown, facet-cut on exterior. Max D fragment ca. 8 cm; Max Th .3 cm.

246. C73.69 Pl. 14
A.P. V–VI, Perimeter Walls, 0
Fragment of rim and side of small jar or other vessel; colorless. Slight iridescence and pitting. Excellent craftsmanship. Upright rim with short projecting lip; nearly vertical side with slight taper downward. On exterior at lip, two cut ridges with a narrow groove dividing them; thickness of rim substantially cut away; considerably thicker side decorated with a partially preserved row of oval facets set side by side. Blown, rotary-polished and facet-cut on exterior. D ca. 7 cm; Max Th rim .3 cm; Max Th side .6 cm.

247. CG 446 Pl. 14
X-E, Building L Exterior NE, 0

Small fragment of side or floor (?) of vessel, possibly from a conical beaker; colorless. Iridescence and slight pitting. Flat fragment with no perceptible curve in any direction. Shape uncertain. On exterior, partially preserved off-set rows of oval facets set side by side in uncertain pattern. Blown, facet-cut on exterior. A Th .3 cm.

Engraved Bowl

248. CF 2325 Pl. 14
A.P. Room 25, I

Fragment of side near rim of bowl or other vessel; colorless. Dulling and pitting. Excellent craftsmanship. On exterior of fragment, a horizontal frieze with an abraded pattern of leaves, stems, and flowers. Above this frieze is a band of two horizontal cut grooves. Possibly cast, cut and abraded on exterior; interior has a fire-polished appearance. Max D fragment ca. 12.5 cm; A Th .3 cm.

Colored Bowls with Pinched Ribs (***Nos. 249–258***)

When first discussed as a group (Pfeffer and Haevernick 1958, 76–91; and Haevernick 1967, 153–166; republished Haevernick 1981, xi–xxviii and 171–179), these small colorful bowls (*zarte Rippenschalen*) were thought to belong to the second and third quarters of the first century A.D., to be concentrated almost exclusively in the Po valley and along the riverine trade routes of central and northwestern Europe, and to have been manufactured at factories operating in northern Italy. An updated survey of known examples now reveals a wholly different situation. In date, the type is much earlier, standing firmly in the first quarter or half of the first century A.D.; it may even have commenced slightly before this. It was devised under Augustus and achieved its greatest popularity under Tiberius and Claudius. Thereafter, examples appear in Neronian, Flavian, and occasionally in second-century contexts, where presumably they represent heirlooms or were redeposited.

In regard to distribution, the bowls occur commonly on sites *throughout* the Roman world, both in the eastern and western Mediterranean. Examples, often in large numbers, are known from central and southern Italy, North Africa, the Iberian Peninsula, Greece and the Aegean, Asia Minor, Cyprus, and the Syro-Palestinian region, as well as from northern Italy and the transalpine provinces. Such a distribution pattern no longer admits readily to a single center of production, in particular one located in the Po valley. Perhaps, as more attention is paid to recognizing the subclasses within the type, regional groupings can be segregated and assigned to appropriate workshops. Suffice it to say at present that these vessels represent the earliest variety of blown, decorated tableware to gain wide currency throughout most parts of the empire.

In color these bowls are reminiscent of most other examples of early blown glass; that is, they primarily appear in deliberately colored, translucent, or opaque fabrics, most often in golden-brown, blue, and purple, but occasionally in such opaque colors as white or light blue. Such colors are associated with the free-blown wares of the first half of the first century and tend to disappear, as noted on pages 81–82 above, shortly after midcentury. Perhaps the naturally colored bluish-green and light green examples, or the far rarer colorless pieces (Vindonissa 1960, pl. 16.1), were manufactured later in time than their brightly colored counterparts. Decoration when present takes the form of opaque white threads wound several times around the rim, shoulder, and occasionally the upper ends of the ribs (**Nos. 254–257**). On some bowls the rim also exhibits a narrow horizontal groove (**No. 250**). Exceptional examples are adorned with splashed decoration in several colors, as is true of two fragmentary pieces from Paestum (see list below).

The shapes of these delicate bowls are highly uniform and exhibit bulbous bodies, off-set, knocked-off rims, and slightly concave bottoms. They closely resemble other contemporary bowls that are both cast and mold-blown. In fact, they bear such an uncanny similarity to certain classes of cast ribbed bowls (here, **Nos. 48–51**) that they may be conscious imitations of the latter. They also resemble certain types of early Romano-Syrian mold-blown wares, especially the colored fluted bowls (here, **Nos. 211–212**). Not surprisingly, all three types are chronologically overlapping, and the similarities among them, despite differences in manufacturing technique, bear witness to this fact (for a brief discussion, see Grose 1982, 28).

Ten examples of these blown bowls (**Nos. 249–258**) were discovered at Cosa during the excavations. Seven are golden-yellow, two blue, and one purple. Among them, **Nos. 249–253** lack applied threads, and except for the occasional cut groove (**No. 250**), as noted above, are otherwise plain. These unadorned examples tend in size to be larger and exhibit more pronounced ribs than most bowls with threads. The decorated pieces, in fact, appear to comprise a somewhat smaller, more fragile group. Whether such differences are meaningful in terms of date or attribution to a particular workshop is uncertain.

Two fragments at Cosa were recovered from dated contexts: **No. 251** from Deposit V, assigned to ca. 25/15 B.C. to ca. A.D. 40/45, and **No. 257** in Deposit VII, belonging to the decade(s) before ca. A.D. 40/45. Such dates are in accord with the many examples from other sites recovered from Tiberian or Claudian levels (see list below). New evidence, however, has also accumulated demonstrating that the type was first devised earlier than previously suspected. Several examples from Magdalensberg 1979, 46, and from Morgantina 1982, 28–29, fig. 11, are considered to be from early Augustan contexts. Moreover, an unpublished example from the Athenian Agora, inv. no. 90 (G 55), is ascribed to a stratum of 50–25 B.C.[48] Although this particular context seems too early for a free-blown vessel of this sort, the type as a whole must have been created around the turn of the millennium, not long after the adoption of free-blowing (see pp. 98–99 above). Moreover, the type seems to have disappeared shortly after midcentury, somewhat earlier than once was thought.[49] The emphasis on color, the polychrome decoration effected by applied threads, and their characteristic shape further corroborate that these bowls belong to the first quarter or half of the first century.

Parallels and Dated Examples of Colored Bowls with Pinched Ribs

Cosa: **Nos. 251** and **257**, Deposits V and VII, respectively dated to ca. 25/15 B.C. to ca. A.D. 40/45, and to the decade before ca. A.D. 40/45.

Pfeffer and Haevernick 1958, reprinted 1981, xvii–xxiv, lists one hundred parallels, with references; Italian examples are on pp. xx–xxi.

Haevernick 1967, reprinted 1981, 172–179, adds many additional pieces, with the Italian examples mostly on p. 175.

Isings 1957, 35–36, Form 17. Discussion with examples.

[48] Information furnished by Gladys D. Weinberg.

[49] For examples, see Isings 1957, 35, who suggests that the bowls appeared "towards the middle of the first century and remained in use until the Flavian period."

Vindonissa 1960, 30–33, nos. 49–59, pls. 4 and 18. Sixty-three examples, of which several are dated in contexts belonging to ca. A.D. 30/40 to as late as A.D. 80.

Magdalensberg 1979, 43–47, nos. 419–467, pls. 12–13 and 21. One fragment from an Augustan context; another from a context terminating ca. A.D. 25. Many parallels cited, including unpublished Italian material.

Pavia Coll. 1983, 78–79, no. 47. Excavated example from Gravellona Lomellina.

Arezzo: Museo Civico. Light green bowl with pronounced ribs and no applied decoration. Unpublished.

Luni I 1973, 469, CM 2499.1, pl. 81.11; KA 376, pl. 215.27 (pronounced ribs); KA 332, pl. 215.34.

Luni II 1977, 278, CM 8246/3, pl. 154.

Ostia II 1970, 38, pl. 8.33a–b and pl. 15, 244a–b.

Paestum: Museo Archeologico. Two fragments with splashed decoration found in the Sanctuary of Bona Dea (University of Michigan–University of Perugia excavations).

Morgantina 1982, 28–29, fig. 11. Two examples among several from the site are illustrated.

Limburg 1971, 80, no. 151, fig. 19.

Kassel Coll. 1967, no. 27, pl. 3. Tall jar or beaker-shaped ribbed bowl similar in form to certain tall, mold-blown fluted beakers.

Tarragona: Pasaje Cobos. Several examples from a context ascribed to the second quarter of the first century A.D. Information furnished by Jennifer Price, University of Durham.

Athenian Agora: Inv. no. 90 (G 55). Small ribbed bowl said to be from a context dated between ca. 50 and 25 B.C.

Tel Anafa (Upper Galilee): Several fragmentary bowls. Unpublished.

Sardis 1980, 12–13, nos. 42–44, pls. 2 and 20.

Colored Bowls with Pinched Ribs

249. C70.302 Pl. 15
S.H. Street M, Sewer, I

Two joining and three non-joining fragments of rim and side of small bowl (profile mostly restorable); golden-yellow. Silvery iridescence and slight pitting. Slightly outsplayed rim with knocked-off edge; side bulges outward at center. On exterior, vertical ribs pinched outward from shoulder and side; ribs extend to junction of side and bottom. Blown and tooled. D ca. 9 cm; Max Th shoulder .3 cm; Min Th .1 cm.

250. CD 1095 Pl. 15
Bas. NEN 1, I

Two joining fragments of rim and side of small bowl; golden-yellow. Slight iridescence and pitting. Shape as in **No. 249**. On exterior, pronounced vertical ribs pinched outward from shoulder and side; also a narrow, horizontal groove at junction of rim and side. Blown, cut and tooled on exterior; some rotary-polishing marks on rim. D ca. 8–9 cm; Max Th shoulder .3 cm; Min Th .1 cm.

251. CE 2116
A.P. Room 16, I Deposit V
ca. 25/15 B.C. to ca. A.D. 40/45

Small fragment of rim (edge missing) and side of small bowl; golden-yellow. Slight iridescence and pitting. Shape and decoration as in **No. 250**. Blown, cut and tooled on exterior. D uncertain; Max Th shoulder .3 cm; Min Th .2 cm.

252. C71.196
V-E, 0

Fragment of rim (edge missing), side, and basal angle of small bowl; blue. Slight iridescence and pitting. Shape as in **No. 249**. On exterior, pronounced vertical ribs pinched outward from shoulder and side; ribs extend to junction of side and bottom. Blown and tooled. D uncertain; Max Th shoulder .3 cm; Min Th .1 cm.

253. C70.627
S.H. Street 5, Sewer, I

Tiny fragment of side near rim of small bowl; golden-yellow. Slight iridescence and pitting. Shape as in **No. 249**. On exterior, the upper portion of a small vertical rib pinched outward from the shoulder; also wheel-polishing marks at junction of rim and side. Blown, rotary-polished and tooled. D uncertain; Max Th shoulder .2 cm.

254. CD 1088 Pl. 15
Bas. ENE 1, II

Fragment of rim and side of small bowl; golden-yellow with opaque white decoration. Slight iridescence and pitting. Slightly outsplayed rim with knocked-off edge; side bulges outward at center. On exterior, narrow, closely set vertical ribs pinched outward from shoulder and side; also two or three horizontal, opaque white threads applied to shoulder with additional threads applied to upper ends of ribs on the body. Blown with applied and tooled decoration. D ca. 7 cm; Max Th shoulder 2 cm; Min Th .05 cm.

255. CD 1089
Bas. ENE 1, I

Fragment of rim (edge missing) of small bowl; golden-yellow with opaque white decorations. Slight silvery iridescence and pitting. Shape and decoration as in **No. 254**. Blown with applied and tooled decoration. D uncertain; Max Th shoulder .25 cm; Min Th .25 cm.

256. C71.194 Pl. 15
V-E, Drain Sediment

Two joining fragments of rim and side of small bowl; purple with opaque white decoration. Black scaly weathering and pitting. Slightly outsplayed rim with knocked-off edge; side bulges outward at center. On exterior, vestiges of narrow ribs pinched outward from shoulder and side; also eight horizontal opaque white threads applied to shoulder and to the upper ends of ribs. Blown with applied and tooled decoration. D ca. 8 cm; Max Th shoulder .2 cm; Min Th uncertain.

257. CE 2048 Pl. 15
A.P. Room 22, II Deposit VII
Decades prior to A.D. 40/45

Small fragment of side of small bowl; golden-yellow. Iridescence and pitting. Shape and decoration as in **No. 254**. Blown with applied and tooled decoration. D uncertain; Max Th shoulder .3 cm; Min Th .04 cm.
Bibliography: Grose 1975, 39, no. 13.

258. C70.577
VII, D-E, 0
Small fragment of side near bottom of small bowl; blue. Slight iridescence and pitting. Convex curving side. On exterior, portions of three vertical ribs pinched from the side. Blown with pinched decoration. A Th .2 cm.

*Vessels with Pinched Decoration: Miscellaneous (**Nos. 259–265**)*

In addition to the ribbed bowls of the previous section, other types or groups of vessels at Cosa also employ pinched decoration, usually small ribs or various projections applied and/or pulled outward from the side or base. **Nos. 260–262**, for example, exemplify the continuation after the mid-first century of this simple method of adorning tablewares. Although their shapes are uncertain, vertical ribs are tooled outward on the body. A small one-handled jug from Pompeii and a ribbed flask, both in the Vatican Museum, illustrate this technique as it was practiced in the Imperial period (Vatican Coll. 1975, 57, nos. 507–508, pl. 21).

More elaborate in conception are ribs that are further ornamented with projecting spikes (**No. 263**). Considerably more common are base-rings comprised of pinched toes (**Nos. 264–265a**), almost always associated with colorless vessels of the second and third centuries A.D., as examples from Ostia and Conimbriga show (Ostia II 1970, 36, no. 27, pl. 8; Conimbriga 1976, 198 and 191, nos. 193 and 195, pl. 41). Although the original forms of the vessels at Cosa cannot now be determined, such bases are associated with a wide variety of shapes—including bowls, beakers, jars, and flasks—as known from western and eastern Mediterranean sites (see discussion in Dura-Europos 1963, 50–51). Similar toed bases are also known from Italy at Luni (Luni I 1973, KA 375, pl. 216) and at Ostia (Ostia II 1970, 36, no. 27, pl. 8; Ostia IV 1977, 26, no. 40, pl. 9; also intact toed beaker in the Antiquarium on the site, unpublished).

The final piece in the section is decorated with a "figure-eight" design, disposed vertically on the body of what appears to be a beaker (**No. 259**). This motif is not especially common, yet it is known from several sites in the western and eastern Mediterranean (see below). The pattern, which is generally repeated several times on the vessel, was achieved by a combination of trailing and tooling as the pincer marks on the Cosan fragment make clear. The same motif is seen on two fragments from Tessin, found in a burial dated ca. A.D. 60/85 (Simonett 1941, 146, nos. 19a–b, fig. 124), and on two fragments now housed in the Museo Civico Archeologico in Bologna (Bologna Coll. 1979, 48–49, no. 34, with additional references). In other instances the figure-eight pattern is arranged horizontally, as is true at Luni (Luni I 1973, KA 330, pl. 216.23); and at eastern sites such as Sardis (Sardis 1980, 20–21, nos. 92–93, pls. 4 and 21) and Dura Europos (Dura Europos 1963, 48–49, nos. 187–192, pls. 5 and 23). The eastern pieces are considered to date to the third century and illustrate the continuing use of this decorative device by Imperial glassmakers. Whether the pattern was disposed vertically or horizontally appears to be of little chronological or stylistic consequence, as implied by examples from Cologne that exhibit both (Fremersdorf 1959, 72, pls. 110–111).

Vessels with Pinched Decoration: Miscellaneous

259. CE 2008 Pl. 15
A.P. Room 15, I
Fragment of side of beaker or other vessel; colorless. Milky weathering and much pitting. Nearly vertical side, tapering downward. On exterior, two vertically set ridges in a figure-eight pattern, pinched from side (or trailed), then tooled. Interior surface uneven. Blown and tooled. D uncertain; A Th .1 cm.

260. C70.573 Pl. 15
S.H. Street 5, Sewer, I
Fragment of side and bottom of small bottle or other vessel; natural light greenish-yellow. Iridescence. Convex curving side; slightly concave bottom. On exterior, portions of three short and unevenly spaced, vertical ribs pinched from side. Blown and tooled. A Th side .25 cm.

261. C70.626
S.H. Street M, 0
Fragment of side and bottom of vessel; natural light green. Iridescence and pitting. Flat bottom. Shape uncertain. On exterior, lower portion of one pinched rib. Blown and tooled. A Th .2 cm.

262. CF 2446 Pl. 15
A.P. Room 25 Exterior, I–II
Two joining fragments of side near bottom of vessel; colorless. Iridescence and corrosion. Convex curving side, with indications that side turns abruptly downward. Shape uncertain. On exterior, three short rounded ribs pinched from side. Blown and tooled. D ca. 8 cm; A Th side .2 cm.

263. C68.662 Pl. 15
Arx W. Slope, 0
Fragment of side of vessel; colorless. Silvery iridescence. Bubbly. Convex curving side. Shape uncertain. On exterior, a partially preserved vertical rib with three projecting spikes pinched from side. Blown and tooled. A Th side .1 cm.

264. CE 2064 Pl. 15
A.P. Room 24, I
Two joining fragments of side and base of small bowl or other vessel; colorless. Iridescence and slight pitting. Convex curving side. At base, numerous toed projections pinched from bottom to provide the vessel with a base-ring. Blown and tooled. D base-ring ca. 3.5 cm; A Th base .2 cm.

265. Additional fragments of toed base-rings.
- **a.** CF 2444, A.P. Room 25 Exterior NW, II. Colorless. Pl. 15.
- **b.** CF 2331, A.P. Room 25 Exterior, I. Colorless.
- **c.** CE 2172, A.P. Room 15, I. Colorless.

Vessels with Applied Prunts or Trails (***Nos. 266–285***)

One of the commonest modes of decorating glass is the application of simple prunts or trailed threads either in the same fabric as the body or in a contrasting color. At Cosa there are numerous Imperial wares that utilize applied threads but only one (**No. 266**) that is decorated with prunts. Its original shape cannot now be determined, but it may have been a beaker similar to one discovered at Vindonissa 1960, 4345, no. 103, pls. 7 and 17 and presumably dating to the first century A.D. In the second and third centuries, this decorative device continued to be used sporadically, as examples from Dura Europos and Sardis make clear (Dura Europos 1963, 51; Sardis 1980, 19–20), but none is known from Cosa. Another bowl with prunts, said to belong to the third century, is of western origin. It is now in the Rheinisches Landesmuseum, Bonn (acc. no. 6662) and is illustrated in Sardis 1980, pl. 3.

Only two of the fragments from Cosa, both beakers (**Nos. 267–268**), employ deliberately colored threads. On each piece a single coil adorns the rim. Neither example is from a dated context, but considering that most colorless Imperial tablewares found at Cosa fall between the late first and mid-third centuries, these pieces probably do as well. On the other hand, the use of dark blue threads (or turquoise blue) was especially favored by glassmakers of the fourth through sixth centuries, and it is possible that **No. 267** actually belongs to the last occupation of the town. A possible parallel for this piece was recovered at Ostia (Ostia IV 1977, 338, no. 370, pl. 49).

The zigzag trailing of **No. 269** belongs to a vessel of uncertain shape and date. Although this design is known from vessels of the fourth century A.D., as preserved in Cologne and the Rhineland (Fremersdorf 1959, 79–81, pls. 131–136), again it is unlikely that the Cosan fragment is that late in date. Moreover, German glasses of this period mostly employ colored threads rather than the decolorized trails found on the fragment at Cosa. **No. 270** preserves a base with a section of an applied thread in the same color as the vessel. Although probably a beaker, the exact form of the vessel and the configuration of the design cannot now be determined.

Tall Beakers with Bogenrippendekor

No. 271 at Cosa (and possibly **Nos. 272–273**) represents a standard group of tall beakers assigned to the mid- to late first century A.D. These are characterized by pronounced vertical ribs arched on both ends (*Becher mit Bogenrippendekor*). The ribs divide the body into oval or oblong sections. Generally the beakers are cylindrical or conical in shape with knocked-off rims and closed cut-out bases. The edge of the rim is generally rotary-polished until smooth, as is suggested by polishing marks found on many examples. This class of beaker is usually decolorized or lightly colored, and the ribs are always in the same fabric as the vessel itself. Although the majority of such vessels closely resemble this description, variations do occur, and these are noted in Vindonissa 1960, 47–48, and Nida-Heddernheim 1974, 25–27. Although **No. 271** can definitely be attributed to this type, the identification of **Nos. 272–273** is less certain owing to their fragmentary condition.

The geographic range of these beakers appears more circumscribed than is usually true of most varieties of drinking cups in the first century. Published examples are found almost exclusively on sites in central and northern Italy, Switzerland, and Germany, although at least one example is known from North Africa (see list below), nor are they found in the eastern Mediterranean. However, this may only reflect the vicissitudes of excavation and publication. Several of the earliest dated examples are beakers from the cemeteries of Tessin, where at least one piece is from a grave dated ca. A.D. 60/75. Most others are reported from Flavian or Trajanic contexts (see list below). Isings 1957, 48, suggests that the type was first devised in Claudian or Claudian/Neronian times, but considering available information, her estimate may be too early. No examples have been found at Cosa in the various deposits of the first half of the first century, nor do they occur at Magdalensberg, abandoned around A.D. 45.

PARALLELS AND DATED EXAMPLES OF TALL BEAKERS WITH *BOGENRIPPENDEKOR*

Simonett 1941, 62, no. 1, fig. 41.1, pl. 12.2, dated ca. A.D. 60–90; 74, no. 3, fig. 56.6, belonging to the last third of the first century; 119, no. 5, fig. 99.5, belonging to the last third of the first century; 122, no. 12, dating around A.D. 100; 169, no. 5, dating to the last third of the century.
Crivelli 1979, unpaged ill., nos. 9 and 13. Two beakers from the necropolis at Tenero, excavated in 1880–1881, as cited in Vindonissa 1960, 48, and attributed to end of the first century A.D.
Castagnole Lanze: As ill. in *Fasti archaeologici* 17, pl. 32, no. 4796.
Mariacher 1966, 10, nos. 12–13, fig. 8. From Zadar.
Aquileia 1968, 54 and 57, nos. 140–141, fig. 6.5; pl. 6.
Luni I 1973, 225, CM 2549/1, pl. 81.13.
Ostia IV 1977, 251, no. 626a–b, pl. 43.626. Probably a beaker of this type.
Vindonissa 1960, 47–48, nos. 110–116, pls. 7 and 19. Examples from Flavian contexts, one from a level with pottery dating around A.D. 70. Six additional parallels from Italy or neighboring areas also cited, as well as one example from the museum at Constantine, Algeria.
Nida-Heddernheim 1974, 25–27, no. 27, pl. 3. Examples from a Flavian/Trajanic stratum. Additional parallels listed.
Isings 1957, 47–48, Form 33. More parallels cited.

Nos. 274–275a preserve the bottoms of naturally colored light green or bluish-green vessels decorated with opaque white spiraling threads. Although these pieces are so fragmentary that complete shapes cannot now be determined, it is possible that they formed the bottoms of ladles, such as those found at Pompeii and now exhibited in the Museo Nazionale, Naples (unpublished), or others, slightly different, from Herculaneum (De Franciscis 1963, 137–139, fig. 1), or one from an undesignated source (Wheaton Coll. 1971, 27–28, no. 52, fig. 31). The color, size, and threading closely resemble those of the Cosan pieces. Other possibilities, however, also present themselves, and the fragments may have belonged to flasks like one from Germany attributed to the first or second century A.D. (La Baume n.d., D 18, pl. 19). Similarly decorated glasses are known from Conimbriga; the fragments there are thought to belong to beakers or drinking cups of some sort (Conimbriga 1976, 195 and 202, nos. 218–224, pl. 42). In the latter instance, they are attributed to beakers of the fourth or fifth century A.D., a period later than almost all material at Cosa. Ostia I 1968, 74, B4.2160, fig. 202a–b, depicts yet another fragmentary bottom with similar threading.

One of the better-preserved bowls at Cosa (**No. 276**) belongs to a group widely distributed in the western and possibly the eastern Mediterranean during the second and third centuries A.D. These bowls make use of colorless glasses that are decorated with horizontal threads. Usually one or more coils are applied to the bottom of the vessel and form a resting platform. The rim is usually upright with a solid, thickened edge. The sides conform to a deep hemispherical shape with an abrupt basal angle and a flat bottom. Quite often the bottom displays a pronounced pontil scar, as is true of **No. 276a**.

Although many such bowls exist, few have been published. From Italy the closest parallel is a shallower bowl, originally found at Zadar and now in the museum at Murano (Mariacher 1966, 15, fig. 15). Also similar is a bowl from Baldock in Britain, dated around A.D. 200 (Charlesworth 1959, 49, pl. II), and others, more fragmentary, from Conimbriga, where such bowls, along with comparable colorless threaded beakers, appear quite routinely (Conimbriga 1965, 80–81, nos. 110–117, pl. 4; cf. Conimbriga 1976, 186 and 190, nos. 167–169, pl. 40). Also similar in many respects is a bowl from Cyprus (Vessberg 1956, 134, fig. 42.28). The pieces cited in Isings 1957, 102–103, Form 85b, more closely resemble beakers rather than large bowls of this sort.

The final prominent group of vessels at Cosa with trailed decoration is a series of colorless beakers, **Nos. 277–283**. These vessels form a homogeneous class of utilitarian drinking cups popular on Roman sites in Italy and the western provinces, possibly elsewhere as well. Rims are usually upright and exhibit a solid, thickened edge. Bodies are slightly conical or cylindrical, and bases (where preserved) are variously treated (slightly concave, closed cut-out, or even with applied circular coils). The colorless threads in horizontal bands of one to six are applied below the rim. For the most part such vessels are delicately blown and unusually thin-walled. Since manufacture of these wares is presumed to have been regional, local variations undoubtedly exist. The geographic distribution and date of these beakers resemble those of the bowls described above. The group seems to represent a typical household ware of the second and third centuries A.D. Two dated examples from Cosa (**Nos. 281** and **283**) confirm such a chronology.

Parallels and Dated Examples of Colorless Cups with Horizontal Thread(s) below the Rim

Cosa: **Nos. 281** and **283**, Deposit IX, belonging to the late first through early third century A.D.

Ostia III 1973, 158, no. 210, pl. 32. Two fragments of similar beakers.

Genty 1972, 69–75. Somewhat different colorless beakers, all with coil base-rings, from sites in France and dated to the late second or third century A.D.

Conimbriga 1976, 186 and 190, nos. 171–177, pls. 40–41. Fragments found in contexts spanning the late first to fifth century. Alarcão, however, compared the fragments to intact vessels known to him from two groups at Mulva, respectively dated to the first half and second half of the second century A.D.

The last two trailed pieces in this catalogue, **Nos. 284–285**, have highly problematic profiles. They are either portions of lids with central knobs or the feet of stemmed vessels. In either case no parallels have yet been located for these fragments. It should be noted, however, that **No. 284** is from Deposit IX and belongs, like most of the decolorized glass at Cosa, to the late first to early third century A.D.

Applied Prunts

266. C72.90 Pl. 15
VI, VII-D, 0

Tiny fragment of side near bottom of vessel; natural yellowish-green. Iridescence and some corrosion. Convex curving side. Shape uncertain, but possibly a beaker. On exterior, a small prunt in the same fabric as the vessel itself. Blown with applied decoration. Max Th side .5 cm.

Applied Threads

267. CE 291 Pl. 16
A.P. II, Room 59, I

Fragment of rim and side of beaker; colorless with blue decoration. Iridescence and pitting. Bubbly. Upright rim with rounded edge; almost vertical, straight side, tapering downward. Probably conical in shape. On exterior, a horizontal thread of blue glass applied 1.6 cm below edge of rim. Blown with applied decoration. D ca. 8 cm; A Th .15 cm.

268. CD 1099 Pl. 16
Bas. ENW 2, III
Fragment of rim and side of small beaker or other vessel; colorless with opaque white decoration. Iridescence and pitting. Inturned rim with thickened, rounded edge; side flares outward below rim. On exterior, a horizontal thread of opaque white glass applied 1 cm below edge of rim. Blown with applied decoration. D ca. 3.4 cm; A Th side .08 cm.

269. CF 2363 Pl. 15
A.P. Room 25A, II Deposit IX
Late first through early third century A.D.
Fragment of side of vessel; colorless. Iridescence and pitting. Side curves diagonally downward. Shape uncertain. On exterior, a zigzag pattern of colorless threads applied between two parallel horizontal threads. Blown with applied decoration. A Th .1 cm.

270. CD 1090 Pl. 15
Bas. ESW 4A, I
Fragment of side and base of vessel; pale purple. Iridescence and pitting. Almost vertical, straight side, tapering downward; slightly concave base; solid cut-out base-ring. On exterior, the lowest portion of V-shaped threads applied in same fabric as vessel. Blown with applied decoration. D base-ring ca. 5 cm; A Th side .2 cm.

271. CE 2104 Pl. 15
A.P. Room 22, I
Three non-joining fragments of rim and side of beaker or other vessel; natural light green. Iridescence and slight pitting. Outsplayed rim with knocked-off edge; nearly vertical but uneven side, tapering downward. On exterior, pronounced ribs applied and drawn from shoulder in the same fabric as the vessel; also a horizontal wheel-cut groove beneath rim. Blown with applied and tooled decoration; rim rotary-polished and cut. D ca. 8 cm; A Th side .1 cm.

272. CF 2388 Pl. 15
A.P. Room 16 Exterior, V
Fragment of side near bottom of vessel; natural pale green. Iridescence. Convex curving side; rounded basal angle. Shape uncertain. On exterior, a pronounced horizontal rib applied around the bottom; from it extend several vertical ribs, all in the same fabric as the vessel. Blown with applied and tooled decoration. A Th .1 cm.

273. CE 2024 Pl. 15
A.P. Room 33, I
Fragment of side of vessel; colorless. Milky weathering and pitting. Crazing on surface. Convex curving side. Shape uncertain. On exterior, a pronounced curved rib and indication of a second

rib, applied in the same fabric as the vessel. Blown with applied and tooled decoration. A Th .1 cm.

274. CG 540 Pl. 16
Curia, SE Room, I
Fragment of side and bottom of ladle or other vessel; natural greenish-blue with opaque white decoration. Dulling and some pitting. Convex curving side; slightly concave bottom. On exterior, an opaque white thread spirals from the center of the bottom and around the sides. Blown with applied decoration. D bottom 6.5 cm; Max Th bottom .5 cm; A Th side .1 cm.

275. Additional fragments similar to **No. 274**.
- **a.** CD 774, Bas. ENE 3, I. Fragment of bottom; natural greenish-blue with opaque white spiraling thread. Pl. 16.
- **b.** CG 581, Curia, SE Room, I. Fragment of bottom; natural greenish-blue with opaque white spiraling thread.
- **c.** CG 56, Curia, Forecourt SE 1, I. Fragment of bottom; natural greenish-blue with opaque white spiraling thread.
- **d.** C69.439, Forum SW, SE Cistern, Sediment Level. Fragment of bottom; natural greenish-blue with opaque white spiraling thread.

276. CF 2396 Pl. 16
A.P. Rooms 16 and 22 Exterior, III–IV
Fourteen joining and four non-joining fragments of rim and side of deep bowl (profile mostly restorable); colorless. Iridescence and slight pitting. Upright rim with thickened, projecting lip; convex curving side. On exterior, a horizontal thread in the same fabric applied 1.1 cm below edge of rim. Blown with applied decoration. H ca. 9 cm; D ca. 13 cm; A Th side .1 cm.

277. CD 1130 Pl. 16
Bas. ENW 2, II
Fragment of rim and side of beaker; colorless. Slight iridescence. Slightly outsplayed rim with thickened, projecting edge; vertical side, expanding outward slightly below rim. On exterior, a narrow horizontal thread in the same fabric applied 1.5 cm below edge of rim. Blown with applied decoration. D ca. 9 cm; A Th side .1 cm.

278. CF 2437 Pl. 16
A.P. Room 26, I–III
Fragment of rim and side of beaker; colorless. Milky weathering and pitting. Upright rim with thickened, projecting lip; almost vertical side. On exterior, a narrow horizontal thread in the same fabric applied .8 cm below edge of rim. Blown with applied decoration. D ca. 9 cm; A Th side .1 cm.

279. C70.623 Pl. 16
S.H. Street M, 0-I
Fragment of rim and side of beaker or other vessel; colorless. Milky weathering and pitting. Upright rim with thickened, round edge; side expands outward below rim. On exterior, a narrow horizontal

thread in the same fabric applied 1.3 cm below edge of rim. Blown with applied decoration. D ca. 9 cm; A Th side .05 cm.

280. CF 2463 Pl. 16
A.P. Room 25, I
Fragment of rim and side of beaker; colorless. Milky weathering and slight pitting. Upright rim with thickened, rounded edge; straight side, tapering downward. On exterior, a narrow horizontal thread in the same fabric applied 1.5 cm below edge of rim. Blown with applied decoration. D ca. 8 cm; A Th side .05 cm.

281. CF 2464 Pl. 16
A.P. Room 25C, III Deposit IX
Late first through early third century A.D.
Fragment of rim and side of beaker; colorless. Milky weathering and slight pitting. Upright rim with thickened, rounded edge; nearly vertical side, tapering downward. On exterior, two narrow horizontal threads in the same fabric applied in a band 1.1 cm below edge of rim. Blown with applied decoration. D ca. 7 cm; A Th side .04 cm.

282. CE 2142 Pl. 16
A.P. Room 16, 0
Fragment of rim and side of beaker; colorless. Slight milky weathering and pitting. Upright rim with thickened, rounded edge; nearly vertical side. On exterior, four narrow horizontal threads in the same fabric applied in a band 1.8–2.5 cm below edge of rim. Blown with applied decoration. D ca. 7 cm; A Th side .1 cm.

283. CF 2465 Pl. 16
A.P. Room 25C, II Deposit IX
Late first through early third century A.D.
Fragment of rim and side of beaker; colorless. Milky weathering and pitting. Upright rim with thickened, rounded edge; vertical side. On exterior, a narrow horizontal thread in the same fabric applied 1.1 cm below edge of rim. Blown with applied decoration. D ca. 7 cm; A Th side .1 cm.

284. CF 2353 Pl. 16
A.P. Room 25 A–B, II Deposit IX
Late first through early third century A.D.
Two joining fragments of vessel; colorless. Iridescence and pitting. Upper surface uneven with convex center and downturned edges. Part of base or lid. On upper side, three threads in the same fabric encircle center. Blown with applied decoration, D ca. 7 cm; A Th .15 cm.

285. CF 2395 Pl. 16
A.P. Room Exterior, IV
Fragment of vessel; colorless. Iridescence and slight pitting. Upper surface uneven with convex center and downturned edges. Part of base or lid. On upper side, two threads in the same fabric encircle center. Blown with applied decoration. D ca. 6 cm; A Th .15 cm.

Vessels with Snake-Thread and Related Appliqués (***Nos. 286–300***)

One of the most inventive decorative motifs achieved by Roman Imperial glassmakers is termed snake-thread, or *Schlangenfadengläser*, owing to the presence of trailed coils resembling stylized snakes or serpentine floral appliqués. Closely related are other vessels decorated with randomly applied curvilinear coils (here possibly **Nos. 292–297**; cf. Fremersdorf 1959, pls. 9, 21–22), or still others with coils in the shapes of birds, or birds with snakes set against a floral background (for examples, see Barag 1969). Although most appliqués are in the same lightly colored or colorless fabrics as the vessels themselves, one series, primarily found at Cologne and in the Rhineland, employs colored threads, usually opaque yellow, opaque white, or translucent blue (for examples, see list below). These designs are found on vessels of all sorts, varying from bowls, beakers, flasks, sprinkler flasks, jugs, and jars, to more unusual shapes, such as saucepans or striking glass slippers fashioned from small bottles (for illustrations, consult Fremersdorf 1959 and the references in the list below). An especially popular shape is a tall cylindrical or U-shaped beaker with a beaded stem and an outsplayed or domical base-ring (here, probably **Nos. 286–299**). This form was also used by other contemporary glassmakers for cut, painted, gilded, and open-work decoration.

Ever since this type was first described in the late nineteenth century (Kisa 1899, 137–147; cf. Kisa 1908, 444–473), controversy has surrounded its place of origin and date, its subsequent dissemination in the eastern and western Mediterranean, and the identification and development of its subclasses. In all these matters, opinion has differed markedly, and no resolution to these questions has yet appeared. Fundamental to the controversy is a distribution pattern thought to center on two geographically disparate regions: on the one hand Cologne and the Rhineland (and to a lesser extent, other provinces of northwestern Europe) and, on the other, the Syro-Palestinian area and its neighboring territories.

In the West, snake-thread glasses tend to be more diverse in shape, and it is customary to find both polychrome and monochrome examples. In the East, fewer forms are encountered, and decoration in most instances is in the same colorless or lightly colored fabrics as the vessels themselves. Also, although based on fewer published sites, the total number of eastern examples seems to be less.

Preliminary studies have also identified stylistic differences in products of the two regions, both in the selection of designs and in the treatment of the applied coils. In the Rhineland realistic "snake" appliqués are often associated with floral motifs or random trailing. By contrast glassmakers in the eastern Mediterranean produced subclasses with birds or birds with snakes (Barag 1969, 55–66). Such subclasses are wholly absent from the Romano-German series. Also, among eastern examples the broad ends of individual coils are often squeezed with a tool, producing a tiny gridiron or checkerboard impression (here **Nos. 286–288**). These are rare on Romano-German vessels, which instead were regularly adorned with parallel tooled indentations on the coils (here **Nos. 289–290**). Finally, in the West there seem to be many more styles of related appliqués that are not present in the East (for examples, vessels with realistically veined leaves; at Cosa, **Nos. 299–300**; cf. Ostia IV 1977, 342, no. 400, pl. 50).

Equally problematic are the place of origin of the style and the inclusive dating of the type and its subclasses. Initially, opinion held that the style first emerged in the eastern Mediterranean, perhaps at Alexandria, then quickly spread to workshops in the Syro-Palestinian region, whence it was carried to Roman glasshouses along the Rhine (Harden in Karanis 1936, 169; Clairmont in Dura Europos 1963, 43–44; and Barag 1969, 65–66, summarize the arguments). Fremersdorf, who more than any other scholar has studied the Rhineland production, at first adhered to an eastern origin but later changed his mind and came to believe that the type was first devised at Cologne under the stimulus of newly arrived glassmakers from the East. Only later was the mature style then copied, he argued, by eastern ateliers (Fremersdorf 1939, 13; cf. Fremersdorf 1959). Fremersdorf first attributed the western production to ca. A.D. 125–200 (Fremersdorf 1955, 243–245) but later suggested that the "standard group of snake-thread vessels appeared only around A.D. 160, and that his Group B dated to the second half of the second century—although this implies (?) that Group A supersedes Group B and therefore belongs to the first half of the century (Fremersdorf 1959, 12, 13–15).

Barag and others, however, have accepted neither his chronology nor his notion of a western origin for the type. Barag 1969, 58, rejects Fremersdorf's interpretation of numismatic evidence for certain tomb groups containing snake-thread wares and concludes that these burials are later than Fremersdorf alleged. According to Barag they belong to the very late second or early third century; that is, they are Severan. In fact, Barag concludes that the type was first invented at the end of the second century and flourished primarily in the third (Barag 1969, 58 and 65–66), conclusions reiterated in Barag 1978, 33–34. There he also ascribes some of the most canonical shapes of snake-thread glasses (U-shaped beaded beakers, for instance) to the third century. Barag also assumes an eastern origin for the type as a whole, but he admits that once the idea had spread to the West, it reached its apogee in the hands of glassmakers along the Rhine.[50]

In regard to date, the examples from Cosa add little to present knowledge. **Nos. 286**, **288**, **292**, **295–296**, and **298–300** are from Deposit IX, but the material that accumulated there spans the late first to early third century A.D. (in addition to some material that potentially is even later). More rewarding but often overlooked is the data on snake-thread and related wares from Gaul, first collected by Morin-Jean 1913, 199–216. He describes several vessels that were recovered from burials and assesses the chronological evidence in each case. In particular, he cites a jug from Cortil-Noirmont and a flask from Cany (Seine-Inferieure). Both came from cemeteries with graves predominantly of the third century (Morin-Jean 1913, 204–207). He even suggests that some varieties of the type come from contexts as late as the fourth century A.D. (Morin-Jean 1913, 212–216).

As for the origin of the style and provenance of individual pieces, the fragments at Cosa are significant in that they demonstrate that snake-thread glasses are as routinely found in Italy as they are in the eastern Mediterranean or farther north in the transalpine provinces. The additional examples discovered at Rome and Ostia (see list below) further confirm the validity of this statement. Indeed, the growing body of evidence from France (Morin-Jean, cited above), Holland (Limburg 1971, 13, no. 24, fig. 2, pl. 1), Britain (Charlesworth 1959, 54, nos. 8–10, fig. 10), and the Iberian Peninsula (Conimbriga 1976, 188 and 191, nos. 183–188, pls. 8 and 41) convincingly argue that the type was standard fare throughout the West under the empire. This in turn suggests that much of the previous discussion regarding the dissemination of the style and its seemingly circumscribed distribution in two regions may be misguided. If this is the case, then emphasis should probably be shifted to the international character of these glasses rather than further argument about the relationship of the Syro-Palestinian and Rhineland production. Like so many of the finer decorated tablewares of the first through third century, these glasses reflect the striking communality of cultural goods throughout the Roman world—even as late as the third century. Either long distance trade was so masterfully organized that goods could be shipped safely from a minimum number of centers, or,

[50] Barag 1969, 65, notes that so few fragments have been recovered from Roman Egypt that there is no longer sound reason to postulate an Alexandrine origin of the style. He also suggests that Rome may have been a center of manufacture.

more probably, decorative styles popular in one part of the empire were quickly emulated elsewhere by local or peripatetic artisans.

At Cosa the dominant shape associated with the group is that of the tall, cylindrical beaker with a beaded stem (probably **Nos. 286–299**). Although only two pieces (**Nos. 293** and **295**) preserve part of a beaded stem and body, forty-three additional but highly fragmentary stems at Cosa (unpublished) indicate the popularity of the shape. All snake-thread fragments at Cosa are colorless, as is the decoration. Most pieces are also affected by white, granular weathering, which causes the glass to appear opaque.

Although the examples are very fragmentary, **No. 286**, and possibly **Nos. 287–288**, may have been decorated in a fashion similar to Barag's "Flower and Bird" group (Barag 1969). These pieces also exhibit a gridiron impression on the ends of the coils. On the other hand, the disposition of the coil of **No. 286** and the presence of parallel tooled indentations closely resemble vessels from the Rhineland (Fremersdorf 1959, pl. 15). **Nos. 289–290** also preserve coils with similar indentations, as do examples from Ostia (Ostia III 1973, 251, nos. 623–624, pl. 43). Despite their fragmentary nature, **Nos. 293–294** may be compared to the randomly trailed snake-thread vessels also known from the Rhineland (cited above), again with a parallel from Ostia (Ostia IV 1977, 342, no. 397, pl. 50). **Nos. 295–297** may form a subclass of their own, a group characterized by pronounced vertical coils.[51] **Nos. 299–300**, with their well-defined leaf appliqués, certainly form a distinct subclass, one that also is known at Ostia (see list below).

PARALLELS AND DATED EXAMPLES OF VESSELS WITH SNAKE-THREAD AND RELATED APPLIQUÉS

Cosa: **Nos. 286**, **288**, **292**, **295–296**, and **298–300**. From Deposit IX, belonging to the late first through early third century A.D.

Ostia II 1970, 251, nos. 623–624, pl. 43.

Ostia IV 1977, 342, no. 400, pl. 50. Colorless fragment with leaf appliqué; also 342, no. 397, pl. 50, a colorless frament with curvilinear arched threads at bottom of a beaded (?) beaker.

Rome: Museo Nazionale Romano. Several fragments. Unpublished.

Rome: Three examples said to be from Rome, now in the Metropolitan Museum of Art, New York, cited in Barag 1969, 60–62, nos. 4, 7–8, fig. 2. The third of these fragments bears a partially preserved Latin inscription. (For another Latin inscription preserved on a vessel possibly related to snake-thread wares, see Gréau Cat. 1903, 169, no. 1212, pl. 226; also two additional inscribed vessels possibly associated with the group in Morin-Jean 1913, 203, fig. 273; also Fremersdorf 1959, 43–44, pl. 24.)

Harden 1934, 50–55. Discussion of type.

Karanis 1936, 168–170, 173, nos. 490 and 492, pl. 16. With discussion.

Fremersdorf 1938, 201–214. General discussion of the type.

Fremersdorf 1959. Numerous examples from Cologne and environs with discussion and parallels.

Morin-Jean 1913, 199–216. Discussion of snake-thread wares with drawings illustrating different varieties of appliqués from European sites, none of which, incidentally, has gridiron impressions on the coils.

Isings 1957, 103, Form 86 and 110, Form 93. These forms describe beaded beakers and ovoid flasks that have in some instances snake-thread or related designs.

[51] Another fragmentary vessel with similar coils is known from Ostia (Ostia III 1973, 251, no. 626, pl. 43) where a base fragment exhibits pronounced threads, arched at the bottom of the vessel. Cf. Fremersdorf 1959, 71–73, pls. 108–109, 113.

Dura Europos 1963, 42–46, nos. 161–178, pls. 5 and 22.
Doppelfeld 1966, 55. Illustrates many of the snake-thread vessels now in Cologne. No. 117, a beaded beaker, is from grave IV at St. Severin and dated to the second half of the second century A.D.
Barag 1969, 55–66. Extensive discussion of the type.
Limburg 1971, 13, no. 24, fig. 2, pl. 1, and 18–19, no. 47, fig. 4.
Nida-Heddernheim 1974, 126–128, nos. 290–292, pls. 18 and 21.
Conimbriga 1976, 188 and 191, nos. 183–188, pls. 8 and 41.
Barag 1978, 33–34. U-shaped beaded beaker from tomb at Hanita, Israel, with discussion of the type.

VESSELS WITH SNAKE-THREAD AND RELATED APPLIQUÉS

286. CF 2305 — Pl. 16
A.P. Room 25A, III — Deposit IX
Late first through early third century A.D.
Four joining fragments of rim and side of conical beaker; colorless. Milky weathering and pitting. Upright rim with thickened, rounded edge; almost vertical, straight side, tapering downward. On exterior, a horizontal thread in the same fabric applied 1.9 cm below edge of rim; snake-thread and other appliqués, also in the same fabric, applied to side. Some appliqués have a cross-hatched pattern impressed into them. Blown with applied and tooled decoration. D ca. 8 cm; A Th .1 cm.

287. CF 2428 — Pl. 16
A.P. Room 25D, II — Deposit IX
Late first through early third century A.D.
Two small, non-joining fragments of side of beaker or other vessel; colorless. Milky weathering and pitting. Nearly straight side, tapering downward. On exterior, snake-thread and other appliqués in the same fabric; some with a cross-hatched pattern impressed into them. Blown with applied and tooled decoration. Max D fragment ca. 8 cm; A Th .12 cm.

288. CF 2427 — Pl. 16
A.P. Rooms 25C, II — Deposit IX
Late first through early third century A.D.
Small fragment of side of beaker or other vessel; colorless. Milky weathering and pitting. Nearly straight side, tapering downward. On exterior, snake-threads and other appliqués as in **No. 287**. Blown with applied and tooled decoration. A Th .1 cm.

289. C72.95 — Pl. 17
IV-B, Column Wedges, 0
Small fragment of side of beaker or other vessel; colorless. Milky weathering and pitting. Nearly straight side, tapering downward. On exterior, snake-thread and other appliqués as in **No. 287**. Blown with applied and tooled decoration. A Th side .1 cm.

290. CE 2144 — Pl. 17
A.P. II, Room 34, II
Small fragment of side of beaker or other vessel; colorless. Milky weathering and pitting. Straight side, tapering downward. On exterior, snake-thread appliqué as in **No. 287**. Blown with applied and tooled decoration. A Th side .08 cm.

291. CE 2206
A.P. Room 15, I
Small fragment of side of vessel; colorless. Milky weathering and pitting. Nearly straight side, tapering downward. On exterior, snake-thread and other appliqués as in **No. 287**. Blown with applied and tooled decoration. A Th side .1 cm.

292. CF 2426 Pl. 17
A.P. Room 25C, II Deposit IX
Late first through early third century A.D.
Two joining fragments of rim and side of conical beaker; colorless. Milky weathering and pitting. Upright rim with thickened, rounded edge; vertical side. On exterior, a horizontal thread in the same fabric applied 1.5 cm below edge of rim; on the side, the upper portion of sharply curving thread, probably part of a snake-thread or related pattern. Blown, with applied decoration. D ca. 7.5 cm; A Th side .1 cm.

293. CF 2309 Pl. 17
A.P. Room 25, I
Fragment of lower side of stemmed beaker or other vessel; colorless. Milky weathering and pitting. Convex curving side; vestiges of a beaded stem. On exterior, a random curvilinear pattern of applied threads in same fabric as the vessel. Blown, with applied decoration. Max D fragment ca. 6.5 cm; A Th side .15 cm.

294. CD 1131 Pl. 17
Bas. ENW 2, III
Fragment of side of beaker or other vessel; colorless. Slight milky weathering and pitting. Nearly straight side, tapering downward. On exterior, a random pattern of applied threads in same fabric as the vessel. Blown, with applied decoration. A Th .1 cm.

295. CF 2425 Pl. 17
A.P. Room 25C, II Deposit IX
Late first through early third century A.D.
Three non-joining fragments of lower side of beaker or other vessel; colorless. Milky weathering and pitting. Convex curving side; probably from a stemmed beaker. On exterior, several uneven but mostly vertical applied threads in the same fabric; several threads have indentations made with a tool. Blown, with applied and tooled decoration. Max D fragment ca. 8 cm; A Th side .2 cm; Max Th lower side .4 cm.

296. CF 2360
A.P. Room 25A, III Deposit IX
Late first through early third century A.D.
Small fragment of side of vessel; colorless. Iridescence and slight pitting. Convex curving side. Shape and decoration similar to **No. 295**. Blown, with applied and tooled decoration. A Th side .15 cm.

297. CE 2105 Pl. 17
A.P. Room 24, III

Small fragment of side of beaker or other vessel; colorless. Slight iridescence and pitting. Convex curving side. Possibly from a stemmed beaker. On exterior, portions of three pronounced vertical threads in the same fabric as the vessel; threads have indentations made with a tool (similar to **Nos. 295–296**). Blown, with applied and tooled decoration. Max D fragment ca. 7 cm; A Th side .15 cm.

298. CF 2364
A.P. Room 25A, II — Deposit IX
Late first through early third century A.D.
Small fragment of side of vessel; colorless. Iridescence and pitting. Straight side, tapering downward. Shape uncertain. On exterior, a pronounced thread in the same fabric as the vessel, set vertically on the side; threads have indentations made with a tool. Blown, with applied and tooled decoration. A Th .1 cm.

299. CF 2308 — Pl. 17
A.P. Room 25A, II — Deposit IX
Late first through early third century A.D.
Fragment of side of beaker or other vessel; colorless. Iridescence and pitting. Straight side, tapering downward. On exterior an appliqué in the shape of a leaf, in the same fabric as the vessel, set vertically on the side; leaf has tooled veins. Blown, with applied and tooled decoration. Max D fragment 6.3 cm; A Th side .08 cm.

300. CF 2429
A.P. Room 25C, II — Deposit IX
Late first through early third century A.D.
Small fragment of side of beaker or other vessel; colorless. Milky weathering and pitting. Straight side, tapering downward, possibly belongs to same vessel as **No. 299**. On exterior, a partially preserved appliqué in the shape of a leaf. Blown, with applied and tooled decoration. A Th side .1 cm.

*Beakers with Indented Decoration (**Nos. 301–306**)*

Yet another conventional technique of adorning blown glass was the formation with a tool of long, vertical indentations around the body of a vessel. Although bowls, jars, jugs, and flasks might be decorated in this manner, the conical or cylindrical beaker was the favored shape to receive four or more indentations that extend from below the rim to near the bottom. In most instances these beakers (*Faltenbecher*) exhibit upright or off-set rims with knocked-off and polished edges. Treatment of bases varies, but closed, cut-out base-rings or simple concave bottoms are the most common. The shape of the bottoms may be square (**No. 304**), round (**No. 305**), or hexagonal (**No. 306**), usually reflecting the number of indentations on the sides. Virtually all examples are colorless or naturally colored. A welcome overview of shapes and sizes based on examples found on Cyprus is provided in Vessberg 1956, 140–141, esp. fig. 44, nos. 25–35, and fig. 45, nos. 13–16. Inspiration for this mode of ornamentation presumably derives from contemporary ceramic "thumb-impressed" beakers, which were especially popular in the mid-first century A.D., as the Romano-Italian "fine ware" pottery at Cosa amply illustrates (Moevs 1973, 238–239, Form 69, dating to Claudian era, possibly earlier).

The six examples from Cosa (**Nos. 301–306**), although highly fragmentary, are typical of other indented beakers found on numerous sites throughout the Mediterranean. When found in dated contexts, these wares usually belong to the mid- to late first century A.D., with some examples continuing into the next century. Although these beakers are often said to begin around the middle of the first century, the earliest securely dated pieces are Flavian, with several examples known from Pompeii (Spinazzola 1928, 228, top shelf). No examples were recovered from Magdalensberg, abandoned ca. A.D. 45. The natural coloration of the group also suggests that they did not become popular until after the middle of the century. The use of indented ornament, of course, continues until late Roman times (see, for example, a series of fourth-century bowls from Aquileia 1968, 173–174, no. 348, pl. 26).

Parallels and Dated Examples of Beakers with Indented Decoration

Ostia I 1968, 73–74, B4.2148, B4.2154–2155, figs. 192–195, with additional references given.
Spinazzola 1928, 228, top shelf. Examples from Pompeii, before A.D. 79.
Simonett 1941, 114–115, fig. 94, pl. 12.1. From Branca grave 1 in Tessin, dated to the second half of the first century A.D.
Bologna Coll. 1979, 48–49, no. 35. Example with additional references.
Pavia Coll. 1983, 85–86, nos. 48–49.
Aquileia 1968, 56–57, nos. 138–139, fig. B.3, pl. 6.
Trier 1977, 74, nos. 311–313, pl. 43.
Nida-Heddernheim 1974, 124–126, nos. 287–289, pl. 19, with discussion.
Vindonissa 1960, 46, nos. 107–109, pls. 7, 19, and 22, with discussion.
Conimbriga 1976, 188–189, and 191; nos. 189–192, pl. 41.
Dura-Europos 1963, nos. 442–443, pl. 10.
Cyprus 1956, 140–141, figs. 44–45, as noted in the text above.
Tipasa 1967, 18–19, 76–84, nos. 144–165, Forms 20–20c.
Isings 1957, 46–47 and 49–50, Forms 32 and 35. Several examples thought to belong to the second half of the second century A.D. are cited, in addition to others known from first-century contexts.

Beakers with Indented Decoration

301. CD 1075 Pl. 17
Bas. ENE 1, I
Fragment of side of beaker; colorless. Slight iridescence and dulling. Outsplayed rim (upper part and edge missing); vertical side, tapering downward to a rounded basal angle. On side, two pronounced vertical indentations made with a tool. Blown and tooled. H fragment 7 cm; A Th .1 cm.

302. CD 1143
Bas. ESW 3, I
Fragment of side of beaker or other vessel; colorless. Slight iridescence and pitting. Vertical side. On side, two pronounced vertical indentations made with a tool, as in **No. 301**. Blown and tooled. A Th .08 cm.

303. CF 2442
A.P. Room 25 Exterior, I

Four non-joining fragments of side of beaker or other vessel; colorless. Slight iridescence and pitting. Vertical side. On side, several vertical indentations made with a tool, as in **No. 301**. Blown and tooled. A Th .1 cm.

304. C69.426 Pl. 17
Forum SW, Drain NW, 0
Fragment of side and bottom of vessel; natural bluish-green. Iridescence and pitting. Four outsplayed sides with broad indentations at their centers; concave bottom. Blown and tooled. D bottom 6 cm; A Th side .08 cm.

305. CE 2025 Pl. 17
A.P. Room 22, I
Bottom with vestiges of sides of vessel; natural light green. Iridescence and pitting. Four outsplayed sides, each with a broad indentation at its center; round bottom with concave underside. Blown and tooled. Max W 6.5 cm; A Th side .1 cm.

306. C70.611 Pl. 17
S.H. Street M, 0-I
Bottom and fragment of sides of vessel; natural light green. Iridescence and pitting. Six vertical sides, each with a broad indentation at its center; hexagonal bottom with concave underside. Blown (possibly into a bottom mold) and tooled. W side 2.6–3 cm; A Th side .08 cm.

ROMAN BLOWN UNDECORATED GLASS

Sometime in the second half (or last quarter) of the first century B.C. the blow pipe was invented, and within a few decades the technology of glassblowing was disseminated around the Mediterranean. Shortly afterward, artisans began to fashion an ever-increasing array of inexpensive table and storage wares until, by the second quarter of the first century, blown glass had become a commonplace in the daily life of the Roman Empire. The economic and social implications of the discovery were immediately felt. Not only was a new industry established in most parts of the empire, but certain age-old industries, especially those dealing with the manufacture of simple ceramics, were adversely affected or displaced altogether.[52]

Contrary to prevailing opinion, however, glassblowing was *not* the invention that first introduced cheap, accessible vessels and articles of glass to the Roman world. This already had been achieved by the casting industry earlier in the first century B.C, and especially by the Romano-Italian factories founded during the course of that century (Grose 1983, 38–40; 45). In fact, the archaeological evidence at Cosa and elsewhere makes it patently clear that casting and other heat-forming methods dominated the craft throughout the first century B.C. and early decades of the next century (Grose 1984, 25–32). Only toward the middle of the reign of Tiberius does the amount of blown glass on archaeological sites begin to exceed the number of cast vessels. Even then, certain cast wares (for

[52] Perhaps the best examples are Romano-Italian thin-walled wares, mostly beakers and bowls, which after decades of production inexplicably disappear under the later Julio-Claudians. In all likelihood this occurred as glass beakers and bowls became the preferred sorts of drinking cups for domestic use; cf. Grose 1977, 9, n. 1.

example, ribbed bowls) retained their popularity for many decades (see pp. 47–49 above), while other types (such as the colorless vessels with broad, overhanging rims) were not introduced until Neronian or possibly Flavian times (see pp. 88–91 above). Not unexpectedly, the cast and blown glasses of the Augustan and Julio-Claudian periods, in certain instances, reveal both imitation and elements of a shared development (see the discussion of blown ribbed bowls, pp. 122–124 above).

Despite the initial prominence of casting methods, however, the discovery of the inflation technique and the proliferation of its products *did* lead to a remarkable and permanent revolution in glass manufacture by the mid-first century A.D. Workshops were established in most every province of the empire, and the material became an assumed fact of life, with multiple uses in the home as well as in architecture and design. Although classical authors of the fifties and sixties A.D. still thought of glass as novel and wondrous, they also remarked on how quickly it had become ordinary and cheap (see references in Grose 1977, 12–16, and Grose 1983, 38–39).

The first center responsible for the invention of blowing is not yet known, but present information suggests that it may well have occurred in the Syro-Palestinian area, where cities such as Sidon are known to have had well-established glass industries (for references and discussion, see Grose 1979, esp. 54–55). Although the evidence is far from conclusive, the earliest published examples of blown vessels are also from this region. For example, a burial at Ein-Gedi in the Judean desert, assigned a date before ca. 40/37 B.C., yielded a blown bottle found with a cast ribbed bowl of the bossed variety (Avigad 1962, 183; cf. Grose 1984, 32, n. 15). More important are several other blown bottles discovered in a stepped bath in the Jewish quarter of the Old City of Jerusalem (Avigad 1972, 200; cf. Grose 1984, 33, n. 17). They were found with numerous cast bowls, hollow glass tubes, and rods, which presumably were the refuse of a nearby workshop. Since the bath had been sealed with a pavement ca. 50/40 B.C., these bottles constitute the earliest known examples of blowing. Close inspection of these bottles, however, reveals that they were not blown on a pipe but were "tube-blown," a technique that probably antedated the invention of standard inflation methods. Nonetheless, they demonstrate that craftsmen in the mid-first century B.C. were not only aware that glass could be blown, but that they were also experimenting with methods to that end (for a fuller discussion of the relationship between tube-blowing and normal blowing, see Grose 1984, esp. 33–34).

Whatever the exact date for the invention of the blow pipe, examples of blown glass on archaeological sites are uncommon until the very close of the first century B.C. or first decade of the following century. This is true on both eastern Mediterranean and Italian sites (for a list of early pieces, see Grose 1977, 16–27). Even Cosa produced one example, a small colored bottle from Deposit IV, dated prior to ca. 25/15 B.C. In almost all instances these earliest blown vessels are either unadorned colored unguent bottles (here **Nos. 362–372**) or colored bowls with pinched ribs and applied threads (here **Nos. 249–258**), types also discussed in Grose 1982, 26–29, and Grose 1984, esp. 32–33.

A survey of the undecorated (or lightly decorated) vessels at Cosa reveals a wide variety of shapes and types, of which most are well known from other sites, both Italian and farther afield. Although most pertain to forms and styles internationally popular in their day, as is true of most classes of the open and closed wares, several appear to be specifically Italian types. Among them are the colored fine wares (pp. 143–145 below) and the bowls with triangular rims (pp. 162–163 below). In date these vessels range from the Augustan period through the middle of the third century A.D., with a few groups as late as the mid-fourth to early fifth century (most notable, the late antique stemmed cups, **Nos. 468–489**). All represent ordinary household items, whether containers or table pieces.

Unfortunately, most are so highly fragmentary that it is difficult to determine complete shapes or assign them to closely dated groups.

Colored Translucent Tablewares and Storage Wares (***Nos. 307–372***)

The blown, colored translucent fine wares discovered at Cosa constitute a diverse group of tablewares that hitherto have not been recognized as forming a distinct class of Roman glass. Because they were manufactured for only a brief time and rarely circulated beyond Italy and a few legionary camps in the West, their pivotal role in the history of glassmaking has been overlooked. Yet they represent an important transitional phase in the changing technologies of the mid-first century A.D. as glassblowing came to illustrate how rapidly glassmakers mastered the intricacies and possibilities of blowing glass. In these vessels one sees a fairly complete panoply of Roman blown glass, expertly made and exhibiting almost every nuance of the glassblower's art.

In form and function the colored fine wares comprise entire dining services with both table pieces and containers. These include plates (**Nos. 307–307a**), dishes (**Nos. 308–314**), bowls of varying profiles and sizes (**Nos. 315–339**), saucepans (**Nos. 340–341**), thin-walled beakers and handleless jars (**Nos. 342–352**), large flasks, jars, and jugs (**Nos. 353–357**), as well as smaller amphoriskoi and bottles (**Nos. 357–372**). Even ladles such as **No. 373** occur. All of these shapes, with their full rounded profiles, illustrate the "natural" forms made possible through blowing. Although a few imitated contemporary cast vessels (compare, for examples, two blue carinated bowls, one cast, the other blown, from Deposit VII, **Nos. 139** and **317**), most represent a wholly new repertoire of shapes, one that would continue to be made (albeit with modifications) for another two centuries or more. No longer would glassmakers habitually copy the forms and decorative devices of stone, metalware, and pottery, as had been true of most core-formed and cast glasses from the late Bronze Age to the advent of the Romano-Italian blowing industry.

Equally significant is the imagination and ingenuity shown by these early glassmakers, who quickly learned to manufacture such a wide array of open and closed shapes. They also rapidly determined most of the basic ways of treating rims and bases. The former can be tubular (or double-tubular, **No. 315**), fire-polished and solid (for example, **No. 307**), or knocked-off and left unfinished (**No. 342**). Bases vary from applied rings (**No. 329**) to solid pads (**No. 309**) and cut-out bases, whether tubular (**No. 307a**), closed (**No. 318**), or open (**No. 354**). More novel is a "special" pad base, as seen on **No. 308**, which is associated almost solely with wares of this sort. Such bases are actually thinly blown paraisons with knocked-off, downturned edges that are attached to the undersides of dishes and bowls. Simultaneously, the same craftsmen also devised strap (**Nos. 315** and **354**), stirrup (**No. 327**), saucepan (**Nos. 340–341**), and vertical hooked handles (**No. 373**). Decoration, however, is usually limited to cut-out ledges on the body (**No. 328**), horizontal cut grooves and abrasions (chiefly on beakers, **Nos. 342**, **344–348**, **350**), and applied blobs or streaks in other colors (**Nos. 354–355**).

Although these workmen produced naturally colored vessels, often in shapes identical to those seen here (compare the colored and naturally colored pieces from Deposit VII; cf. Grose 1975, esp. 38–46), the colored fine wares can still be discussed as a separate entity owing to their use of brilliantly colored fabrics. The colors employed are both transparent and opaque. The most common are a rich transparent blue and an emerald green, but purple (**No. 354**), peacock blue (**No. 356**), and deep blue-green (aquamarine, **No. 358**) also occur. Although Cosa yielded no examples in opaque fabrics, they are known from Magdalensberg, Tarragona, and other sites (see list below) and include pieces in opaque white, light blue, and blue-green. These colors, of course, are precisely the same as those used in the production of cast transparent and opaque fine wares (pp. 76–82 above). Clearly this is not a chance occurrence. Either the glassblowers imitated the color palette of the casting industry, or both cast and blown fine wares were fabricated in the same factories or cluster of factories.

Among vessels described in this section most can be attributed to a common source manufacturing many different shapes. This is demonstrated by the presence of so many plates, dishes, bowls, beakers, and even jars, jugs, bottles, and ladles found together in Deposit VII, a storeroom of a shop destroyed ca. A.D. 40/45 (see list below). Regardless of form, the overall homogeneity of these pieces is self-evident. Only the splashed wares (**Nos. 354–355**), the thin-walled beakers (**Nos. 343–351**), and some of the small unguent bottles (**Nos. 362–372**) can be singled out as perhaps standing apart from the standard series of colored fine wares, yet even they partake of the same rich colors and are of the same era (for further discussion, see pp. 76–82 above). The precise location of the factories is unknown, but based on the distribution of the type, it undoubtedly was centered in Italy. Once again, one is reminded of Strabo's emphasis on the innovative color technology that he associated with the glass industry based at Rome (*Geog.* 16.2.25; cf. p. 77 above).

More assured is the date of the group. At Cosa (Deposits V, VII, and VIII), as at Pompeii, Magdalensberg, and Tarragona, specimens are assigned mainly to the second quarter of the first century A.D. (see list below), with the earliest examples from Augustan contexts. Excavations at the Regia in the Forum Romanum, for example, yielded an emerald green dish comparable to **Nos. 308–310** from a context of the last quarter of the first century B.C./first decade of the following century (Grose 1977, 18–21, no. 10, figs. 1–4). This would suggest that the colored fine wares were among the first blown products to have been devised once blowing had been introduced into Italy in a serious way. Although evidence for their disappearance is scanty, it seems that they were no longer in vogue after the mid-first century A.D., a period coinciding with the general abandonment of deliberately colored fabrics (whether cast, mold-blown, or free-blown) and the subsequent rise of naturally colored and colorless glasses (see pp. 81–82 above).

Parallels and Dated Examples of Undecorated Colored Translucent Tablewares and Storage Wares

Cosa: **No. 364** (small bottle). Deposit IV, prior to ca. 25/15 B.C; **Nos. 360**, **371–372** (small bottles, Deposit V, ca. 25/15 B.C. to ca. A.D. 40/45; **Nos. 307–311e**, **315–319a**, **320**, **323**, **333**, **337**, **342**, **351–354**, **358**, **373**, Deposit VII, decade prior to A.D. 40/45; **No. 344**, Deposit VII, prior to ca. A.D. 50/55. See also Grose 1977 for the publication of all glass vessels found in Deposit VII.

Grose 1977, 18, no. 10, fig. 2. Rim fragment of an emerald green dish from the excavations of the Regia in the Forum Romanum, Rome.

Pompeii: Museo Nazionale, Naples. Several small blue bowls similar to **Nos. 320–324** at Cosa; also a purple bowl with off-set rim like **No. 317** or **318** and a special pad base like **No. 308** at Cosa. Unpublished.

Grose 1977, fig. 3a. Emerald green dish from a Tiberian deposit at the House of Sallust at Pompeii.

Paestum: Museo Archeologico. Several fragments of vessels, including dishes such as **No. 375**, from the Sanctuary of Bona Dea, University of Michigan–Perugia University excavations. Unpublished.

Simonett 1941, 115–117. Branca, grave 2, a grave with several comparable vessels: blue dish (no. 1); spherical blue flask (no. 7) akin to **No. 353** at Cosa; a purple dove-shaped unguent container (no.

8); a large blue, one-handled jar (no. 9); a small opaque white bowl (no. 14); and a two-handled jar with splashed decoration, opaque white blobs on a greenish body (no. 15). According to Berger in Vindonissa 1960, chart opp. p. 93, the grave dates to the first half of the first century, probably the first quarter of the century.

Magdalensberg 1979. The site, abandoned by A.D. 45, yielded many examples of blown, colored fine wares, but in the catalogue they are dispersed by shape and therefore difficult to locate without additional profile drawings and color plates. Nonetheless, presumed examples include the following: 82, no. 613, pl. 4 (blue dish); 84, nos. 624 and 627, pl. 4 (blue and blue-green dishes); 58, no. 490 (blue bowl); 65, no. 515, pl. 3 (blue-green bowl with cut-out ledge below rim); 64, nos. 510–511, pl. 3 (purple and opaque white bowls with cut-out ledges); 51, no. 471, pls. 3 and 13 (purple bowl with special pad base). Not found at Cosa but known from Magdalensberg and Tarragona are sack-shaped beakers, mostly in opaque colors, such as 99–103, nos. 762–792, or taller beakers with off-set rims, such as an opaque white example, 94–95, no. 729, pls. 5 and 14 (Isings Form 4).

Pasaje Cobos, Tarragona: Several translucent and opaque colored fine wares similar to those at Cosa, including bowls with special pad bases. Information furnished by Dr. Jennifer Price, University of Durham.

Morin-Jean 1913, 33, fig. 9.18, illustrates a pad base like that of **No. 308** but does not cite actual examples of such vessels known to him, presumably from sites in Gaul.

*Colored Translucent Plates and Dishes (**Nos. 307–314**)*

307. CF 2432A — Pl. 18
A.P. Room 25 Exterior, II–III

Four joining fragments of rim, side, and floor of large plate (profile mostly restorable); blue. Iridescence and pitting. Slightly outsplayed rim with solid, thickened edge; straight side, tapering diagonally downward; uneven floor. Blown. H rim 3.5 cm; D ca. 26 cm; A Th .15 cm. Found with base-ring **No. 307a**; however, the diameter of the base-ring appears too large to belong to this plate.

a. CF 2432B, A.P. Room 25 Exterior, II–III. Fragment of floor and base-ring of large plate; blue. Double tubular base-ring cut out from bottom. Possibly **No. 307**. Blown and tooled. D base-ring ca. 18 cm. Pl. 18.

308. CE 1954 — Pl. 18
A.P. Room 22, II — Deposit VII
Decade prior to ca. A.D. 40/45

Thirty-two joining fragments of rim, side, floor, and base-ring of dish (profile restorable); blue. Portions of vessel restored in araldite, an epoxy resin. Iridescence and pitting. Upright rim with solid, thickened edge, nearly vertical side; flat floor; thin pad base with knocked-off edge. Blown. H 3 cm; D 17 cm; D base ring 7.7 cm; A Th .1 cm.

Bibliography: Grose 1975, 38, no. 9; Grose 1983, 44.

a. CE 2145, A.P. Room 22, II, Deposit VII, decade prior to ca. A.D. 40/45. Two joining fragments of rim and side of dish identical to **No. 308**, and joining one of three pad bases (**Nos. 311c–e**). Blown. D ca. 16 cm.
Bibliography: Grose 1975, 38, no. 9A.

309. CE 2146 and CE 1958 Pl. 18
A.P. Room 22, II Deposit VII
Decade prior to ca. A.D. 40/45

Two joining fragments of rim and side, and three joining fragments of floor and base-ring of dish (profile restorable); blue. Iridescence and pitting. Upright rim with solid, thickened edge; nearly vertical side; uneven floor, sloping upward at center; thick pad base with short, rounded edge. Blown. H 3 cm; D ca. 16 cm; D base-ring 9.5 cm; A Th side .1 cm.
Bibliography: Grose 1975, 38, no. 9B.

310. CE 1965 Pl. 18
A.P. Room 22, II Deposit VII
Decade prior to ca. A.D. 40/45

Six joining and nine non-joining fragments of rim, side, floor, and base-ring (edge missing) of dish (profile restorable); emerald green. Iridescence and pitting. Tubular rim, folded outward; nearly vertical side; highly uneven floor, noticeably convex above base-ring; pad base as in **No. 308**. Blown. H 3 cm; D ca. 15.3 cm; D base-ring ca. 6.6 cm; A Th .25 cm.
Bibliography: Grose 1975, 39, no. 10; Grose 1977, 20, figs. 3–4.

311. Additional rim and base fragments from similar dishes, all from A.P. Room 22, II, Deposit VII, decade prior to ca. A.D. 40/45.

- **a.** CE 1991. Numerous non-joining fragments of rim and side; blue. Tubular rim folded outward. Joins one of three pad bases, **Nos. 311c–e** below. D ca. 16 cm.
  Bibliography: Grose 1975, 39, no. 10A.
- **b.** CE 2147. Numerous non-joining fragments of rim and side; blue. Tubular rim, folded outward. Joining one of three pad bases, **Nos. 311c–e** below. D ca. 16 cm.
  Bibliography: Grose 1975, 39, no. 10B.
- **c.** CE 1955. Four joining fragments of floor and base-ring; blue. Very uneven floor; pad base similar to **No. 308**. D base-ring 7.2 cm.
  Bibliography: Grose 1975, 39, nos. 10A–B.
- **d.** CE 1956. Seven joining fragments of floor and base-ring; blue. Very uneven floor; pad base similar to **No. 308**. D base-ring ca. 7 cm.
  Bibliography: Grose 1975, 39, nos. 10A–B.
- **e.** CE 1957. Three joining fragments of floor and base-ring; blue. Very uneven floor; pad base as in **No. 308**. D base-ring ca. 7.7 cm.
  Bibliography: Grose 1975, 39, nos. 10A–B.

312. CD 1136
Bas. ENE 3, I

Two joining fragments of rim and side of dish; blue. Iridescence and pitting. Tubular rim, folded outward; nearly vertical side, tapering downward to a rounded basal angle. Shape probably similar to **No. 310**. Blown and tooled. D ca. 15 cm; A Th .1 cm.

313. Additional fragments of tubular rims from colored dishes similar in shape to **No. 310**.

- **a.** CD 1135, Bas. ENE 3, I. Fragment of rim and side; emerald green. D ca. 16 cm. Pl. 18.
- **b.** CD 1154, Bas. NW 3 Exterior, I. Fragment of rim and side; emerald green. D ca. 15–16 cm.

c. C70.631, S.H. Street M, 0. Fragment of rim and side; emerald green. D ca. 15–16 cm.
d. C68.710, Forum SW, 0. Fragment of rim and side; emerald green. D ca. 15–16 cm.
e. C66.872, V-D, E, 0. Fragment of rim; emerald green. D ca. 15–16 cm.
f. C69.459, Forum SW, I. Fragment of rim; blue. D uncertain.
g. CE 2171B, A.P. Room 29, I. Fragment of rim; emerald green. D uncertain.

314. CD 1155 Pl. 18
Bas. Nave, 0
Fragment of rim and side of dish or other vessel; dark blue-green (aquamarine). Iridescence and pitting. Upright rim with solid, thickened edge; below edge of rim, a rounded horizontal ridge projects outward from side; vertical side; rounded basal angle. Blown and tooled. D ca. 13 cm; A Th .1 cm.
a. CE 2122, A.P. Room 17, I. Fragment of rim and side similar in shape to **No. 314**; dark blue-green (aquamarine). D uncertain; A Th .1 cm.
b. CE 2133. A.P. Room 4, 0. Fragment of rim and side similar in shape to **No. 314**; dark blue-green (aquamarine). D uncertain; A Th .1 cm.

*Colored Translucent Bowls (**Nos. 315–339**)*

315. CE 1992 Pl. 18
A.P. Room 22, II Deposit VII
Decade prior to ca. A.D. 40/45
Three joining fragments of rim, side, and handle (incomplete) of large bowl (possibly a footed bowl); blue. Iridescence and pitting. Outsplayed rim with double tubular edge, folded outward; almost vertical side; side may possibly have expanded outward below lower handle attachment; three-ribbed handle, oval in cross section, attached below rim, presumably twisted. Blown, with drawn and tooled handle. D ca. 16 cm; A Th side .25 cm.
Bibliography: Grose 1975, 39, no. 11.

316. CE 2045 Pl. 18
A.P. Room 22, II Deposit VII
Decade prior to ca. A.D. 40/45
Three non-joining fragments of rim and side of large bowl or other vessel; purple. Enamel weathering and pitting. Off-set, upright rim with solid, thickened edge; carinated side. Shape probably similar to **No. 317** below. Blown. D ca. 16 cm; A Th .1 cm.
Bibliography: Grose 1975, 39, no. 12B.

317. CE 1959 Pl. 18
A.P. Room 22, II Deposit VII
Decade prior to ca. A.D. 40/45
Three joining and seven non-joining fragments of rim, side, and base of bowl (profile restorable); blue. Iridescence and pitting. Off-set, upright rim with solid, thickened edge; carinated side; flattened, cut-out base-ring. Blown and tooled. H 4.8 cm; D ca. 12 cm; D base-ring 5.6 cm; A Th side .15 cm.
Bibliography: Grose 1975, 39, no. 12; Grose 1983, 44.

318. CE 2036A and B Pl. 18
A.P. Room 22, II Deposit VII
Decade prior to ca. A.D. 40/45.

Two joining and two non-joining fragments of rim and side of small bowl; dark blue-green (aquamarine). Iridescence and pitting. Inturned rim with tubular edge, folded outward; side bulges outward above basal angle. Blown. D ca. 11 cm; A Th .1 cm.
Bibliography: Grose 1975, 40, no. 14.

319. Additional fragments of rims of colored, carinated bowls similar in shape to **Nos. 316–318**.

- **a.** CE 1987, A.P. Room 22, II, Deposit VII, decade prior to ca. A.D. 40/45. Six fragments of rim and side; blue. D ca. 12 cm.
- **b.** CE 2173, A.P. Room 15, I. Fragment of rim and side; blue. D uncertain.

320. Description of **No. 320** (illustration on pl. 19) is missing.

321. C66.870 Pl. 19
V-D, SE Balk, 0 L
Two non-joining fragments of rim and side of small bowl; emerald green. Iridescence and pitting. Inturned rim with tubular edge, folded outward; side expands outward below rim. Blown. D ca. 10 cm; A Th side .1 cm.

322. C72.56 Pl. 19
A.P. Room 15, I
Fragment of rim and side of small bowl; blue. Iridescence and pitting. Inturned rim with tubular edge, folded outward; side expands outward below rim. Blown. D ca. 8 cm; A Th .08 cm.

323. CE 2148 Pl. 19
A.P. Room 22, II Deposit VII
Decade prior to ca. A.D. 40/45
Two joining fragments of rim and side of small bowl; blue. Iridescence and pitting. Inturned rim with tubular edge, folded outward; almost vertical side; rounded basal angle. Blown. D ca. 6.8 cm; A Th .1 cm.
Bibliography: Grose 1975, 40, no. 15.

324. Additional fragments of rims of small colored bowls similar in shape to **Nos. 320–323**.

- **a.** C68.712, S.H., 0-I. Fragment of rim and side; emerald green. D ca. 10 cm. Pl. 19.
- **b.** CF 2449, A.P. Room 25 Exterior, II. Fragment of rim; emerald green. D ca. 8 cm.
- **c.** C69.457, Forum SW, I. Fragment of rim and side; emerald green. D ca. 8 cm.
- **d.** C69.458, Forum SW, I. Fragment of rim and side; emerald green. D ca. 8 cm.
- **e.** C66.871, V-D, E, I. Fragment of rim and side; emerald green. D ca. 8 cm.

325. CE 1971 Pl. 19
A.P. Room 15, I
Two non-joining fragments of rim and side of small bowl or beaker; blue. Iridescence and pitting. Nearly horizontal rim with solid, thickened edge; convex curving side. Blown. D ca. 9 cm; A Th .1 cm.

326. CF 2412 Pl. 19
A.P. Room 16 Exterior, V

Fragment of rim and side of small bowl; blue. Iridescence and pitting. Outsplayed rim with solid, thickened edge; straight side, tapering downward. Blown. D ca. 9 cm; A Th .1 cm.

327. C70.613 Pl. 19
S.H. Street M, 0-I
Two joining fragments of rim, side, and handle of bowl; blue. Iridescence and pitting. Short, outsplayed rim with solid, thickened edge; nearly vertical side, tapering downward; stirrup handle with projecting thumb stalls attached at rim, presumably one of two such handles. Blown, with drawn and tooled handle. D ca. 10 cm; A Th side .08 cm; H handle 3.8 cm.

328. CE 1970 Pl. 19
A.P. Room II, I–II
Three joining fragments of side of bowl or other vessel; blue. Iridescence and pitting. Body encircled by a flattened tube, cut-out from the side; convex curving side. Shape uncertain. Blown and tooled. Max D body ca. 8.5 cm; A Th .1 cm.

329. C66.815 Pl. 19
S.H. Street M, 0-I
Fragment of side, floor, and base-ring of bowl or other vessel; emerald green. Iridescence and pitting. Convex curving side; concave floor; outsplayed base-ring with rounded edge. Blown and tooled. D base-ring ca. 7 cm; Max Th floor .4 cm.

330. C66.812 Pl. 19
S.H. Street M, 0-I
Three joining fragments of side, floor, and base-ring of bowl or other vessel; blue. Iridescence and pitting. Convex curving side; slightly concave floor; outsplayed base-ring with rounded edge. Blown. D base-ring ca. 6.4 cm; Max Th floor .4 cm.

331. C69.408 Pl. 19
Forum SW, I
Fragment of side, floor, and base-ring of bowl or other vessel; blue. Iridescence and pitting. Shape and technique of manufacturing as in **No. 330**, except that the upper side of the floor is slightly convex. D base-ring 5.2 cm; Max Th floor .4 cm.

332. C72.135
VI/VII-D, 0
Fragment of side, floor, and base-ring of small bowl or other vessel; emerald green. Iridescence and pitting. Shape and technique of manufacturing as in **No. 329**. D base-ring ca. 4 cm; Max Th floor .2 cm.

333. CE1988 Pl. 19
A.P. Room 22, II Deposit VII
Decade prior to ca. A.D. 40/45
Fragment of side, floor, and base-ring of small bowl or other vessel; blue. Iridescence and pitting. Shape and technique of manufacturing as in **No. 329**. D base-ring 3.9 cm; Max Th floor .15 cm.
Bibliography: Grose 1975, 40, no. 17.

334. C66.811 Pl. 19
S.H. Street M, 0-I

Fragment of side, floor, and base-ring of small bowl or other vessel; emerald green. Iridescence and pitting. Shape and technique of manufacturing as in **No. 329**. D base-ring 3.1 cm; Max Th floor .2 cm.

335. Additional fragments of colored base-rings similar to **Nos. 329–334**.

a. CD 1168, Bas. ENW 2, III. Blue. D base-ring ca. 5.1 cm.
b. C70.584, Forum SATR, 0. Emerald green. D base-ring ca. 5 cm.
c. C69.452, S.H., 0. Emerald green. D base-ring ca. 5 cm.

336. CE 1167 Pl. 19
A.P. Room 22, II Deposit VII
Decade prior to ca. A.D. 40/45

Fragment of side, floor, and base-ring of bowl or other vessel; dark blue-green (aquamarine). Iridescence and pitting. Convex curving side; uneven, slightly convex floor; pad base as in **No. 308**. Blown. D base-ring ca. 4.8 cm; Max Th floor .7 cm.

337. CE 2171 Pl. 19
A.P. Room 22, II Deposit VII
Decade prior to ca. A.D. 40/45

Fragment of side, floor, and base-ring (edge missing) of bowl or other vessel; blue. Iridescence and pitting. Shape and technique of manufacturing as in **No. 336**. D base-ring ca. 3.5 cm; Max Th floor .7 cm.
Bibliography: Grose 1975, 40, no. 18.

338. CD 1166 Pl. 19
Bas. ENW 3, I

Fragment of side, floor, and base-ring (edge missing) of small bowl or other vessel; dark blue-green (aquamarine). Iridescence and pitting. Shape and technique of manufacturing as in **No. 336**. D base-ring ca. 3.5 cm; Max Th floor .6 cm.

339. C70.660
S.H. Street M, 0

Fragment of side, floor, and base-ring (edge missing) of small bowl or other vessel; blue. Iridescence and pitting. Convex curving side; concave floor; distinctive pad base as in **No. 308**. Blown and tooled. D base-ring uncertain; Max Th floor .6 cm.

*Colored Translucent Saucepans (**Nos. 340–341**)*

340. C70.612 Pl. 20
S.H. Street M, 0-I

Fragment of handle (incomplete) and rim of saucepan; blue. Iridescence and pitting. Outsplayed rim with solid, thickened edge; nearly vertical side; horizontal handle attached to top of rim. Blown, with drawn glass handle in same fabric. D ca. 12 cm; A Th side .1 cm.

341. C72.44 Pl. 20
A.P. Room 13, Cesspool

Handle and fragment of rim and side of saucepan; emerald green. Iridescence and pitting. Outsplayed rim with solid, thickened edge; nearly vertical side, tapering downward; horizontal handle attached to top of rim. On either side of the handle, a thin, hemispherical projection folded outward at right angles to the rim; end of handle folded under. Blown, with drawn and tooled handle. D ca. 10 cm; A Th side .05 cm; L handle 7.6 cm; W handle at end 3 cm.
Parallel: See also naturally colored fragments of similar saucepans, **Nos. 423–423b**.

*Colored Translucent Beakers (**Nos. 342–352**)*

342. CE 2041 Pl. 20
A.P. Room 22, II Deposit VII
Decade prior to ca. A.D. 40/45
Four joining and six non-joining fragments of rim, side (incomplete), and bottom of tall, conical beaker (profile restorable); purple. Enamel weathering and pitting. Off-set, upright rim with knocked-off edge; nearly vertical side, tapering downward to a rounded basal angle; kicked bottom. On exterior, rotary-polishing marks and several horizontal bands of wheel-cut lines below rim and around body; rotary-polishing marks on interior. Blown and rotary-polished. Restored H 12 cm; D ca. 8 cm; D bottom 4.5 cm; A Th side .08 cm.
Bibliography: Grose 1975, 40, no. 19.

343. CD 1156 Pl. 20
Bas. ENE 3, I
Fragment of rim and side of conical beaker, yellow. Iridescence and pitting. Off-set, upright rim with knocked-off edge; nearly vertical side, tapering downward. On exterior, rotary-polishing marks below edge of rim. Blown and rotary-polished. D ca. 8 cm; A Th .1 cm.

344. CD 1157 Pl. 20
Bas. ENE 3, III Deposit VIII
Prior to ca. A.D. 50/55
Two joining fragments of rim, side, and bottom of hemispherical beaker (profile mostly restorable); blue. Iridescence and pitting. Inturned rim with knocked-off edge; convex curving side; indications of a slightly concave bottom. On interior, several horizontal rotary-polishing marks around body; on exterior, two horizontal grooves: one below edge of rim, the other just above the junction of the side and bottom. Blown, rotary-polished on rim and around interior; exterior cut. Restored H 5 cm; D ca. 10 cm; A Th .15 cm.

345. CE 2185 Pl. 20
A.P. Room 15, I
Fragment of rim and side of beaker; lime green. Iridescence, pitting, and discoloration. Slightly inturned rim with knocked-off edge; straight side, tapering diagonally downward. On exterior, a horizontal groove around center of body; elsewhere, horizontal rotary-polishing marks. Blown, rotary-polished on rim and exterior; exterior cut. D ca. 9 cm; A Th .08 cm.

346. C69.415 Pl. 20
Forum SW, I
Two joining fragments of rim and side of beaker; purple. Iridescence and pitting. Upright rim with knocked-off edge; convex curving side. On exterior, a thin, horizontal groove immediately below

edge of rim; two additional grooves in a band around the center of the body. Blown (?), rotary-polished on rim and on both surfaces; exterior cut. D ca. 9 cm; A Th side .1 cm.

347. CD 1158 Pl. 20
Bas. ENE 3, I
Fragment of rim and side of beaker; light blue. Iridescence and pitting. Upright rim with knocked-off edge; straight side, tapering upward. On exterior, a thin, horizontal groove below edge of rim. Blown, rotary-polished on rim; cut on exterior. D ca. 7 cm; A Th .08 cm.

348. CD 1159 Pl. 20
Bas. ENE 3, I
Fragment of rim and side of beaker; green. Iridescence and pitting. Upright rim with knocked-off edge, almost vertical side. On exterior, a horizontal groove below edge of rim; rotary-polishing marks on interior below rim. Blown, rotary-polished on rim and on both surfaces; cut on exterior. D ca. 8 cm; A Th .15 cm.

349. C70.632 Pl. 20
S.H. Street M, Sewer
Fragment of rim and side of beaker; blue. Silvery iridescence and pitting. Upright rim with knocked-off edge; almost vertical side, slightly inturned. On exterior, a horizontal band of rotary-polishing marks below edge of rim. Blown, rotary-polished on rim and on exterior. D ca. 8 cm; A Th .15 cm.

350. CE 2019 Pl. 20
A.P. Room 21, IV
Two non-joining fragments of rim and side of beaker; golden-yellow. Iridescence and pitting. Outsplayed rim with thickened edge; convex curving side. On interior, a thin, horizontal groove immediately below edge of rim; on exterior, three horizontal grooves in a band around the center of the body. Blown, cut on both surfaces; rotary-polished on rim. D ca. 8 cm; Max Th side .1 cm.

351. CE 2149 Pl. 20
A.P. Room 22, II Deposit VII
Decade prior to ca. A.D. 40/45
Fragment of rim and side of beaker or jar; emerald green. Iridescence and pitting. Outsplayed rim with solid, thickened edge; side expands outward below rim. Blown. D ca. 7 cm; A Th .08 cm.
Bibliography: Grose 1975, 40, no. 16.

352. CE 2150
A.P. Room 22, II Deposit VII
Decade prior to ca. A.D. 40/45
Fragment of rim and side of beaker or jar; emerald green. Iridescence and pitting. Shape and techniques of manufacturing as in **No. 351**. D ca. 7.5 cm; A Th .08 cm.
Bibliography: Grose 1975, 40, no. 16A.

Colored Translucent Flasks, Jugs, and Jars (***Nos. 353–357***)

353. CE 1961 Pl. 21
A.P. Room 22, II Deposit VII
Decade prior to ca. A.D. 40/45

Twenty-five joining and four non-joining fragments of side and bottom of large spherical flask; light blue. Slight iridescence and pitting. Spherical body; concave bottom. Blown very thin. H body 9.5 cm; D body ca. 10 cm; D bottom 5 cm; A Th .05 cm.
Bibliography: Grose 1975, 40, no. 20.

354. CE 2043 Pl. 21
A.P. Room 22, II Deposit VII
Decade prior to ca. A.D. 40/45

Four joining and fifteen non-joining fragments of neck (upper portion and rim missing), one handle, side, and base of a two-handled jar (profile mostly restorable); purple with opaque white decoration. Iridescence and pitting. Cylindrical neck; piriform body; open cut-out base; strap handle, oval in cross section, presumably one of two. On exterior, unmarvered blobs of opaque white glass, applied to surface to achieve the final shape (hence elongated on neck, more rounded in body). Blown and tooled, with drawn glass handle and applied decoration. Restored H 20 cm; Restored D 6.5 cm; D base-ring ca. 6 cm; A Th side .13 cm. The lower third of the vessel, known from photographs and drawings, is on loan to the sons of Marchesa Maria Rita San Felice Guglielmi.
Bibliography: Grose 1975, 42, no. 22.

355. CE 1972 Pl. 21
A.P. Room 21A, II

Eight joining fragments of neck, shoulder, and side of large jar or jug; light blue with opaque white decoration. Iridescence and pitting. Cylindrical neck; straight shoulder flared diagonally downward. Precise shape uncertain. On exterior, a wide band of opaque white glass applied diagonally across shoulder and neck. Blown with applied decoration. D neck ca. 4.5 cm; A Th neck .08 cm.

356. CE 2107 Pl. 21
A.P. Room 21, 0

Six non-joining fragments from lower part of side of large jar, jug, or other vessel; peacock blue. Silvery iridescence and pitting. Convex curving side. Precise shape uncertain. On exterior, four parallel, thin threads applied in same fabric around body; threads pointed in cross section. Blown, with applied decoration. Max D fragment ca. 15 cm; A Th side .03 cm.

357. CD 1161
Bas. ENW 3, I

Handle and indication of side of large jar or jug; peacock blue. Iridescence and pitting. Convex curving side; strap handle with lower attachment intact. Shape uncertain, but probably a two-handled jar similar to **No. 354**. Blown, with drawn glass handle. H handle 7.4 cm; Max W handle at point of attachment 1.9 cm.

Colored Translucent Thick-Walled Bottles and Amphoriskoi (***Nos. 358–361***)

358. CE 1966 Pl. 21
A.P. Room 22, II Deposit VII
Decade prior to ca. A.D. 40/45

Five joining fragments of rim, neck, and body of small bottle; blue. Iridescence and pitting. Upright rim with projecting, rounded lip; cylindrical neck; shoulder curves outward, then downward to form a cylindrical body. On upper side of rim, a narrow tooled groove. On exterior, a tool mark at junction of neck and shoulder. Blown and tooled. H fragment 6.5 cm; D 2 cm; D body 3.5 cm; A Th side .3 cm.
Bibliography: Grose 1975, 42, no. 21; Grose 1983, 44.

359. CG 375 Pl. 21
Curia, SE Room, 0

Fragment of side and bottom (incomplete) of small bottle; blue. Iridescence and pitting. Ovoid or cylindrical body, tapering downward to a rounded basal angle; slightly concave bottom. Blown. A Th .3 cm.

360. CE 2063 Pl. 21
A.P. Room 16, I Deposit V
ca. 25/15 B.C. to ca. A.D. 40/45

Fragment of side and bottom (tip missing) of amphoriskos; dark blue. Iridescence and pitting. Side curves diagonally downward to a pointed or rounded tip. Blown. Max D fragment 4 cm; A Th side .3 cm.

361. C69.409 Pl. 21
S.H. Room 11, I

Fragment of side and bottom (tip missing) of amphoriskos; dark blue. Iridescence and pitting. Straight side, tapers diagonally downward to a pointed or rounded tip. Blown. Max D side 2.5 cm; A Th side .3 cm.

Colored Translucent Small Thin-Walled Bottles (***Nos. 362–372***)

362. C70.303 Pl. 22
S.H. Street M, Sewer

One non-joining and five joining fragments of rim, neck, body, and bottom of small bottle (profile restorable); golden yellow. Silvery iridescence and pitting. Outsplayed rim with uneven, fire-polished edge; cylindrical neck, tapering upward; shoulder flares outward to form bulbous body; slightly concave bottom. On exterior, a thin, horizontal groove at junction of neck and shoulder; two additional grooves in a band around the greatest diameter of the body. Blown, cut on exterior. H 6.3 cm; D rim 2.2 cm; D body 5.6 cm; A Th side .05 cm.
Bibliography: Grose 1982, 44; Grose 1984, 440.

363. CE 2108 Pl. 22
A.P. Room 10, II

Two joining and two non-joining fragments of rim, neck, and side of small bottle; yellow. Silvery iridescence and pitting. Outsplayed rim with uneven, fire-polished edge; uneven cylindrical neck, tapering upward; shoulder flares diagonally outward from neck; piriform body. Blown. D 2.2 cm; H neck 1.5 cm; A Th side .04 cm.

364. C68.600 Pl. 22
S.H. Room 4, I Deposit IV
Prior to 25/15 B.C.

Fragment of rim, neck, and shoulder of small bottle; golden-yellow. Silvery iridescence and pitting. Horizontal rim with edge polished to a point; cylindrical neck, tapering downward; shoulder flares sharply outward from neck. Blown and polished. D 1.7 cm; H neck 2 cm; A Th .04 cm.

365. CD 72 Pl. 22
Bas. EW, I

Fragment of rim, neck, and shoulder of small bottle; blue. Iridescence and pitting. Outsplayed rim with uneven, fire-polished edge; short, funnel-shaped neck; shoulder curves diagonally outward from neck, forming ovoid (?) body. Blown. D 1.7 cm; H neck .8 cm. A Th .04 cm.

366. C71.207 Pl. 22
VI-D, SW Perimeter Trench, 0

Two joining fragments of rim (edge missing) and neck of small bottle; blue. Iridescence and pitting. Outsplayed rim; cylindrical neck, tapering upward. Blown. D uncertain; H neck 2.6 cm; A Th .05 cm.

367. CE 2189 Pl. 22
A.P. Room 15, I

Fragment of neck of small bottle; dark blue. Iridescence and pitting. Narrow cylindrical neck, bulging slightly at center. On exterior, a tooling mark at junction of neck and side. Blown and tooled. Max D neck .5 cm; H neck 1.9 cm; A Th .1 cm.

368. CE 2188
A.P. Room 16A, I

Tiny fragment of rim and neck of small bottle; blue. Iridescence and pitting. Outsplayed rim with uneven, fire-polished edge; short, cylindrical neck. Blown. D ca. 1.4 cm; H neck 1.2 cm; A Th .04 cm.

369. CE 2187
A.P. Room 21A, II

Tiny fragment of rim of small bottle; blue. Iridescence and pitting. Outsplayed rim with fire-polished edge. Blown. D ca. 2 cm; A Th .06 cm.

370. CE 1139 Pl. 22
Bas. ENE 1, I

Small fragment of rim of small bottle; blue. Iridescence and pitting. Outsplayed rim with solid, thickened edge; cylindrical neck. Blown. D ca. 2.7 cm; A Th .1 cm.

371. CE 2201
A.P. Room 16, I Deposit V
ca. 25/15 B.C. to ca. A.D. 40/45

Small fragment of neck and side of small bottle; yellow. Iridescence and pitting. Cylindrical neck; side flares diagonally outward from neck, as in **No. 365**. Blown. A Th .05 cm.

372. CE 2017 Pl. 22
A.P. Room 16, I Deposit V
ca. 25/15 B.C. to ca. A.D. 40/45
Small fragment of side and bottom of small bottle; yellow. Iridescence and pitting. Ovoid body; slightly concave bottom. Blown. Max D body 3 cm; A Th .04 cm.

*Colored Translucent Ladle (**No. 373**)*

373. CE 1962 Pl. 21
A.P. Room 22, II Deposit VII
Decade prior to ca. A.D. 40/45
Five joining fragments of bowl and handle (tip missing) of ladle (profile restorable); blue. Iridescence and pitting. Bowl of ladle has a sharply inturned rim with a thickened, rounded edge; rim off-set from convex curving side; concave bottom, convex on interior; vertical handle with a hooked end. Blown, with a drawn glass handle; middle of handle flattened with a tool. H 11.5 cm; D bowl 5.8 cm; H bowl 2.7 cm; A Th side .2 cm.
Bibliography: Grose 1975, 42, no. 23; Grose 1984, 44.

*Naturally Colored and Colorless Tablewares (**Nos. 374–489**)*

*Various Naturally Colored Plates, Dishes, and Bowls (**Nos. 374–391**)*

374. CE 2154 Pl. 22
A.P. Room 22, II Deposit VII
Decade prior to ca. A.D. 40/45
Five joining fragments of rim, side, and floor of dish (profile mostly restorable); natural bluish-green. Iridescence and pitting. Slightly outsplayed rim with solid, thickened edge; vertical side; flat floor, sloping obliquely to the outside. Blown. H ca. 3.2 cm; D ca. 16 cm; A Th .1 cm. Restored with pad base **No. 374a**.
Bibliography: Grose 1975, 42, no. 24.

- **a.** CE 2151B, A.P. Room 22, II, Deposit VII, decade prior to ca. A.D. 40/45. Two non- joining fragments of floor and base-ring of dish; natural bluish-green. Pad base with short, rounded, slightly outsplayed base-ring. D base-ring ca. 9 cm; A Th floor .1 cm. Restored with **No. 374**. Pl. 22.
  Bibliography: Grose 1975, 44, 25B.

375. CE 2153C Pl. 22
A.P. Room 22, II Deposit VII
Decade prior to ca. A.D. 40/45
Two joining and five non-joining fragments of rim, side, and floor of dish (profile mostly restorable); natural bluish-green. Slight iridescence. Inturned tubular rim, edge folded outward; slightly inturned side; uneven floor, convex at center. Blown. H ca. 2.8 cm; D ca. 16 cm; A Th .15 cm. Restored with pad base **No. 375a**.
Bibliography: Grose 1975, 52, no. 25.

a. CE 2049, A.P. Room 22, II, Deposit VII, decade prior to ca. A.D. 40/45. Fragment of floor and base-ring of dish; natural bluish-green. Pad base with uneven floor and rounded, slightly outsplayed base-ring. D base-ring ca. 8.5 cm; A Th floor .15 cm. Restored with **No. 375**.

376. CE 2153A
A.P. Room 22, II — Deposit VII
Decade prior to ca. A.D. 40/45

Six joining and four non-joining fragments of rim and side of dish; naturally bluish-green. Iridescence and pitting. Shape and technique of manufacture as in **No. 375**. D ca. 19 cm.
Bibliography: Grose 1975, 44, no. 25A.

a. CE 2151A, A.P. Room 22, II, Deposit VII, decade prior to ca. A.D. 40/45. Three joining fragments of pad base and base-ring; natural bluish-green. D base-ring 10 cm. Possibly belongs to same vessel as **No. 376**.
Bibliography: Grose 1975, 44, nos. 25A–B.

377. CE 2153B
A.P. Room 22, II — Deposit VII
Decade prior to ca. A.D. 40/45

Two joining and four non-joining fragments of rim and side of dish; natural bluish-green. Iridescence and pitting. Shape and technique of manufacture as in **No. 375**. D ca. 16 cm.
Bibliography: Grose 1975, 44, no. 25B.

378. CE 2151C
A.P. Room 22, II — Deposit VII
Decade prior to ca. A.D. 40/45

Fragment of pad base and base-ring of dish; natural bluish green. Iridescence and pitting. Shape and technique of manufacture as in **No. 375a**. D uncertain.
Bibliography: Grose 1975, 44, nos. 25A–B.

379. CE 2033 — Pl. 22
A.P. Room 22, II — Deposit VII
Decade prior to ca. A.D. 40/45

Five joining fragments of rim, side, and base of small bowl (profile restorable); natural bluish-green. Slight iridescence and pitting. Inturned rim with tubular edge, folded outward; side expands outward, then downward, forming a hemispherical body; pad base with rounded, outsplayed base-ring. Blown. H 5.9 cm; D 10 cm; A Th .1 cm.
Bibliography: Grose 1975, 44, no. 26.

380. Fragments of seven rims and seven base-rings, presumably from seven different bowls, similar in shape and technique of manufacture to **No. 379**. These fragments are also from A.P. Room 22, Deposit VII, decade prior to ca. A.D. 40/45.

a. CE 2153D. Tubular rim fragment; natural bluish-green. D ca. 14 cm.
b. CE 2153E. Tubular rim fragment; natural bluish-green. D ca. 13 cm. Pl. 22.
c. CE 2153F. Tubular rim fragment; natural bluish-green. D ca. 12.5 cm.
d. CE 2053. Tubular rim fragment; natural bluish-green. D ca. 12 cm.
e. CE 2153G. Tubular rim fragment; natural bluish-green. D ca. 12 cm. Pl. 22.
f. CE 2151E. Tubular rim fragment; natural bluish-green. D ca. 11 cm. Pl. 22.

g. CE 2153I. Tubular rim fragment; natural bluish-green. D ca. 8.5 cm. Pl. 22.
h. CE 2151J. Base fragment; natural bluish-green. D base ca. 6.5 cm.
i. CE 2151D. Base fragment; natural bluish-green. D base ca. 5.5 cm. Pl. 23.
j. CE 2151H. Small base fragment; natural bluish-green. D base ca. 5 cm.
k. CE 2151F. Base fragment; natural bluish-green. D base ca. 4.5 cm; Pl. 23.
l. CE 2153H. Base and base-ring; natural bluish-green. D base-ring ca. 4.2 cm. Pl. 23.
m. CE 2151G. Base and base-ring; natural bluish-green. D base-ring ca. 4 cm. Pl. 23.
n. CE 2151I. Small base fragment; natural bluish-green. D base uncertain.

381. CE 2034 — Pl. 22
A.P. Room 22, II — Deposit VII
Decade prior to ca. A.D. 40/45
Three joining fragments of rim (edge missing), side, and base of small bowl (profile mostly restorable); natural bluish-green. Iridescence and slight pitting. Slightly outsplayed rim rises vertically from tubular but flattened cut-out ledge; convex curving side; distinctive pad base with unfinished, slightly outsplayed edge. Blown. H ca. 7 cm; D ca. 9 cm; A Th side .2 cm.
Bibliography: Grose 1975, 44, no. 27.

382. CE 2195A–I
A.P. Room 16, I — Deposit V
ca. 25/15 B.C. to ca. A.D. 40/45
Three fragments of rim and six fragments of bases (or base-rings) of dishes or bowls similar to **No. 375** or **Nos. 379–380**, respectively.
a. Tubular rim fragment; natural bluish green. D ca. 15 cm.
b. Tubular rim fragment; natural bluish-green. D ca. 14 cm.
c. Tubular rim fragment; natural bluish-green. D ca. 13 cm.
d. Base and base-ring fragment; natural bluish-green. D ca. 6 cm.
e. Base and base-ring fragment; natural bluish-green. D ca. 5.5 cm.
f. Base and base-ring fragment; natural bluish-green. D ca. 5 cm.
g. Base and base-ring fragment; natural bluish-green. D ca. 5 cm.
h. Base and base-ring fragment; natural bluish-green. D uncertain.
i. Base and base-ring fragment; natural bluish-green. D uncertain.

383. CD 1141 — Pl. 23
Bas. ENE 1, III — Deposit VIII
Prior to ca. A.D. 50/55
Base and base-ring of small bowl; natural bluish-green. Iridescence and pitting. Pad base with outsplayed base-ring. Shape of bowl probably similar to **No. 379**. Blown. D base-ring 5.5 cm; A Th floor .2 cm.

384. CD 1153A and B — Pl. 23
Bas. ENE 3, III — Deposit VIII
Prior to ca. A.D. 50/55
Two non-joining fragments of sides and base-rings of bowl(s); natural bluish-green. Iridescence and pitting. Convex curving sides; pad bases with outsplayed base-rings. Shape of bowl or bowls probably similar to **No. 379**. Blown. D base-rings ca. 6 cm.

385. CE 2003A Pl. 23
A.P. Room 15, I
Three joining fragments of rim and side of bowl; natural pale greenish-yellow. Slight iridescence and pitting. Outsplayed tubular rim, edge folded outward; almost vertical side, tapering downward. Blown. D ca. 15 cm; A Th side .1 cm.

386. CE 2003B Pl. 23
A.P. Room 15, I
Two joining fragments of side and base of bowl; natural pale yellowish-green. Slight iridescence and pitting. Convex curving side; uneven, slightly convex floor; pad base with solid, outsplayed base-ring. Blown. D base-ring 5.5 cm; A Th side .1 cm. This base was restored with **No. 385**.

387. C70.607 Pl. 23
S.H. Street M, 0-I
Fragment of side and base of bowl; natural bluish-green. Slight iridescence and pitting. Convex curving side; flat but uneven floor; pad base with outsplayed base-ring. Blown. D base-ring 5.2 cm; A Th side .2 cm.

388. C66.822 Pl. 23
A.P. Room 22, I
Two joining fragments of side and base-ring of bowl; natural pale yellowish-green. Iridescence and pitting. Convex curving side; concave floor; tall, outsplayed base-ring applied to bottom of vessel. Blown. D base-ring 5.1 cm; A Th side .2 cm.

389. C66.881 Pl. 23
S.H. Street M, 0-I
Base and base-ring of bowl or other vessel; natural light green. Slight iridescence and pitting. Uneven convex bottom; pad base with outsplayed base-ring. Blown. D base-ring 7 cm; A Th floor .25 cm.

390. C70.662 Pl. 23
Forum SE, Trench 70D, I
Base and base-ring of bowl or other vessel; dark greenish-yellow. Slight iridescence. Uneven, slightly concave floor; pad base with rounded outsplayed base-ring. Numerous oblique tooling marks on exterior of base-ring. Blown and tooled. D base-ring 5.3 cm; A Th floor .25 cm.

391. CF 2351A Pl. 23
A.P. Room 25 Exterior, NW, I
Fragment of side and base of bowl or other vessel; natural bluish-green. Iridescence and pitting. Convex curving side; tall conical pad base. Max D base 6.5 cm; A Th side .15 cm.

*Bowls with Applied Fillets (**Nos. 392–402**)*

Surprisingly plentiful at Cosa are a series of bowls (and possibly plates), usually fairly shallow but of varying diameter, that are ornamented with glass fillets attached in pairs at opposite sides of the rim (**Nos. 392–402**). The fillets are actually ribbed coils tooled while hot into alternating ridges and depressions. Usually they are symmetrically designed with the greatest width at the center. At times the coils act as handles (and may in fact resemble counterparts in silver); at other times they are purely decorative. Mostly they are in the same fabric as the vessel itself.

The rims of these bowls are uniformly tubular or double-tubular (**No. 395**), and sides generally proceed diagonally downward to closed cut-out base-rings or, more rarely, applied base-rings. Most pieces are either colorless or lightly colored, generally with a greenish or greenish-yellow tinge.

In distribution most published examples are known from the eastern Mediterranean, where they appear to be especially common on Cyprus and on Syro-Palestinian sites (see list below). For this reason their presence at Cosa was initially unexpected. Yet it is clear from the number of examples found at the town (as well as others from Italian contexts; see list below) that these bowls were abundantly available in Italy, possibly throughout the western Mediterranean. Like so much early Imperial glassware, they were widely disseminated or imitated. Their apparent absence at Vindonissa, Fishbourne, and other transalpine settlements, however, may indicate that they rarely circulated in those regions.

Based on current knowledge, these bowls were introduced sometime in the second half of the first century A.D., probably after midcentury. They do not occur at Magdalensberg, abandoned around A.D. 45; nor do they seem to exist at Pompeii and Herculaneum, destroyed in A.D. 79. The earliest examples are from burials in Canton Ticino attributed only generally to the mid-first century or to the latter half of that century. Later examples are assigned a Hadrianic–Antonine date in Isings 1957, 59, Form 43, or are said to belong to the late second century A.D., as at Tipasa (see list below). In all likelihood most examples fall in the last quarter of the first and continue well into the second century.

Parallels and Dated Examples of Bowls with Applied Fillets

Ostia II 1970, 250, no. 616, pl. 93. Fairly deep bowl.

Luni II 1977, 280, CM 3106, pl. 155.9, a dish; 284, CM 9967, pl. 156.13, a deep bowl; also 552, K 1059, pl. 285.14.

Simonett 1941, 85, fig. 69.1, pl. 10.I.1. Neronian to Vespasianic grave, which accords with Berger's redating in Vindonissa 1960, chart opp. p. 92, to the second half of the first century A.D.

Ulrich 1914, 539–540. Examples from a burial at Bellinzona attributed to the mid-first century A.D.

Murano, Museo Vetrario: Acc. nos. 3641, 3534, and 6512, as cited by Auth in Newark Coll. 1976, 95. Presumably the vessels are from Zadar (Croatia) or from Italy.

Simoni 1963, 41–43, fig. 3. Shallow bowl from a burial at Salò dated to the first century A.D.

Doppelfeld 1966, pl. 84, upper photo. Without find-spot, but possibly from Cologne, as is most glass in the catalogue.

Alarcão 1968, nos. 33–34. Several examples from Portuguese museums, possibly of western origin.

Tipasa 1967, 94, no. 195, pl. 10.4. Bowl from Tipasa found in a burial with a coin of Marcus Aurelius or Lucius Verus.

Buechner 1960, no. 16. Fragment of a bowl from Tarrha, Crete.

Sardis 1980, 21–22, nos. 94–99, pls. 4 and 21.

Jalame (Israel): Several fragmentary examples. Unpublished. Information furnished by Gladys D. Weinberg.

Barag 1962, 210, fig. 4. A precisely dated example from the Cave of Horror, a site destroyed and abandoned ca. A.D. 132–135.

Wadi ed-Daliyeh, Israel 1974, 104, pl. 39.4. Bowl dates to first third of the second century.

Samaria-Sebaste III 1957, 415, fig. 96.5. With additional references to examples from the eastern Mediterranean.

Vessberg 1956, 133, fig. 42.16–17, and 135, fig. 43.5–6. Cypriot examples, with additional references.
Isings 1957, 59, Form 43. Additional references cited.

Both Isings 1957, 58–59, Forms 42 and 43, and Vessberg 1956, 133 and 135, divide examples of the type between two groups based on shape (one broad and shallow, the other smaller and deeper). This has not been done in this catalogue because, regardless of the precise form, the basic conception of these bowls and their decoration are identical.

A most unusual, broad shallow bowl, said to be from Sidon and now in Berlin, is in a dark blue fabric with opaque white fillets; Berlin Coll. 1976, 61, no. 115. D 17.5 cm.

Bowls with Applied Fillets

392. CD 1182 Pl. 24
Bas. ENE 1, I
Fragment of rim and side of large, circular (or oval) plate or bowl; colorless with greenish-yellow tinge. Slight iridescence and pitting. Outsplayed tubular rim, folded outward. Attached to edge of rim, a short, symmetrically designed ribbed coil, presumably one of two. Blown, with drawn and tooled fillet. D ca. 20 cm; A Th side .1 cm.

393. Description of **No. 393** (illustration on pl. 24) is missing.

394. C66.878 Pl. 24
S.H. Street M, 0-I
Fragment of rim of plate or bowl; colorless. Slight iridescence. Outsplayed tubular rim, folded outward. Attached on top of rim, a ribbed coil, partially preserved, presumably one of two. Blown, with drawn and tooled filet. D ca. 19 cm; A Th side .1 cm.

395. CF 2473 Pl. 24
IX-E, Forum East Corner, NE Passage, I
Fragment of rim (edge missing) and side of plate or bowl; colorless with greenish-yellow tinge. Slight iridescence. Partially preserved tubular rim. Attached to edge of rim, a ribbed coil, partially preserved, presumably one of two. Blown, with drawn and tooled fillet. D ca. 18 cm; A Th side .1 cm.

396. CD 1184 Pl. 24
Bas. ENE 2, I
Fragment of rim and side of plate or bowl; colorless with slight greenish-yellow tinge. Slight iridescence. Outsplayed tubular rim, folded outward; side tapers diagonally downward. Attached to edge of rim, a ribbed coil, partially preserved, presumably one of two. Blown, with drawn and tooled fillet. D ca. 18 cm; A Th side .1 cm.

397. CD 1183 Pl. 24
Bas. ENE 2, I
Fragment of rim and side of plate or bowl; colorless with greenish tinge. Slight iridescence. Narrow, outsplayed tubular rim, folded outward; side tapers downward. Attached to underside of rim, a symmetrically designed ribbed coil, partially preserved, presumably one of two; preserved end of coil folded abruptly outward from body. Blown, with drawn and tooled fillet. D ca. 18 cm; A Th side .1 cm.

398. C66.876 Pl. 24
S.H. Street M, 0-I
Fragment of rim and side of plate or bowl; colorless with slight greenish-yellow tinge. Iridescence and pitting. Outsplayed rim with solid thickened edge; convex curving side. Attached to edge of rim, a narrow ribbed coil, partially preserved, presumably one of two. Blown, with drawn and tooled fillet. D ca. 18 cm; A Th side .1 cm.

399. C72.140 Pl. 24
S.H. Street M, 0
Two joining fragments of rim and side of plate or bowl; colorless. Silvery iridescence. Outsplayed rim with solid thickened edge; side tapers diagonally downward. Attached to edge of rim, a narrow ribbed coil, partially preserved, presumably one of two. Blown, with drawn and tooled fillet. D ca. 18 cm; A Th side .1 cm.

400. C66.837 Pl. 24
S.H. Street M, 0-I
Fragment of rim and side of bowl; colorless with greenish tinge. Slight iridescence. Horizontal tubular rim, folded outward; side appears to expand outward beneath rim. Attached to edge of rim a symmetrically designed ribbed coil, mostly preserved, presumably one of two. Blown, with drawn and tooled fillet. D ca. 16 m; A Th side .1 cm.

401. C66.777 Pl. 24
S.H. Street M, 0-I
Fragment of rim of plate or bowl; colorless with greenish-yellow tinge. Slight iridescence. Outsplayed tubular rim, folded outward. Attached to edge of rim, a symmetrically designed ribbed coil, mostly preserved, presumably one of two. Blown, with drawn and tooled fillet. D ca. 16 cm; A Th side .1 cm.

402. CD 1181 Pl. 24
Bas. ENE 1, I
Fragment of rim and side of bowl; colorless. Slight iridescence and discoloration. Outsplayed tubular rim, folded outward; almost vertical side. Attached to edge of rim, a ribbed coil, partially preserved, presumably one of two. Blown, with drawn and tooled fillet. D ca. 14 cm; A Th side .1 cm.

*Bowls with Triangular Rims (**Nos. 403–413**)*

Virtually unique to the excavations at Cosa is a series of bowls (**Nos. 404–413**) and possibly a plate (**No. 403**) of differing diameters but otherwise enjoying a uniform character. These vessels all possess upright rims with thick, solid edges, triangular in cross section. The walls are delicately blown and taper diagonally downward to a slightly concave bottom (**No. 408**). Diameters vary between 23 and 11 cm, with at least one piece as large as 30 cm (**No. 403**). In color these bowls are either pale bluish-green (**Nos. 405**, **408**, **410**, and **412**) or colorless (all other examples). In most instances, however, blowing spirals, elliptical bubbles, and horizontal striations obscure the clarity of the glass.

Although a few examples of the type have been published elsewhere, the type itself has never been recognized or described. Based on its distribution, it appears to have been confined to Italy and may well be a regional Italian tableware of the Imperial period. Aside from Cosa and its port, it is known from the excavations at Luni, Villa Potenza, Ostia, and possibly the Veneto (see list below). The only intact piece is now in the Rothschild Collection at Waddesdon Manor (Great Britain), but its provenance is unknown. The only dated examples are from Cosa (and possibly the bowl from the Veneto), where six fragments were recovered from Deposit IX, spanning the late first to early third century A.D.

Parallels and Dated Examples of Bowls with Triangular Rims

Cosa: **Nos. 403**, **406**, and **409–412**, Deposit IX, late first through third century A.D.

Villa Potenza 1971, 415, no. 48, fig. 11.4 and possibly 11.10. The brevity of the description does not allow a positive identification.

Luni II 1977, 285, CM 5419/8 and CM 5777, pl. 156.16 and 21.

Ostia IV 1977, 340, nos. 381–382, pl. 49; 341, no. 389, pl. 50. Possibly others on pls. 49–50 as well.

Calvi 1975, 207, fig. 8. An oval colorless bowl from the Veneto now in the Museo di Brescia, found in a tomb tentatively assigned to the early second century. Based on the photograph, this piece *may* possess a rim that is triangular in cross section, but it differs from those at Cosa in that it is oval and exhibits a true base-ring of some sort.

Waddesdon Manor Coll. 1977, 70–71, no. 10. Although this bowl (H 3.8 cm; D 14.3 cm) is ascribed to the fourth century and perhaps to Syria, the evidence from Italy now suggests that it probably is western in origin and belongs to the second or third centuries.

Bowls with Triangular Rims

403. CF 2449B — Pl. 25
A.P. Room 25D, III — Deposit IX
Late first through early third century A.D.

Fragment of rim and side of large plate or bowl; colorless. Iridescence, milky weathering, and pitting. Numerous horizontal striations on exterior surface. Horizontal rim with downturned, pointed edge; rim notably thicker at junction with side; convex curving side. Blown. D ca. 30 cm; A Th side .15 cm.

404. C72.138 — Pl. 25
IV-B, Column Wedges, 0

Six non-joining fragments of rim and side of plate or bowl; colorless. Iridescence, milky weathering, and pitting. Numerous horizontal striations on exterior surface. Upright rim, triangular in cross section, with an almost pointed edge; straight side, tapering diagonally downward. Blown. D ca. 23 cm; A Th side .1 cm.

405. CF 2390 — Pl. 25
A.P. Room 16 Exterior, V

Three joining and two non-joining fragments of rim and side of plate or bowl; pale bluish-green, possibly incompletely decolorized. Numerous horizontal striations on exterior surface. Upright rim, triangular in cross section, with almost pointed edge; straight side, tapering diagonally downward. Blown. D ca. 23 cm; A Th side .08 cm.

406. CF 2467 — Pl. 25
A.P. Room 25C, III — Deposit IX
Late first through early third century A.D.

Fragment of rim and side of plate or bowl; colorless. Silvery iridescence and pitting. Numerous horizontal striations on exterior surface. Outsplayed rim, triangular in cross section, with almost pointed edge; straight side, set obliquely to rim. Blown. D ca. 22 cm; A Th side .06 cm.

407. CE 2209 Pl. 25
A.P. Room 15, I
Two joining fragments of rim and side of bowl; colorless. Silvery iridescence and pitting. Numerous horizontal striations on exterior surface. Upright rim, triangular in cross section, with almost pointed edge; convex curving side. Blown. D ca. 20 cm; A Th side .05 cm.

408. CD 1175 Pl. 25
Bas. ENW 2, II
Two joining and three non-joining fragments of rim and side, and fourteen joining and non-joining fragments of side and bottom of bowl (profile mostly restorable); colorless with pale bluish-green tinge. Silvery iridescence and slight pitting. Numerous horizontal striations on exterior surface. Upright rim, triangular in cross section, with almost pointed edge; straight side, tapering diagonally to bottom; slightly concave bottom. Blown. Restored H 5 cm; D ca. 17 cm; A Th side .08 cm.

409. CF 2468A Pl. 25
A.P. Room 25B, III Deposit IX
Late first through early third century A.D.
Fragment of rim and side of dish or bowl; colorless. Silvery iridescence and pitting. Numerous horizontal striations on external surface. Outsplayed rim, triangular in cross section, with almost pointed edge; straight side, set obliquely to rim. Blown. D ca. 16.5 cm; A Th side .08 cm.

410. CF 2468B Pl. 25
A.P. Room 25 B, III Deposit IX
Late first through early third century A.D.
Four non-joining fragments of rim and side of bowl; colorless with bluish-green tinge. Silvery iridescence and pitting. Numerous horizontal striations on external surface. Vertical rim, triangular in cross section, with almost pointed edge; straight side, tapering diagonally downward. Blown. D ca. 15 cm; A Th side .06 cm.

411. CF 2468C Pl. 25
A.P. Room 25B, III Deposit IX
Late first through early third century A.D.
Fragment of rim and side of bowl; colorless. Silvery iridescence and pitting. Numerous horizontal striations on external surface. Vertical rim, triangular in cross section, with almost pointed edge; straight side, tapering diagonally downward. Blown. D ca. 13 cm; A Th side .06 cm.

412. CF 2358 Pl. 25
A.P. Room 25B, III Deposit IX
Late first through early third century A.D.
Two joining fragments of rim and side of bowl; pale bluish-green, possibly incompletely decolorized. Silvery iridescence and pitting. Numerous horizontal striations on exterior surface; also many

elliptical bubbles set horizontally in the glass. Vertical rim, triangular in cross section, with almost pointed edge; straight side, tapering diagonally downward. Blown. D ca. 13 cm; A Th side .07 cm.

413. C72.137B Pl. 25
S.H. Street N, 0
Fragment of rim and side of bowl; colorless. Silvery iridescence and pitting. Numerous horizontal striations on outside surface. Slightly inturned rim, triangular in cross section, with almost pointed edge; straight side, tapering diagonally downward. Blown. D ca. 11 cm; A Th side .05 cm.

*Various Colorless Plates, Dishes, and Bowls (**Nos. 414–422**)*

This final section devoted to free-blown plates, dishes, and bowls comprises a disparate group of glasses dating to the second or early third century A.D. All are colorless (or almost so), and most exhibit numerous elliptical bubbles or blowing spirals. Diameters are fairly large, but it is usually difficult to reconstruct complete shapes owing to the fragmentary nature of the material. For the same reason it is difficult to draw exact comparisons with fragments or intact vessels from other sites. Five examples (**Nos. 415**, **417**, **419**, **419a**, and **420**), representing each of the three principal rim varieties, are from Deposit IX and belong to the late first through early third centuries, although most objects in the deposit are of the second century.

The largest pieces are broad plates (**Nos. 414–415**) with straight sides and rims proceeding diagonally downward. Diameters are substantial, varying between 30 and 32 cm. **Nos. 416–418** are large dishes with solid, fire-polished rims, probably with slightly concave bottoms rather than base-rings. The last five pieces (**Nos. 419–422**) have characteristic hook-shaped rims with thick, downturned edges.

Although these vessels do not comprise a closely related type in the same sense as the plates and bowls of the preceding two sections, they nevertheless seem to typify the ordinary colorless tablewares of the mid-Imperial period.

Parallels and Dated Examples of Colorless Plates, Dishes, and Bowls

Cosa: **Nos. 415**, **417**, **419–419a**, and **420**. Deposit IX, late first through early third century A.D.
Ostia IV 1977, 340, no. 379, pl. 49. Rim similar to **Nos. 414–415** at Cosa.
Conimbriga 1976, 206–207, no. 280, pl. 45. Another rim but from a smaller version of **Nos. 414–415**.
Alarcão 1965, 78–79, no. 107, pl. 4; also 120–124, nos. 205–225, pls. 8–9. Additional rim fragments from Conimbriga similar to **Nos. 414–415**.
Tipasa 1967, 90, no. 181, pl. 10.9. Yet another vessel with a rim like **Nos. 414–415**, but this one preserving a closed cut-out base-ring.
Vasa 1958, 48, fig. 19a. Another potential parallel for **Nos. 414–415** from Cyprus and said to date to the third century A.D.
Luni II 1977, CM 2917, pl. 155.12; CM 3004, pl. 155.17; and CM 7392/1, pl. 156.22. Rims similar to **Nos. 419–422**.
Ostia IV 1977, 340, nos. 374–375, pl. 49. Rims of vessels possibly similar to **Nos. 419–422**.
Tipasa 1967, 86, no. 169, pl. 9.1. An intact bowl with a rim roughly similar to **Nos. 419–419a**.
Sardis 1980, 62, no. 393, pl. 24. Although disposed vertically on the vessel, this rim fragment is hooked not unlike **Nos. 420–422**. It, however, illustrates the longevity of such rims because the piece is said to date to the fifth or sixth century A.D., a period much later than the examples at Cosa.

Colorless Plates, Dishes, and Bowls

414. Description of **No. 414** (illustration on pl. 26) is missing.

415. Description of **No. 415** (illustration on pl. 26) is missing.

416. Description of **No. 416** (illustration on pl. 26) is missing.

417. CF 2469 Pl. 26
A.P. Room 25B, III Deposit IX
Late first through early third century A.D.
Two non-joining fragments of rim and side of bowl; colorless. Silvery iridescence and pitting. Numerous horizontal striations on exterior surface. Upright rim with solid, rounded edge; convex curving side. Blown. D ca. 23 cm; A Th side .06 cm.

418. CF 2470 Pl. 26
A.P. Room 16 Exterior, III–V
Fragment of rim and side of bowl; colorless. Iridescence and pitting. Numerous horizontal striations on exterior surface. Upright rim with solid, rounded edge; almost straight side, tapering diagonally downward. Blown. D ca. 19 cm; A Th side .1 cm.

419. CF 2450A Pl. 26
A.P. Room 25D, III Deposit IX
Late first through early third century A.D.
Two non-joining fragments of rim and side of bowl; colorless. Iridescence and pitting. Horizontal rim with solid, downturned edge; convex curving side. Blown. D ca. 24 cm; A Th side .1 cm.

a. CF 2350, A.P. Room 25D, III, Deposit IX, late first through early third century A.D. Fragment of rim and side similar in shape to **No. 419**. D ca. 21.5 cm; A Th side .05 cm. Pl. 26.

420. CF 2496 Pl. 26
A.P. Room 25C, III Deposit IX
Late first through early third century A.D.
Fragment of rim and side of plate or bowl; colorless. Silvery iridescence and pitting. Numerous horizontal striations on exterior surface. Outsplayed rim with uneven, solid thickened edge; edge slightly downturned; straight side tapers diagonally downward. Blown. D ca. 23 cm; A Th side .06 cm.

421. CF 2507
Temple B (Forum), Courtyard, 0
Fragment of rim and side of plate or bowl; colorless. Silvery iridescence. Shape of rim and side similar to **No. 420**. Blown. D ca. 20–22 cm; A Th side .05 cm.

422. CF 2508
Temple B (Forum), Courtyard 0
Fragment of rim and side of plate and bowl; colorless. Silvery iridescence and pitting. Shape of rim and side similar to **Nos. 420** and **421**. Blown. D ca. 18–20 cm; A Th side .05 cm.

Naturally Colored Saucepans (***No. 423***)

The naturally colored saucepans (**Nos. 423–423b**), though fragmentary, presumably resemble the forms of their colored counterparts, **Nos. 340–341**, which are noted in an earlier section, pages 143–145 above. When intact, they probably were similar in form to the examples illustrated from Pompeii in Spinazzola 1928, pl. 222; northern Italy in Pavia Coll 1983, 38, 43, and 48, pl. 3, with references; Tipasa in Tipasa 1967, 94, pl. 10.1; Cyprus in Vessberg 1956, 137, fig. 43.21–22, where they are called ladles; and others noted in Isings 1957, 92, Form 75b. In date the bluish-green saucepans are associated with the second half of the first century A.D. but continue into the second and possibly the third century. If a handle fragment found at Magdalensberg 1979, 165, no. 1265, pl. 10, is from a saucepan, then the shape must have been devised before ca. A.D. 45, when the camp was abandoned.

Naturally Colored Saucepan

423. C72.93 Pl. 23
S.H. Street M, 0
Fragment of rim, side, and handle (incomplete) of saucepan; natural bluish-green. Iridescence and pitting. Outsplayed rim with solid thickened edge; straight, almost vertical side; wide horizontal handle attached to top of rim. Blown, with drawn glass handle. D ca. 11 cm; A Th side .2 cm.

a. C66.879, S.H. Street M, 0-I. End portions of saucepan handle; natural bluish-green.
b. C66.880, V-D, E, I. End portion of saucepan handle; natural bluish-green.
See also colored fragments of similar saucepans, **Nos. 340–341**.

Beakers and Handleless Jars (***Nos. 424–467***)

Among the earliest forms of blown glass—and perhaps the most ubiquitous—is the beaker or drinking cup. The popularity of the form undoubtedly arose due to the translucency or transparency of glass itself. This particular attribute enabled ordinary people to observe for the first time the sparkle and color of the liquids held within a vessel—a novelty ever since taken for granted. Moreover, glass vessels could easily be cleaned and repeatedly reused; they never became impregnated as did pottery with the flavors and residues of the substance they held. It is therefore scarcely surprising to discover an abundant number of free-blown beakers among the earliest finds of blown glass on virtually all Roman archaeological sites, including Cosa.

Almost from the outset of their production, beakers appeared in four basic shapes or variations thereof. These include the so-called Hofheim cup decorated with horizontal grooves (**Nos. 424–435**), the short, unadorned sack-shaped beaker or handleless jar (**Nos. 437–438** and possibly others), and a more diverse series of taller conical or bulbous beakers (**Nos. 439–444** and **448–462**). Virtually all of these forms emerged by the second quarter of the first century A.D., although excavated examples of the Hofheim cups are already present in the Augustan Age. Not surprisingly, these cups also disappear earlier than the other shapes. Examples of all forms, however, are to be found on sites in both the eastern and western Mediterranean and attest to the international character of most types of early Imperial blown glass.

Unfortunately, at Cosa, most examples are highly fragmentary, and once again it is difficult to determine complete profiles or to draw exact parallels with pieces from other sites. It is also impossible to distinguish clearly between those vessels that served solely as drinking cups and others that may have been handleless (and possibly lidded) jars or containers (see "Sack-Shaped Beakers," p. 169 below).

Hofheim Cups

The earliest recognizable group of blown beakers, as noted above, are the Hofheim cups named after the Claudian fort where more than thirty examples were found and first identified (Ritterling 1912, 365–366, fig. 93.2 and 4, pl. 38). Although a few may have been cast, the majority are free-blown, as is also true of most examples discovered at Cosa, Magdalenberg, Vindonissa, and Camulodunum. For the most part these beakers exploit the clarity of natural bluish-green or light green glass, although some known from other sites (see list below) utilize the same intentional colors (especially blue, purple, golden-brown, and dark green) characteristic of so much cast and blown glass of the early first century A.D. Far fewer pieces are decolorized. Those that are, however, may be counted among the earliest known Roman glasses to initiate the shift at midcentury from colored to colorless fabrics (for example, **Nos. 425–426** and **431–432** at Cosa; cf. Fishbourne 1971, 345–346, nos. 50–51, fig. 137, pl. 27).

The shapes of these cups are fairly standard. Most possess slightly inturned rims with knocked-off and polished edges. Bodies are usually hemispherical, and the bottoms are either slightly concave or kicked. Also probably related to the group are pieces that are more cylindrical (**No. 433**) or conical (**Nos. 431–432**) but that in all other respects conform to the type. Decoration takes the form of pronounced grooves often arranged in horizontal bands around the body.[53] Their presence is reminiscent of similar grooves found on such contemporary cast vessels as linear-cut and ribbed bowls or the translucent and opaque fine wares.

The earliest examples of Hofheim cups come from Magdalensberg 1979, 40, where several pieces are assigned to late Augustan contexts. Elsewhere the vast majority of examples appear in Tiberian–Claudian levels, with additional examples from Neronian or even Flavian strata (see list below). The same chronological pattern is reflected on Samothrace in the eastern Mediterranean, where such beakers are attributed to graves of the second quarter of the first century A.D. At Cosa **Nos. 426** and **430** are from Deposit V (ca. 25/15 B.C. to ca. A.D. 40/45), a period wholly consistent with the evidence found at other sites.

Parallels and Dated Examples of Hofheim Cups

Cosa: **Nos. 426** and **430**, Deposit V, dated to ca. 25/15 B.C. to ca. A.D. 40/45.

Luni I 1973, 468–469, CM 2499/2 and CM 2499/3; 778–779, KA 313, pl. 216.6.

Luni II 1977, 277, CM 5271 and CM 7310/28, pl. 154.12–13, with others cited, some of which are from strata dated to A.D. 40/50 or to A.D. 50–65/75.

Magdalensberg 1979, 37–43, nos. 329–418, pls. 2, 12, and 20. Colored and naturally colored examples, with discussion and an exhaustive list of parallels. At Magdalensberg most dated pieces are from Tiberian to Claudian contexts, but with several earlier examples attributed to Augustan levels.

Vindonissa 1960, 43–45, esp. nos. 98–101, pls. 7 and 17. No. 100 is from a Claudian–Neronian context; no. 101 is from a Tiberian/early Claudian level. Many parallels cited.

Ritterling 1912, 365–366, figs. 93.2 and 4, pl. 38. Examples from Hofheim from the Claudian period.

Camulodunum 1947, 302–303, nos. 68–75, pl. 88. Dated beakers from Periods IV–VI, ca. A.D. 49–65.

Fishbourne 1971, 344–345, nos. 46–48 and 50–51, fig. 139, pl. 27. The earliest examples are in levels dated ca. A.D. 43/55.

[53] Certain examples of the shape are more elaborately decorated and include pieces with painted scenes or applied blobs of polychrome glass. Others are cased, and some of these are cut in a cameo technique. For examples, see the references cited in Isings 1957 and Grose 1978, as noted in the list of parallels below. It can be assumed that such cups are contemporary with other vessels decorated in the same ways. Since most examples use deliberately colored fabrics, they can be assigned an Augustan or Julio-Claudian date.

Tarragona (Pasaje Cobos): Several examples attributed to the second quarter of the first century A.D. Information furnished by Jennifer Price, University of Durham. Unpublished.
Samothrace 1967, 45, no. 36, fig. 36. Beaker attributed to the second quarter of the first century A.D.
Vessberg 1956, 133, fig. 42.25. A colorless beaker possibly belonging to this group.
Isings 1957, 28–30, Form 12. Many colored and naturally colored examples listed; also others more highly decorated in cased, cameo, and painted techniques. Dated examples range from the Augustan through Julio-Claudian periods, with pieces from later contexts also cited.
Grose 1978, 78, fig. 14. A colored Hofheim beaker decorated with applied blobs of polychrome glass as well as cut grooves.

Sack-Shaped Beakers

Invented somewhat later than the Hofheim cups, the sack-shaped beakers or jars were already in production in the second quarter of the first century A.D. They appear to have enjoyed wide circulation around the Mediterranean basin but seem less common in the transalpine provinces of Europe. Although most published examples from the West are ascribed to the first century A.D., examples from Cyprus and the eastern Mediterranean are attributed to contexts as late as the third century. None of the examples at Cosa (**Nos. 437–438**) is dated, but **No. 438** was recovered from Forum SW, I, an area containing considerable quantities of material belonging to the early first century A.D. (see pp. 14, 55 above).

These vessels are characterized by a short, sack-shaped body and an off-set, often unfinished rim. Usually they are undecorated except for an occasional band of polishing marks. Generally they are either colorless or naturally colored. Some probably served as drinking vessels, while others were used as jars, which during the second century A.D. were occasionally capped with painted glass lids (Cyprus 1956, 1982, fig. 56.2–3 and 5).

Parallels and Dated Examples of Sack-Shaped Beakers

Luni II 1977, 549 and 551, K2336 and K2140, pl. 285.9. Two fragmentary examples of this group.
Magdalensberg 1979, 43, nos. 410–418, pl. 2. These fragments presumably predate the abandonment of the site, ca. A.D. 45. They are incorrectly catalogued with beakers of the Hofheim variety.
Aquileia 1968, 68–69, 73–74, nos. 176–179, pl. C.9.
Pompeii: Museo Nazionale Archeologico, Naples. Inv. no. 13466. Pre-A.D. 79. Unpublished.
Cyprus 1956, 139 and 198, fig. 44.9–22. Numerous examples, mostly attributed to the second and third centuries A.D. Several are lidded jars rather than drinking vessels.
Samothrace 1967, 45–46, no. 37, fig. 37. An example attributed to the first half of the first century A.D.
Tarragona (Pasaje Cobos): Several examples attributed to the second quarter of the first century A.D. Information furnished by Jennifer Price, University of Durham. Unpublished.

Taller Beakers

Nos. 439–444 (and possibly others) are rims from taller, cylindrical or conical beakers that are well documented on Italian, European, and eastern Mediterranean sites. In fact, the shape is so basic to blown glass that it is difficult to discern any significant development over time that might be helpful in distinguishing regional groupings. Fabrics are mostly colorless, and decoration of any kind is rare. None of the examples at Cosa is dated, but similar pieces from Magdalensberg, Tarragona, and Samothrace (see list below) belong to the first half of the first century A.D. They demonstrate that the earliest beakers of this shape are contemporary with those already discussed. Later examples, as cited in Isings 1957, 45 and 48–49, are attributed to the third or even fourth century A.D.

Parallels and Dated Examples of Taller Beakers

Luni II 1977, 555, J 781/3 and K 2255, pl. 286.10 and 13.

Samothrace 1967, 46–47, nos. 39–45, figs. 39–45. All examples are attributed to the first half of the first century A.D.

Tarragona (Pasaje Cobos): Several examples dated to the second quarter of the first century A.D. Information furnished by Jennifer Price, University of Durham. Unpublished.

Vessberg 1956, 144, fig. 45.4 and 9. These represent possible parallels to **Nos. 439–444** found at Cosa but are assigned dates in the second and third centuries A.D.

Isings 1957, 45, Form 30, and 48–49, Form 34. These two forms are thought to have begun in the second half of the first century A.D. (but they continue to the fourth century); both provide possible parallels for the examples at Cosa.

Among beakers at Cosa **No. 451** is unusually well preserved. In shape it closely approximates Isings Form 34 (Isings 1957, 48–49). Although its body is considerably more bulbous, like Form 34 it possesses an off-set, unfinished rim and a closed cut-out base. According to Isings, such beakers commenced production in the second half of the first century A.D. and continued well into the third century. A fragment from Camulodunum 1947, 303, no. 77, pl. 88, found in a level dated to ca. A.D. 49–61, however, may also possess this shape, which would indicate that the form actually began by midcentury, even if its greatest popularity only developed later. Since similar vessels appear at Pompeii (Isings 1957, 48–49), they definitely were being produced during the sixties and seventies A.D. At Cosa **Nos. 452–459** may originally have also been part of similar vessels.

Aside from the shapes mentioned above, other pieces at Cosa are equally typical of blown Imperial beakers found at many locations around the Mediterranean. **No. 455**, which dates to the decade before ca. A.D. 40/45, belongs to a well-known group of vessels with short cylindrical or rounded bodies and inturned rims with solid, thickened edges. Other examples dated to the second quarter of the first century have been recovered from Tarragona (unpublished) and Magdalensberg 1979, 164, no. 1229, pl. 10. A similar chronology also holds true for **Nos. 446–447a**, bluish-green beakers (or small bowls) with cut-out ledges below the rim. Two at Cosa (**Nos. 447–447a**) are from Deposit VII (decade prior to ca. A.D. 40/45), and this date is further confirmed by similar finds from the second quarter of the first century from Tarragona (unpublished) and Magdalensberg 1979, 65, no. 515, pl. 3. Apparently, the cut-out ledge was a specialty of the Romano-Italian industry of the early first century (see also **Nos. 328** and **381**, bowls with comparable ledges).

Finally, **Nos. 463–467** illustrate various ways of handling the bases of beakers. They range from simple concave bottoms to open or closed cut-outs, or even pads with pronounced, solid bumps on the interior (**No. 464**). Bases of this type are associated with beakers manufactured from the second quarter of the first century onward. Of these, **No. 464** is perhaps the most puzzling technically and aesthetically, but it is well documented at several sites, including Samothrace 1967, 46, no. 41, fig. 41; Magdalensberg 1979, 102, no. 767, pl. 5; Camulodunum 1947, 303, no. 77, pl. 88; and Tarragona (unpublished). In all cases the pieces are ascribed to the first half or the second quarter of the first century A.D.

Beakers and Handleless Jars

424. CE 2062 Pl. 27
A.P. Room 15, I
One non-joining and eight joining fragments of rim and side of beaker (profile mostly restorable); colorless with light green tinge. Slight iridescence and pitting. Lightly inturned rim, knocked-off and rotary-polished; slightly inturned side curves inward at bottom. On exterior of body, three horizontal grooves and two polishing marks. Blown, rotary-polished and cut. H ca. 8.5 cm; D ca. 9 cm; A Th .2 cm.

425. CE 2004 Pl. 27
A.P Room 15, I
Two joining fragments of rim and side of beaker (profile mostly restorable); natural light green, possibly incompletely decolorized. Iridescence and pitting. Inturned rim, knocked-off and rotary-polished; inturned side expands outward, then curves inward at bottom. On exterior, six horizontal grooves. Blown, rotary-polished and cut. H ca. 8.6 cm; D ca. 8 cm; A Th .2 cm.

426. CE 2133
A.P. Room 16, I Deposit V
ca. 25/15 B.C. to ca. A.D. 40/45
Fragment of lower side of beaker; natural bluish-green, possibly incompletely decolorized. Iridescence and pitting. Convex curving side. Shape probably similar to **Nos. 424–425**. On exterior, a band of two horizontal grooves. Blown, rotary-polished and cut. Max D fragment ca. 8–9 cm; Max Th side .3 cm.

427. C69.424 Pl. 27
Forum SW, I
Two non-joining fragments of rim and side of beaker; colorless with greenish-yellow tinge. Iridescence and pitting. Slightly inturned rim; knocked-off and rotary-polished; slightly inturned side. On exterior, three horizontal grooves. Blown, rotary-polished and cut. D ca. 7 cm; A Th .15 cm.

428. CD 1169 Pl. 27
Bas. ENW 2, II
Fragment of rim and side of beaker; natural pale bluish-green. Slight iridescence and pitting. Inturned rim, knocked-off and rotary-polished; side curves outward and downward. On exterior, one horizontal groove; numerous faint polishing marks. Blown, rotary-polished and cut. D ca. 6 cm; A Th .2 cm.

429. CE 837 Pl. 27
A.P. Room 11, I
One non-joining and two joining fragments of rim and side of beaker; natural light bluish-green. Iridescence and pitting. Off-set rim, knocked-off and rotary-polished; side expands outward, then turns in at bottom. On exterior, three narrow horizontal grooves. Blown, rotary-polished and cut. D ca. 8 cm; A Th .2 cm.

430. CE 2136 Pl. 27
A.P. Room 16, I Deposit V
ca. 25/15 B.C. to ca. A.D. 40/45
Fragment of rim and side of beaker; natural bluish-green. Iridescence and pitting. Off-set rim, knocked-off and rotary-polished; side expands outward, then turns in at bottom. On exterior, three narrow horizontal grooves in same configuration as **No. 429**. Blown, rotary-polished and cut. D ca. 8 cm; Max Th .2 cm.

431. CD 1174A Pl. 27
Bas. ESW 3, I
Fragment of rim and side of beaker; colorless. Iridescence and pitting. Upright rim, knocked-off and rotary-polished; straight side, tapering diagonally downward. On exterior, three narrow horizontal grooves. Blown, rotary-polished and cut. D ca. 8 cm; A Th .15 cm.

432. CE 2008B Pl. 27
A.P. Room 12, II
Fragment of rim and side of beaker; colorless. Iridescence and pitting. Upright rim, knocked-off and rotary-polished; straight side, tapering diagonally downward. On exterior, two bands of three narrow, horizontal grooves. Blown, rotary-polished and cut. D ca. 10 cm; A Th .18 cm.

433. C68.495 Pl. 27
Arx W. Slope, 0
Six joining fragments of rim, side, and bottom of beaker (profile restorable); colorless with greenish tinge. Upright rim, knocked-off and rotary-polished; straight, nearly vertical side, tapering downward; slightly concave bottom. On exterior, five horizontal grooves. Blown, rotary-polished and cut. H 4.8 cm; D ca. 8 cm; A Th .15 cm.

434. C66.825 Pl. 27
S.H. Street M, 0-I
Two non-joining fragments of rim and side of beaker; colorless. Slight iridescence and pitting. Off-set rim, knocked-off and rotary-polished; almost vertical side, curves inward at bottom. On exterior, three horizontal grooves. Blown, rotary-polished and cut. D ca. 10 cm; A Th .2 cm.

435. CD 1171 Pl. 27
Bas. Exterior W, I
Fragment of side and bottom of beaker; natural light green. Iridescence and pitting. Straight side, tapering diagonally downward; slightly concave bottom. On exterior, one horizontal groove. Blown and cut. Max D ca. 8 cm; Max Th .25 cm.

436. CE 2207 Pl. 27
A.P. Room 16, I Deposit V
ca. 25/15 B.C. to ca. A.D. 40/45
Two non-joining fragments of rim and side of beaker; natural pale blue, possibly incompletely decolorized. Iridescence and pitting. Slightly outsplayed rim, knocked-off and rotary-polished; nearly vertical side turns inward near bottom. On exterior, a narrow horizontal groove below rim. Blown, rotary-polished and cut. D ca. 8 cm; A Th .12 cm.

437. CF 2466 Pl. 27
A.P. Room 25 Exterior, II
Fragment of rim and side of beaker; colorless with bluish-green tinge. Iridescence and pitting. Off-set rim, knocked-off and rotary-polished; side expands outward, then turns abruptly inward toward bottom. On exterior of body, several horizontal wheel-cut lines. Blown and rotary-polished. D ca. 8 cm; A Th .2 cm.

438. C69.423 Pl. 27
Forum SW, I
Fragment of rim and side of beaker; colorless with distinct bluish-green tinge. Iridescence and pitting. Off-set rim, knocked-off and rotary-polished; inturned side bulges outward, then curves inward at bottom. On exterior, horizontal wheel-cut lines below rim and around body. Blown and rotary-polished. D ca. 7 cm; A Th .12 cm.

439. CD 1174B Pl. 28
Bas. ESW 3, I
Two joining fragments of rim and side of conical beaker; colorless with distinct bluish-green tinge. Iridescence and pitting. Off-set rim, knocked-off and rotary-polished; straight, almost vertical side, tapering downward. On exterior, horizontal wheel-cut lines below rim. Blown and rotary-polished. D ca. 10 cm; A Th .18 cm.

440. CE 2028 Pl. 28
A.P. Room 22, I
Two joining fragments of rim and side of tall beaker; colorless with distinct bluish-green tinge. Iridescence and pitting. Outsplayed rim, knocked-off and rotary-polished; almost vertical, straight side, tapering downward. On exterior, horizontal wheel-cut lines below rim. Blown and rotary-polished. D ca. 7.2 cm; A Th .09 cm.

441. CD 1172B Pl. 28
Bas. ENE 3, I
Fragment of rim and side of conical beaker; colorless with light green tinge. Iridescence and pitting. Off-set rim, knocked-off and rotary-polished; straight side, tapering downward. On exterior, horizontal wheel-cut lines below rim. Blown and rotary polished. D ca. 9 cm; A Th .08 cm.

442. C72.136 Pl. 28
IV-B, TA, 0
Fragment of rim and side of conical beaker; colorless. Iridescence and pitting. Outsplayed rim with rounded thickened edge; straight side. Tapering diagonally downward. Blown. D ca. 7.5 cm; A Th .09 cm.

443. CD 1172A Pl. 28
Bas. ENE 3, I
Fragment of rim and side of conical beaker; colorless with distinct light-green tinge. Iridescence and pitting. Off-set rim, knocked-off and rotary-polished; straight side, tapering diagonally downward. On exterior, horizontal wheel-cut lines below rim and around body. Blown and rotary-polished. D ca. 9 cm; A Th .1 cm.

444. CE 2031 Pl. 28
A.P. Room 22, I
Two joining fragments of rim and side of conical beaker; colorless with distinct greenish-yellow tinge. Iridescence and pitting. Off-set rim, knocked-off and rotary-polished; straight side, tapering diagonally downward. Blown and rotary-polished. D ca. 9 cm; A Th .1 cm.

445. CE 2164 Pl. 28
A.P. Room 22, II Deposit VII
Decade before ca. A.D. 40/45
Fragment of rim and side of beaker; natural bluish-green. Iridescence and pitting. Inturned rim with solid, thickened edge; side expands outward slightly. Blown. D ca. 6 cm; A Th .11 cm.
Bibliography: Grose 1975, 45, no. 31.

446. CD 1188 Pl. 28
Bas. Exterior NW 3, I
Fragment of rim and side of beaker or other vessel; natural bluish-green. Iridescence and pitting. Outsplayed rim with solid thickened edge; below rim, a tubular ledge cut-out from side; straight side, tapering downward. Blown and tooled. D ca. 11 cm; A Th .12 cm.

447. CE 2054 Pl. 28
A.P. Room 22, II Deposit VII
Decade prior to ca. A.D. 40/45
Two non-joining fragments of rim and side of beaker or other vessel; natural bluish-green. Slight iridescence and pitting. Outsplayed rim with rounded edge; below rim, a closed tubular ledge cut-out from side; straight side, tapering downward. Blown and tooled. D ca. 8 cm; A Th .1 cm.
Bibliography: Grose 1975, 45, no. 33.

- **a.** CE 2161, A.P. Room 22, II, Deposit VII, decade prior to ca. A.D. 40/45. Fragment of rim and side identical to **No. 447**, possibly from the same vessel, except that its thickness is .25 cm.
  Bibliography: Grose 1975, 45, no. 33A.

448. C68.688 Pl. 28
Forum ENW, 0
Fragment of rim and side of beaker; natural light green. Slight iridescence and pitting. Slightly inturned rim, knocked-off and rotary-polished; side expands slightly outward, then curves inward. On exterior, horizontal wheel-cut line below rim. Blown and rotary-polished. D ca. 10 cm; A Th .2 cm.

449. CD 1170 Pl. 28
Forum ENW 2, III
Fragment of rim and side of beaker; colorless. Iridescence and pitting. Upright rim with rounded, inturned edge; straight, almost vertical side. Blown. D ca. 8 cm; A Th .07 cm.

450. CE 2163 Pl. 28
A.P. Room 22, II Deposit VII
Decade prior to ca. A.D. 40/45

One non-joining and three joining fragments of rim and side of beaker; colorless. Iridescence and pitting. Vertical rim, knocked-off and rotary-polished; side curves outward, then downward. On exterior, horizontal wheel-cut lines below rim. Blown and rotary-polished. D ca. 7 cm; A Th .1 cm. *Bibliography:* Grose 1975, 44, no. 28.

451. C70.568 Pl. 28
S.H. Street 5, Sewer
Five joining fragments of rim, side, and base of beaker (profile restorable); colorless. Iridescence, pitting, and brown scaly weathering. Sharply off-set rim, knocked-off and rotary-polished; convex curving side; closed cut-out base-ring. On exterior, horizontal wheel-cut lines below rim and around body. Blown and rotary-polished. H 9 cm; D ca. 8 cm; A Th .08 cm.

452. CE 2169 Pl. 28
A.P. Room 22, II Deposit VII
Decade prior to ca. A.D. 40/45
Small fragment of rim and side of beaker; colorless. Slight iridescence and pitting. Sharply off-set rim, knocked-off and rotary-polished; side curves outward and downward. On exterior on shoulder, a narrow horizontal wheel-cut groove. Blown, rotary-polished and cut. D ca. 8 cm; A Th .14 cm. *Bibliography:* Grose 1975, 45, no. 30.

453. CF 2356 Pl. 28
A.P. Room 25A, IV
Fragment of rim and side of beaker or jar; colorless. Silvery iridescence and pitting. Sharply off-set rim, knocked-off and rotary-polished; straight side, expands diagonally outward. On exterior, horizontal wheel-cut lines around body. Blown and rotary-polished. D ca. 7 cm; A Th .07 cm.

454. CE 2056 Pl. 29
A.P. Room 22, II Deposit VII
Decade prior to ca. A.D. 40/45
Two joining fragments of rim and side of beaker or jar; colorless with distinct bluish-green tinge. Slight iridescence and pitting. Off-set rim, knocked-off and rotary-polished; convex curving side expands outward. On exterior, horizontal wheel-cut lines below rim and around body. Blown and rotary-polished. D ca. 8 cm; A Th .2 cm.
Bibliography: Grose 1975, 45, no. 29.

455. CD 1106 Pl. 29
Bas. ENE, I
Fragment of rim and side of beaker or jar; colorless. Iridescence and pitting. Sharply off-set rim, knocked-off and rotary-polished; convex curving side expands outward. On exterior below rim, horizontal wheel-cut lines. Blown and rotary-polished. D ca. 8.7 cm; A Th .08 cm.

456. C72.42 Pl. 29
H.T. (House of Quintus Fulvius), Room 10, 0
Two joining fragments of rim and side of beaker or jar; colorless. Iridescence and pitting. Off-set rim, knocked off and rotary-polished; convex curving side expands outward. Blown. D ca. 8 cm; A Th .11 cm.

457. C66.823 Pl. 29
S.H. Street M, 0-I
Fragment of rim and side of beaker; natural bluish-green. Iridescence and pitting. Off-set rim, knocked-off and rotary-polished; straight, nearly vertical side. On exterior, horizontal wheel-cut lines below rim and around body. Blown and rotary-polished. D ca. 6.5 cm; A Th .15 cm.

458. C65.803 Pl. 28
V-D, Casupola, I
Fragment of rim and side of beaker; colorless with distinct bluish-green tinge. Slight iridescence and pitting. Off-set rim, knocked-off and rotary-polished; side expands slightly outward, then downward. Blown. D ca. 7 cm; A Th .1 cm.

459. CD 1173 Pl. 29
Bas. ENE 1, I
Fragment of rim and side of beaker; colorless with greenish tinge. Iridescence and pitting. Off-set rim, knocked-off and rotary-polished; nearly vertical side curves inward toward base. On exterior, horizontal wheel-cut lines below rim and around body. Blown and rotary-polished. D ca. 6.5 cm; A Th .09 cm.

460. C70.582 Pl. 29
S.H. Street M, 0-I
Three joining fragments of rim and side of beaker or jar; natural bluish-green. Slight iridescence and pitting. Upright rim with solid edge, folded inward; side tapers diagonally downward, then turns outward. On interior beneath edge of rim, a horizontal groove caused by the folding of rim. Blown. D ca. 7.5 cm; A Th .1 cm.

461. C68.687 Pl. 29
Arx W. Slope, 0
Fragment of rim and side of beaker; colorless with bluish-green tinge. Slight iridescence and pitting. Short off-set rim, knocked-off and polished; side tapers slightly outward. Blown. D ca. 5.5 cm; A Th .1 cm.

462. C66.875 Pl. 29
V-D, SE Balk, 0
Fragment of rim and side of beaker; natural light green. Slight iridescence and pitting. Slightly outsplayed rim with solid, rounded edge; almost vertical side, tapering downward. On exterior, a distinct horizontal wheel-cut groove below rim. Blown and rotary-polished. D ca. 7 cm; A Th side .2 cm.

463. CE 1990 Pl. 29
A.P. Room 22, II Deposit VII
Decade before ca. A.D. 40/45
Fragment of side and base of beaker; natural bluish-green. Iridescence and pitting. Convex curving side; closed cut-out base-ring; concave on underside. Blown. D base-ring 5 cm; A Th side .1 cm.
Bibliography: Grose 1975, 45, no. 32.

464. C70.605 Pl. 29
S.H. Street M, 0-I
Fragment of side and base of beaker; natural bluish-green. Iridescence and pitting. Convex curving side; closed cut-out base-ring flat on underside but with pronounced, solid convex bump on upper side. Blown. D base-ring 4.4 cm; A Th side .1 cm.

465. C70.581 Pl. 29
S.H. Street 5, 0-I
Fragment of side and bottom of beaker; colorless with greenish-yellow tinge. Iridescence and pitting. Nearly vertical side turns outward at top of fragment; bottom slightly concave. On exterior, bands of horizontal wheel-cut lines. Blown and rotary-polished. D bottom 6 cm; A Th side .06 cm.

466. C70.580 Pl. 29
S.H. Street M, 0-I
Two joining fragments of side and bottom of beaker; colorless. Iridescence and pitting. Straight side, tapering downward; bottom slightly concave. Blown. D bottom 5 cm; Max Th side .2 cm.

467. C66.828 Pl. 29
S.H. Street M, 0-I
Fragment of side and bottom of beaker; colorless. Iridescence and pitting. Side curves downward to bottom; bottom slightly concave. Blown. D bottom ca. 5 cm; Max Th side .2 cm.

Beaded Stemmed Cups (***Nos. 468–474***)

468. CD 1100 Pl. 30
Bas. ENW 2, I
Fragment of side, beaded stem, and base of stemmed cup; colorless. Iridescence and pitting. Convex curving side; solid beaded stem; outsplayed base with solid, thickened edge; base uneven but concave on underside. On exterior of side above stem, a horizontal thread in the same fabric. Blown and tooled, with applied thread. D base-ring ca. 7.5 cm; A Th side .2 5 cm.

469. C68.659 Pl. 30
S.H. Garden, 0-B
Fragment of side, beaded stem, and base (incomplete) of stemmed cup; colorless. Iridescence, milky weathering, and pitting. Convex curving side; solid beaded stem; outsplayed base. On exterior of side above stem, a horizontal thread in the same fabric. Blown and tooled, with applied thread. Max D fragment 5 cm; A Th side .3 cm.

470. CF 2497A Pl. 30
A.P. Room 25C, II Deposit IX
Late first through early third century A.D.
Fragment of side and beaded stem of stemmed cup; colorless. Iridescence, milky weathering, and pitting. Convex curving side; solid beaded stem; indications of outsplayed base. On exterior of side above stem, a horizontal thread in the same fabric. Blown and tooled, with applied thread. Max D fragment 4 cm; Max Th side .5 cm.

- **a.** CF 2497B, A.P. Room 25C, II, Deposit IX. Late first through early third century A.D. Two joining fragments of side and beaded stem of stemmed cup; colorless.
- **b.** CF 2497C, A.P. Room 25C, II, Deposit IX. Late first through early third century A.D. Beaded stem with vestiges of side of stemmed cup; colorless.
- **c.** CF 2497D, A.P. Room 25C, II, Deposit IX. Late first through early third century A.D. Beaded stem with vestiges of side of stemmed cup; colorless.

471. C69.419 Pl. 30
Forum ENE, 0
Fragment of side, beaded stem, and base (incomplete) of stemmed cup; colorless with greenish-yellow tinge. Slight iridescence and pitting. Convex curving side; solid beaded stem; outsplayed base. Blown and tooled. Max D fragment 5 cm; Max Th side .3 cm.

472. CF 2339 Pl. 30
A.P. Room 25, I
Fragment of side and beaded stem of stemmed cup; colorless. Slight iridescence and pitting. Convex curving side; hollow beaded stem; indications of outsplayed base. Blown and tooled. Max D fragment 5 cm; Max Th side .4 cm.

473. CF 2498 Pl. 30
A.P. Room 25, I
Fragment of side and vestiges of beaded stem of stemmed cup; colorless. Iridescence and pitting. Convex curving side; upper portion of beaded stem preserved. On exterior of side, two bands of horizontal wheel-cut lines. Blown, tooled and rotary-polished. Max D fragment 5.5 cm; Max Th side .3 cm.

474. CD 1128 Pl. 30
Bas. EW, I
Fragment of side of beaded stemmed cup; colorless. Iridescence, milky weathering, and pitting. Convex curving side. On exterior of side, two horizontal grooves in a band. Blown and cut. Max Th fragment 5.5 cm; Max Th side .3 cm.

*Various Stemmed Cups (**Nos. 475–489**)*

475. C70.593 Pl. 30
S.H. Street 5, 0-I
Fragment of side, solid stem, and base (incomplete) of stemmed cup or other vessel; natural bluish-green. Slight iridescence and pitting. Convex curving side; cylindrical solid stem; outsplayed base. Blown. Max D fragment 5.7 cm; Max Th side .3 cm.

476. CD 1087 Pl. 30
Bas. ENE 1, II
Stemmed base and fragment of side of stemmed cup; natural greenish-yellow. Iridescence and pitting. Convex curving side; cylindrical solid stem, tapering downward; outsplayed base with solid thickened edge; concave on underside. Blown. D base 4.8 cm; A Th side .2 cm.

477. C69.454 Pl. 30
Forum SW, I

Fragment of rim and side of small stemmed cup; natural pale yellow. Iridescence and pitting. Upright rim with solid, thickened edge, folded inward; straight side, tapering diagonally downward. Blown. D ca. 7.5 cm; A Th side .08 cm.

478. CG 577 Pl. 30
Curia, SE Room, 0
Fragment of rim and side of small stemmed cup; natural pale yellowish-green. Silvery iridescence and pitting. Upright rim with solid, thickened edge, folded inward; straight side, tapering diagonally downward. Blown. D ca. 7.5 cm; A Th side .08 cm.

479. CE 1997 Pl. 30
A.P. Room 12-a-5, Cesspool, I
Stemmed base and fragment of side of small stemmed cup; natural pale yellow. Silvery iridescence and pitting. Straight, almost vertical side, tapering downward to flat floor; hollow stem; outsplayed base with tubular edge; pontil scar blocks lower portion of stem. Blown and tooled. D base-ring 3.7 cm; A Th side .05 cm.

480. C69.400A Pl. 30
Forum Cistern (SW Extension) Deposit X
Below fallen vaulting
Fourth through mid-fifth century A.D.
Stemmed base and fragment of floor of small stemmed cup; natural pale yellowish-green. Iridescence and pitting. Nearly flat floor; hollow stem; outsplayed base with tubular edge; pontil wad obstructs lower portion of stem. Blown and tooled. D base-ring 4 cm; A Th floor .1 cm.

481. C69.400B Pl. 30
Forum Cistern (SW Extension) Deposit X
Below fallen vaulting
Fourth through mid-fifth century A.D.
Stemmed base and fragment of side of small stemmed cup; natural yellowish-green. Iridescence and pitting. Convex curving side; hollow stem with pronounced bubble of glass at upper end; nearly flat, outsplayed base with tubular edge; pontil wad obstructs lower end of stem, causing vessel to wobble on its base. Residual metal "scale" from pontil adheres to underside of wad. Blown and tooled. D base-ring 4.85 cm; A Th side .18 cm.

482. C72.43 Pl. 30
A.P. Room 13, I
Stemmed base and fragment of floor of small stemmed cup; natural pale green. Iridescence and pitting. Pronounced bubble of glass at upper end of hollow stem; outsplayed base-ring, conical in shape, with tubular edge; pontil wad obstructs lower end of stem. Blown and tooled. D base-ring 4.8 cm; A Th side .1 cm.

483. CB 1149
Capitolium Exterior, 0
Stemmed base and fragment of side of small stemmed cup (deformed by heat); natural bluish-green.

Iridescence and pitting. Hollow stem; outsplayed base-ring with tubular edge. Blown and tooled. D base-ring 4.2 cm; A Th floor .12 cm.

484. C69.402 Pl. 30
Forum SW, 0
Stemmed base and floor of unique, small stemmed cup; natural pale yellow. Iridescence and pitting. Undulating floor rests on three loops (one broken) of drawn glass that are inserted into a dome-shaped base with a solid, thickened edge; pontil wad at center of underside of base. Blown and tooled. D base-ring 3.7 cm; A Th floor .1 cm.

485. CD 308 Pl. 30
Bas. N 7, I
Stemmed base of small stemmed cup; natural pale yellow. Iridescence and pitting. Hollow stem; outsplayed base with tubular edge; pontil wad obstructs lower end of stem. Blown and tooled. D base-ring 4.2 cm.

486. C69.400C Pl. 30
Forum Cistern (SW Extension) Deposit X
Below fallen vaulting
Fourth through mid-fifth century A.D.
Fragment of base of small stemmed cup; natural pale yellow. Iridescence and pitting. Outsplayed base with tubular edge. Blown and tooled. D base-ring ca. 4 cm.

487. C68.657A Pl. 30
Forum ENE, 0
Stemmed base of small stemmed cup; natural pale yellow. Iridescence and pitting. Hollow stem; outsplayed base with tubular edge. Blown and tooled. D base-ring 3.9 cm; A Th floor .1 cm.

488. CG 386 Pl. 30
Curia SE, I–II
Fragment of base and stem of small stemmed cup; natural pale yellow. Iridescence and pitting. Hollow twisted stem; flat base with tubular edge. Blown and tooled. D base-ring ca. 3.9 cm; A Th floor .1 cm.

489. CG 382 Pl. 30
Curia, II
Stemmed base of small stemmed cup; natural pale yellow. Iridescence and pitting. Hollow stem; outsplayed base with tubular edge. Blown and tooled. D base-ring 2.9 cm; A Th floor .1 cm.

*Naturally Colored and Colorless Storage Wares (**Nos. 490–574**)*

The sections that follow publish a representative sampling of the naturally colored and decolorized blown storage wares and containers found at Cosa. Although few of the fragments retain a continuous profile, they nonetheless illustrate the presence at the site of many of the standard kinds of such vessels belonging to the early to mid-Imperial periods. Included are wide-mouthed jars (**Nos. 490–493**), large and medium-sized flasks and jugs (**Nos. 494–523**), an oil flask (**No. 524**), mold-blown prismatic bottles (**Nos. 525–548**), and a variety of small unguent bottles (**Nos. 549–573**). Within each group there usually are two obvious divisions, with one set of containers dating to the early to mid-first century A.D. and occurring in natural bluish-green or light green fabrics, in contrast to others of the late first through third century A.D. favoring colorless or very lightly colored glasses. In all but a few instances these storage wares are plain and unadorned, except, of course, for the molded design found on prismatic bottles.

*Wide-Mouthed Jars (**Nos. 490–493**)*

Among containers that became popular in Italy almost from the beginning of glassblowing were large storage jars of two principal forms. One is free-blown with a bulbous body and a broad tubular (or double tubular) rim, either upright (**No. 490**) or flattened (**Nos. 491–492**). Some specimens possess handles, and others were originally fitted with knobbed lids. They are often referred to as *ollae* (see Isings 1957, 81, Form 62). At Cosa one or both of these types is represented by several rims (**Nos. 490–492**), of which the first two are from Deposit VIII and are dated before ca. A.D. 50/55. Predictably, both are bluish-green, as is so much of the earlier blown utilitarian glassware on the site. At Magdalensberg such wide-mouthed vessels, presumably ollae, have also been recovered from contexts of the first half of the first century and are said to have begun in the first quarter of the century, a date that may be too early based on what is known from other excavations (see list below). By A.D. 79, however, they were present at Pompeii and Herculaneum in large numbers and were evidently a staple of the glassmakers' craft (again, see references that follow). According to Isings both kinds of large storage jars continued to be fabricated throughout the second century A.D. and possibly into the third century as well.

Parallels and Dated Examples of Wide-Mouthed Jars

Cosa: **Nos. 491–492**, Deposit VIII, prior to ca. A.D. 50/55.

Pompeii: A Square jar dating before A.D. 79 and typical of many unpublished examples from Pompeii now in the Museo Archeologico, Naples.

Bologna Coll. 1979, 88–89, no. 84. A square jar, handled, with additional references.

Magdalensberg 1979, 153–156, nos. 1155–1157 and 1160, pl. 9. Several dated examples of ollae attributed as early as the first quarter of the first century A.D. Numerous references cited.

Bologna Coll. 1979, 71–88. Various shapes of ollae, with additional references.

Aquileia 1968, 88–92, pls. F and I, and 15.2–3. Additional forms of ollae, some lidded, with two examples in purple and blue glass with applied blobs of opaque white (nos. 228–229).

Fishbourne 1971, 355–356, nos. 79 and 81, fig. 141, with a brief discussion of wide-mouthed jars.

Wide-Mouthed Jars

490. CD 1162A — Pl. 31
Bas. ENE 4, II — Deposit VIII
Prior to ca. A.D. 50/55

Fragment of rim and side of wide-mouthed jar; natural bluish-green. Iridescence and pitting. Slightly inturned rim with pronounced tubular edge, folded outward and downward; side almost vertical. Blown and tooled. D ca. 16 cm; A Th side .15 cm.

491. CD 1162B Pl. 31
Bas. ENE 4, II Deposit VIII
Prior to ca. A.D. 50/55

Fragment of rim and side of wide-mouthed jar; natural bluish-green. Iridescence and pitting. Inturned rim with uneven and flattened tubular edge, folded outward, then inward; side expands diagonally outward below rim. Blown and tooled. D ca. 14 cm; A Th side .2 cm.

492. CD 1190 Pl. 31
Bas. ENE 2, I

Fragment of rim of wide-mouthed jar; colorless with greenish tinge. Iridescence and pitting. Nearly horizontal, flat tubular rim, folded outward, then inward, and with thickened edge turned under. Blown and tooled. D ca. 16 cm; A Th rim .12 cm.

493. CD 1189 Pl. 31
Bas. ENE 1, I

Fragment of rim, side, and handle (incomplete) of jar or other vessel; natural yellowish-green. Iridescence and pitting. Outsplayed rim with solid, thickened edge; below edge, two closed tubular ledges cut out from side; side tapers downward; upper thumb-rest of handle attached to edge of rim. Blown and tooled, with drawn glass handle. D ca. 9.5 cm; A Th side .08 cm.

*Large Flasks and Jugs (**Nos. 494–503**)*

An assortment of rims and bases documents the existence at Cosa of large flasks and handled jugs, which in profile are associated with everyday storage wares known from a great many archaeological sites in Italy and around the Mediterranean but less frequently in central and northwestern Europe. Although few at Cosa come from dated contexts, most appear in such well-known forms that their chronological framework has been established from dated examples unearthed elsewhere. It can also be assumed that the bluish-green pieces (here **Nos. 496**, **498**, and **500**) belong to the first half of the first century A.D., whereas the colorless examples are mostly confined to late first, second, and third centuries. **Nos. 497**, **499**, and **502–503** are from Deposit IX and definitely belong to the second or early third century A.D. Since **Nos. 494–495** also exhibit the same rim as **No. 497** (see below), they too can be assigned a mid-Imperial date.

Although it is difficult to determine complete shapes from these fragments, parallels for most exist at numerous sites in both the western and eastern Mediterranean. This again illustrates the international character of so much of the blown glass of the early and mid-Imperial period. Either artisans (or their products) freely moved throughout the empire, or forms and fashions created in one region were quickly copied at other centers in the Roman world. For example, **Nos. 494–495** preserve funnel-shaped necks, outsplayed rounded rims, and a characteristic open cut-out ledge below the rim. Such profiles are also found as far east as Dura-Europos (Dura-Europos 1963, 104–105), where comparanda from Syro-Palestinian excavations are also noted. Such rims, however, are also found as far west as Conimbriga in the Iberian Peninsula (Conimbriga 1976, 196–197, nos. 233–234, pl. 43). For another Italian parallel, see Ostia I 1968, 71, no. 162, pl. 8.

No. 497 is similar in that it has a protruding ledge beneath the rim. In this case, however, the ledge may actually be a wide applied coil, not unlike others known from Dura-Europos, Sardis, Conimbriga, or elsewhere, and usually ascribed to large, cylindrical bottle-jugs or two-handled bottles (Isings 1957, 156–158, Forms 126 and 127); cf. Dura-Europos 1963, 117–119, nos. 584–587, pl. 13; Sardis 1980, 85, nos. 633–634, pl. 27; and Conimbriga 1976, 196–197, nos. 233–234, pl. 43. Since the vessels cited in these publications, however, are thought to have been manufactured in the third century A.D., and continued in production during the fourth and fifth centuries, they are later chronologically than the examples from Deposit IX at Cosa and may not furnish appropriate parallels.

Bluish-green jugs such as **No. 498** (and possibly **No. 496**) are in many respects comparable to Isings Form 52 (Isings 1957, 69–70), a shape that is first found in Claudian–Neronian levels but that remained in vogue throughout much or all of the first century, as finds from Pompeii and Herculaneum attest. **No. 499**, a thick, uneven colorless base, was recovered from Deposit IX and is therefore no later than the third century A.D., despite the fact that the base is reminiscent of others customarily associated only with the late third through fifth centuries A.D. (see, for example, Isings Form 120 b; Isings 1957, 150–152). Isings's jug is typical of a group found in late Roman and early Byzantine contexts around the Mediterranean (cf. Sardis 1980, 66, no. 444, pl. 24, for a similar fragmentary base from a late antique level).

The thinly blown, long-necked flasks with trefoil spouts (**No. 500**), out-turned rim attachments (**No. 501**), or rolled rim (**Nos. 502–503**) are also characteristic products of the later Imperial Roman industry. **No. 501** finds a close parallel at Ostia (Ostia II 1970, 29, nos. 8a–b, pl. 7). The rolled rims of **Nos. 502–503** are probably an innovation of the second century, as is suggested by the date of the deposit in which they were found.

Large Flasks and Jugs

494. CF 2373 — Pl. 31
A.P. Room 25 Exterior, I

Six joining fragments of rim, neck, and shoulder of large flask; colorless. Silvery iridescence and pitting. Outsplayed funnel-shaped rim with solid edge; below it, an open cut-out ledge encircles the neck; cylindrical neck, tapering upward; shoulder flares abruptly outward, then turns downward. Blown and tooled. D 8.5 cm; A Th shoulder .15 cm.

495. CF 204 — Pl. 31
Temple B (Forum), Exterior SE, II

Fragment of rim and neck of large flask; colorless with greenish tinge. Iridescence and pitting. Outsplayed funnel-shaped rim with solid edge; below it, an open cut-out ledge encircles the neck; vestiges of a cylindrical neck. Blown and tooled. D ca. 8 cm; A Th neck .2 cm.

496. CE 2002 — Pl. 31
A.P. Room 15, I

Five joining fragments of neck and shoulder and four joining fragments of side and base of large flask or jug; natural bluish-green. Slight iridescence and pitting. Cylindrical neck, tapering upward; shoulder flares abruptly outward, then curves downward; convex curving side; kicked base with applied, solid base-ring. Blown. Max D neck 7 cm; D base-ring 10.5 cm; A Th side .1 cm.

497. CF 2369 — Pl. 31
A.P. Room 25A, III — Deposit IX
Late first through early third centuries A.D.

Fragment of rim, neck, and handle (incomplete) of large jug; colorless. Silvery iridescence and pitting. Outsplayed funnel-shaped rim with solid edge; below it, a solid, horizontal ledge encircles the neck; cylindrical neck, tapering upward; upper portion of ribbed handle attached below rim. Blown and tooled, with drawn glass handle. D ca. 7.8 cm; A Th neck .2 cm.

498. C67.72 Pl. 31
S.H. Street M, I
Two joining fragments of rim, neck, shoulder, and handle of large jug; natural bluish-green. Iridescence and pitting. Outsplayed rim, folded downward, then upward; cylindrical neck appears to bulge at center; shoulder flares outward; a two-ribbed strap handle with pronounced thumb-rest is attached at the rim and on the shoulder; a coil of glass with six upright pinched protrusions (three on either side of handle) applied to the rim. Blown, with tooled coil and handle. D 5.5 cm; A Th side .1 cm.

499. CF 2367 Pl. 31
A.P. Room 25A, III Deposit IX
Late first through early third century A.D.
Fragment of side and base of large flask or jug; colorless. Silvery iridescence and pitting. Side expands outward from base; uneven, concave base with tall, solid base-ring, outsplayed from side. Blown and tooled. D base-ring ca. 10.5 cm; A Th side .15 cm.

500. CD 1096 Pl. 32
Bas. ENW 2, II
Two non-joining fragments of rim and neck of large flask or jug with trefoil spout; natural bluish-green. Iridescence and pitting. Outturned rim in trefoil shape with tubular edge, folded inward; long cylindrical neck; no handle attachment preserved. Blown and tooled. Preserved H neck 15 cm; Max W rim ca. 8 cm; A Th neck .08 cm.

501. CD 1191 Pl. 32
Bas. Exterior W, I
Rim (incomplete) and fragment of neck of large flask or jug; colorless with greenish tinge. Silvery iridescence and pitting. Outsplayed rim with edge turned downward, then diagonally upward; cylindrical neck; two vertical, pinched protrusions, set side by side, attached to rim; no handle attachment preserved. Blown and tooled. D 6 cm; A Th side .08 cm.

502. CF 2342 Pl. 32
A.P. Room 25D, III Deposit IX
Late first through early third century A.D.
Rim (incomplete) and two joining fragments of neck of large flask; natural light green. Iridescence and pitting. Outsplayed rim with rolled edge, folded inward; cylindrical neck, tapering slightly upward. Blown. D 5.5 cm; A Th side .1 cm.

503. CF 2328 Pl. 32
A.P. Room 25D, III
Late first through early third century A.D.
Five joining fragments of rim and neck of large flask; colorless with greenish tinge. Iridescence, corrosion, and discoloration. Outsplayed rim with rolled edge, folded inward; cylindrical neck,

tapering upward, with tooled constriction at junction of neck and shoulder. Blown. H neck 15 cm; D 4.6 cm; A Th side .1 cm.

Medium-Sized Flasks and Jugs (**Nos. 504–523**)

Given the fragmentary state of the vessels in this section, it is not possible to distinguish the precise form of each piece, but all represent medium-sized containers, and most were originally handleless flasks of one sort or another. The majority are natural bluish-green in color and belong to the first half of the first century A.D., as the numerous examples from Deposits V, VII–VIII, and from Forum SW, I, demonstrate. Rims and bodies, insofar as they are preserved, are typical of hundreds of examples known from other Romano-Italian sites—including the somewhat later pieces from excavations at Pompeii and Herculaneum. The shapes involved, however, are sufficiently basic to the glassblower's craft that many may have continued with little change throughout the second and even the third century A.D.

Parallels and Dated Examples of Medium-Sized Flasks and Jugs

Cosa: **Nos. 512** and **514**, Deposit V, ca. 25/15 B.C. to ca. A.D. 40/45; **Nos. 509–510**, Deposit VII, decade prior to ca. A.D. 40/45; **No. 507**, Deposit VIII, prior to ca. A.D. 50/55.

Magdalensberg 1979, 125–127, nos. 978–1008, pl. 7. Handleless flasks of Isings Form 16; 145–148, nos. 1056–1057 *inter alia*, pl. 9, which resemble (except for the presence of handles) **Nos. 515–516** at Cosa. All examples date before ca. A.D. 45.

Tarragona (Pasaje Cobos): Examples of handleless, naturally colored flasks and medium-sized jugs of the second quarter of the first century A.D. Unpublished. Information furnished by Jennifer Price, University of Durham.

Charlesworth 1965, 233, no. 13, fig. 7, pl. 13. Handleless flask from Pompeii, pre-A.D. 79, and typical of many others from the town, now in the Museo Archeologico, Naples.

Aquileia 1968, 57–59, no. 144, pls. B.6 and 12.7. Bluish-green handleless flask.

Later in date than the majority of bluish-green flasks and jugs are five other medium-sized vessels that also were probably parts of flasks but more unusual in shape. Two examples preserve funnel-shaped rims that taper to a tooled constriction at the junction of the neck and body (**Nos. 519–520**). The other three, all bottoms (**Nos. 521–523**), are probably from amphora-shaped flasks and resemble several from Aquileia (Aquileia 1968, 131, nos. 262–263, pls. 20 and L.3, with parallels from Pompeii cited). At Aquileia these containers are ascribed to the second half of the second century or third century A.D. If the pieces at Cosa are not from flasks, however, they could possibly be the bases of late antique or medieval lamps, such as several from Italy now in the Vatican Collection and described as "early medieval" (Vatican Coll. 1975, 96–97, nos. 877–879, pl. 61). Considering how little glass at Cosa belongs to the period between the third and thirteenth centuries A.D., it nonetheless seems more reasonable to assume that these fragments represent flasks of a mid-Imperial date.

Medium-Sized Flasks and Jugs

504. CF 2377 Pl. 32
A.P. Room 25 Exterior, III

Rim, neck, and fragment of shoulder of flask; natural bluish-green. Iridescence and pitting. Outsplayed rim, folded outward, then diagonally upward; cylindrical neck; shoulder flares diagonally outward. Blown. D 3.7 cm; A Th shoulder .1 cm.

505. C69.432 Pl. 32
Forum SW, I

Fragment of rim, neck, and shoulder of flask; natural bluish-green. Iridescence and pitting. Outsplayed rim, folded outward, then inward; cylindrical neck; shoulder flares diagonally outward. Blown. D ca. 3.4 cm; A Th shoulder .2 cm.

506. CD 1151 Pl. 32
Bas. ENE 3, III Deposit VIII
Prior to ca. A.D. 50/55

Rim and fragment of neck of flask; natural bluish-green. Iridescence and pitting. Outsplayed rim, folded outward, then inward; cylindrical neck, bulging slightly at center. Blown. D 3.2 cm; A Th neck .18 cm.

507. CD 1150 Pl. 32
Bas. ENE 3, III Deposit VIII
Prior to ca. A.D. 50/55

Fragment of neck and shoulder of flask; natural bluish-green. Iridescence and pitting. Cylindrical neck bulging slightly at center; shoulder curves diagonally outward from tooled constriction at junction of shoulder and neck. Blown. D neck 2.8 cm; A Th neck .2 cm.

508. C69.453B Pl. 32
VIII-E, F, 0

Fragment of rim and neck of flask; natural light green. Iridescence and pitting. Horizontal rim, folded outward and inward to form open tubular edge; cylindrical neck. Blown. D ca. 3.2 cm; A Th neck .1 cm.

509. CE 2046 Pl. 32
A.P. Room 22, II Deposit VII
Decade prior to ca. A.D. 40/45

Rim, neck, and vestiges of shoulder of flask; natural bluish-green. Iridescence and pitting. Outsplayed rim with solid, thickened edge; cylindrical neck, bulging slightly at center; shoulder flares diagonally outward from tooled constriction at junction of shoulder and neck. Blown. D ca. 2.8 cm; A Th neck .13 cm.
Bibliography: Grose 1975, 45, no. 34.

510. CE 2156
A.P. Room 22, II Deposit VII
Decade prior to ca. A.D. 40/45

Fragment of rim and neck of flask; natural bluish-green. Iridescence and pitting. Outsplayed rim with solid, thickened edge; cylindrical neck, bulging slightly at center. Blown. D 2.8 cm; A Th .13 cm.
Bibliography: Grose 1975, 45, no. 34A.

511. C68.736
Forum ENW, 0
Fragment of rim and neck of flask; natural bluish-green. Iridescence and pitting. Outsplayed rim with rounded, thickened edge; cylindrical neck, tapering upward. Blown. D ca. 2.8 cm; A Th neck .1 cm.

512. CE 2194 Pl. 32
A.P. Room 16, I Deposit V
ca. 25/15 B.C. to ca. A.D. 40/45
Fragment of neck and shoulder of flask; natural bluish-green. Iridescence and pitting. Cylindrical neck with slight tooled constriction at junction of neck and shoulder; shoulder curves diagonally outward. Blown. D neck 2 cm; A Th neck .1 cm.

513. C69.455 Pl. 32
Forum SW, I
Two joining fragments of rim (edge missing) and neck of flask; natural light green. Iridescence and pitting. Outsplayed funnel-shaped rim; long narrow cylindrical neck with tooled constriction at junction of neck and shoulder. Blown. D neck 1.5 cm; A Th neck .1 cm.

514. CE 2193 Pl. 32
A.P. Room 16, I Deposit V
ca. 25/15 B.C. to ca. A.D. 40/45
Fragment of rim, neck, and vestiges of shoulder of flask; natural pale green. Iridescence and pitting. Short, horizontal rim, probably rotary-polished on edge; short cylindrical neck; shoulder flares diagonally outward. Blown. D 1.9 cm; A Th neck .2 cm.

515. C66.839 Pl. 32
S.H. Street M, 0-I
Rim, neck, and fragment of shoulder of flask; natural light green. Iridescence and pitting. Outsplayed rim, folded outward, then diagonally upward; short, cylindrical neck, bulging slightly at center; faint tooled constriction at junction with shoulder; horizontal shoulder turns abruptly downward; no handle attachment in evidence. Blown. D 3.5 cm; A Th shoulder .1 cm.

516. CD 1138 Pl. 32
Bas. ENW 2, III
Fragment of rim, neck, and shoulder of flask; natural bluish-green. Iridescence and pitting. Outsplayed rim, folded outward, then diagonally upward; short cylindrical neck, tapering upward; horizontal shoulder turns abruptly downward; no handle attachment in evidence. Blown. D ca. 1.2 cm; A Th shoulder .15 cm.

517. CE 2030 Pl. 32
A.P. Room 22, II Deposit VII
Decade prior to ca. A.D. 40/45
Two joining fragments of rim, neck, shoulder, and upper handle attachment (incomplete) of jug; natural bluish-green. Iridescence and pitting. Outsplayed rim, folded outward then diagonally

upward; cylindrical neck expands at shoulder; upper part of handle attached below rim. Blown, with drawn glass handle. D ca. 2.6 cm; A Th neck .1 cm.
Bibliography: Grose 1975, 46, no. 36.

518. CD 1192 Pl. 32
Bas. Trib. Vault, I
Rim, fragment of neck, and upper handle attachment (incomplete) of jug; natural bluish-green. Iridescence and pitting. Outsplayed rim, folded downward, then upward; short cylindrical neck turns abruptly outward at shoulder; upper part of handle attached below rim. Blown, with drawn glass handle. D 2.8 cm; A Th neck .1 cm.

519. CD 1097 Pl. 33
Bas. ENE 2, I
Fragment of rim, neck, and side of flask; colorless. Iridescence and pitting. Off-set rim, knocked-off and rotary-polished; funnel-shaped neck with tooled constriction at junction with side; side curves diagonally downward. Blown. D 4 cm; A Th side .1 cm.

520. C70.592 Pl. 33
S.H. Street M, 0-I
Fragment of rim (edge missing), neck, and side of flask; colorless. Iridescence and pitting. Rim probably off-set; funnel-shaped neck with tooled constriction at junction with side; side expands diagonally downward. Shape similar to **No. 519**. Blown. D ca. 3.5 cm; A Th side .1 cm.

521. CF 2403 Pl. 33
Temple B (Forum), Cella, 0
Fragment of side and bottom of amphora-shaped flask or other vessel; natural light green. Iridescence and pitting. Conical shaped body terminating with a convex bottom. Blown. A Th side .1 cm.

522. CA 665 Pl. 33
Capitolium, Area 4C, 0
Fragment of side and bottom of amphora-shaped flask or other vessel; natural light green. Iridescence and pitting. Conical shaped body terminating with a convex bottom, similar to **Nos. 521** and **523**. Blown. A Th side .1 cm.

523. CF 2402 Pl. 33
Temple B (Forum), Pavement Fill, I
Fragment of side and bottom of amphora-shaped flask or other vessel; natural light green. Iridescence and pitting. Conical shaped body terminating with a convex bottom, similar to **Nos. 521–522**. Blown. A Th side .1 cm.

Spherical Oil Flasks (***No. 524***)

Another shape to emerge among the varieties of blown glass before the mid-first century A.D. was that of the simple oil flask that mimicked ceramic and metal aryballoi with ring or "dolphin" handles on opposite sides of the rim. These flasks, such as **No. 524** at Cosa, vary in size but all exhibit spherical bodies with slightly concave bottoms, short necks, and broad applicator rims. Their association with strigils in several burials strongly suggests that they were regularly employed as containers of olive oil (see Isings and other references below). Among intact examples, many still preserve metal stoppers and metal carrying handles or chains, inserted through the glass handles (see especially Morin-Jean 1913, 84–85, figs. 88–89). Although most examples of such glass flasks have been found in Italy or the provinces of the western Mediterranean and Europe, they are also known from eastern sites such as Karanis 1936, 252–253, no. 773, pl. 20, and Dura-Europos 1963, 113 and 115, no. 565, pl. 34.

These flasks were devised at least as early as Claudian–Neronian times and seem to have retained their popularity as late as the third or fourth century A.D. Accordingly, they represent one of the longest-lived forms of blown glass. The earliest published example is from Carnuntum, where it is said to be from a Claudian level. For it and later examples, see the references in the list below, especially Fishbourne 1971; Isings 1957; and Fremersdorf 1958, 28–31, pls. 30–31, 36–39, and 41.

PARALLELS AND DATED EXAMPLES OF SPHERICAL OIL FLASKS

Luni I 1973, 477, CM 921, pls. 82.14 and 109.4.
Ostia I 1968, 77, fig. 234 a–b. Said to be from a context of the mid-third century A.D.
Aquileia 1968, 26–28, nos. 14–15, pl. 2, with additional references.
Vindonissa 1960, 75 and 77, no. 197, pls. 13 and 20.
Lombardy Cat. 1964, 36, nos. 5–5a, pl. 3. Flasks from the Museo Civico at Brescia and another from Torricella (Mantova), now in the Museo Archeologico at Mantova; the latter is said to have been found with coins of Tiberius and Flavian emperors.
Fishbourne 1971, 356–357, nos. 82–82a, fig. 141, pl. 28.
Nida-Heddernheim 1974, 30–35, nos. 30–49, pl. 4. With extensive discussion and many parallels cited. All examples from Flavian or second-century contexts.
Conimbriga 1976, 170, no. 83, pl. 37. An example from a Claudian context.
Dura-Europos 1963, 113 and 115, no. 565, pl. 34.
Vessberg 1956, 208, fig. 50.33–36. Examples from Cyprus.
Karanis 1936, 252, pl. 20, Class XII, Group A. Examples from Egypt.
Isings 1957, 78–81, Form 61. Among the earliest oil flasks cited is one from a Claudian context at Carnuntum.

SPHERICAL OIL FLASK

524. C70.621 Pl. 33
VI-D, E, 0

Two joining fragments of rim, neck, shoulder, and one handle of a two-handled globular oil flask; natural pale yellowish-green. Silvery iridescence and pitting. Outsplayed rim, folded outward, then obliquely inward; short cylindrical neck; globular body; small ring handles attached beneath rim and on shoulder. Blown, with drawn glass handles. D 4.5 cm; A Th shoulder .25 cm.

Prismatic and Mercury Bottles (***Nos. 525–548***)

Mold-blown prismatic bottles (also known as square bottles) require little introduction, being among the most familiar types of Roman Imperial glassware. They were devised shortly after the invention of mold-blowing in order to store, preserve, display, and transport liquids or other commodities. They must have been especially popular in household settings, and several are actually included in a relief sculpture depicting a home interior on the inside of a stone sarcophagus now in the Leiden Museum (*Mnemosyne* 3, 1935–1936, 88 ff.). At times they were reused as cinerary urns in much the same way as the wide-mouthed jars, or ollae, discussed earlier.

In shape, most prismatic bottles are square in horizontal cross section (**Nos. 539–544**), although some are rectangular. Others are multi-sided with as many as thirteen, as few as three sides (see, respectively, an unpublished example in the Toledo Museum of Art [23.975] and one from Pompeii illustrated in Morin-Jean 1913, 60, fig. 36). In addition, there exist a separate series of cylindrical or barrel-shaped mold-blown bottles, but no examples of these types were found at Cosa. Also related in technique and function are the so-called Mercury bottles discussed in the next section, as well as the class of mold-blown jars noted on pages 107–108 above.

The prismatic bottles at Cosa, as elsewhere, were all blown into body-molds that imparted flat, vertical walls, usually thicker at the center than near the corners. The bottom-molds quite often made use of simple geometric designs (**Nos. 541–545**), or even figural motifs (not represented at Cosa). Rims are uniformly large and can be classified into three or four common shapes (for examples, see **Nos. 525–532**). Handles, whether one or a pair, are usually of the "hip" variety with pronounced vertical ribbing, as is true of **No. 531**. The overall sizes of these containers are remarkably diverse, and when examined, their capacities fail to correspond to any standard units of Roman liquid measures (for discussion, see Charlesworth 1966, 29).

Comparable studies of the geometric patterns and decorative motifs on the underside of these bottles also reveal little attempt at uniformity or standardization, which otherwise might aid us in attributing certain groups of bottles to particular regions, workshops, or dates. In fact, although most of the geometric designs are repeatedly employed—concentric circular ridges, raised dots, and right-angled bars in the corners—variations are exceedingly numerous, as the discussion of Charlesworth 1966 and Alarçao 1975 both emphasize. The two inscribed examples at Cosa appear in Latin, as would be expected in Italy. Of these, **No. 540** preserves the lowest two lines of a four-lined inscription bearing the letters ALEX/NDI, probably the Greek name ALEXANDROS in the Latin genitive using a ligature for AND and curiously lacking the R. This name is known from other bottles but always in the Greek alphabet (Trowbridge 1930, 120, with references; cf. *CIL* 15.2.1.7001). The other inscribed bottle (**No. 539**) is more complete and can easily be restored. The full inscription in three lines reads EX OFFICINA TITIENI YACINTHI. Although this individual is not otherwise identified in the literary or epigraphical sources, the name is known from a second inscribed bottle with a similar (but not identical) inscription found at Rome and now in the collection of the Antiquarium Comunale, Rome (cf. Kisa 1908, 930, no. 39; cf. *CIL* 15.2.1.6997).

It is unclear when prismatic bottles were first introduced. Czurda-Ruth in Magdalensberg 1979, 136, argues that some examples are present in late Augustan levels. This, however, seems doubtful, if only because the technique of mold-blowing is with little doubt a later invention (see pp. 98–99 above). Also, the pieces she cites are so fragmentary that it is difficult to equate them with the normal shapes of square bottles. The fact that none carries relief decoration on the bottom and the pieces appear to be commensurate in size with standard prismatic bottles further detracts from her assertion. On the other hand, the evidence from Deposit VII at Cosa (supplemented by the material from Magdalensberg) does suggest that by the forties A.D. small bottles were indeed being blown into four-sided body-molds and that these vessels also used bottom-molds with simple relief decoration (at Cosa, see **Nos. 545–547**). In all likelihood, these vessels preceded by only a few years the mass production of the larger, more standard versions of these wares. Once devised, these containers enjoyed tremendous popularity from Neronian times to the end of the first century A.D.

They are also discovered in securely dated deposits of the second century, as late as the reign of Hadrian (Charlesworth 1966, 29). Thereafter, they seem to have fallen from favor and were replaced with other types of large and free- and mold-blown bottles, almost always cylindrical, such as Isings Forms 89–90 (Isings 1957, 106–108).

Parallels and Dated Examples of Prismatic Bottles

Cosa: Nos. **534–535**, **537–538**, **545–547**, Deposit VII, decade prior to ca. A.D. 40/45; **No. 536**, Deposit IX, late first through early third century A.D.

Magdalensberg 1979, 131–136, nos. 1017–1018, 1020–1021, pl. 7, with numerous fragmentary rims and handles thought to belong to prismatic bottles but, ironically, none with decorated bottoms. Numerous parallels cited on pp. 133–135.

Vindonissa 1960, 78–82, pls. 13 and 21, with numerous parallels listed for square, rectangular, hexagonal, and cylindrical prismatic bottles. The earliest dated examples cited are Claudian.

Camulodunum 1947, 306, no. 98, pl. 88. The earliest fragments that seem to be from prismatic bottles are assigned to Periods II to IV, ca. A.D. 43/44–61.

Fishbourne 1971, 361–364, nos. 94–105, figs. 143–144. All pieces are highly fragmentary.

Nida-Heddernheim 1974, 67–81. Discussion of the various shapes of prismatic and cylindrical bottles, both free-blown and mold-blown, with comparanda listed.

Samothrace 1967, 47, no. 46, fig. 46. A bottle attributed to a Claudian burial.

Tipasa 1967, 14–15 and 48–52, nos. 55–64, pls. 4–5. Examples from North Africa.

Vessberg 1956, 148–149, and 200–201; fig. 47, nos. 1–5. Examples from Cyprus, with mention of others from Egyptian and Syro-Palestinian contexts.

Isings 1957, 63–67, Form 50 a and b. Additional references listed.

As is evident from the parallels cited above, most examples of prismatic bottles are reported from Italian and western Mediterranean communities, with a lesser number known from the eastern Mediterranean. The presence of Greek inscriptions on some pieces, however, suggests that some of these vessels must have been manufactured in the eastern Roman provinces, and not exclusively in the West. Actual sites producing these wares have yet to be discovered, but the bottom-mold of a bottle was found at Cologne (see p. 98, n. 29 above).

Mercury Bottles

A variation of the standard prismatic bottle is the mold-blown Mercury bottle, actually a handleless flask, usually colorless or natural bluish-green, such as the single example found at Cosa (**No. 536**). Like prismatic bottles, they were blown into four-sided body-molds and made use of a decorated bottom-mold. Mercury bottles take their name from a relief representation of that god depicted on the undersides of many of these containers. Usually the god holds a purse in one hand and is accompanied by a ram, rooster, or tortoise. In some instances Latin letters appear in the corners of the bottom, and a few pieces even bear Latin inscriptions on the side of the flasks as well (see Isings 1957, 100–101, Form 84; cf. Kisa 1908, 780 ff., 926ff., and 940; also Morin-Jean 1913, 66–71). In shape, the bottles are square in horizontal cross section with a distinct downward taper. Necks are long and narrow, with right-angled shoulders and broad, flat rims with a small aperture, not unlike the rims of the oil flasks discussed above. Most Mercury flasks vary between 12 and 20 cm in height, but taller examples, up to 40 cm, are known in the collection of the Museo Nazionale Romano (unpublished).

Previous studies of Mercury bottles have ascertained that they belong primarily to the second or third centuries A.D. (see discussion in Isings 1957, 100). Their distribution is identifiably Italian and western, but at least one bottle, albeit a variant of the group, was found in a burial at Samothrace (see list below). The purpose of these distinctive yet oddly shaped containers is uncertain, but their limited capacities and applicator rims show that they contained liquids of some sort. One bottle discovered at Clemarais in France bears the inscription FIRMI HILARI AD TYLOSIM AROMATICUM, an indication that the contents were an aromatic ointment, possibly for medicinal purposes (cf. Aquileia 1968, 57).

Parallels and Dated Examples of Mercury Bottles

Luni I 1973, 478, CM 1768/1. An unillustrated fragment of a Mercury bottle.

Vatican Coll. 1975, 70–71, nos. 682–691, pls. 30–31. Examples presumably from Rome or environs.

Bologna Coll. 1979, 68, nos. 61–62, with parallels to others known from northern Italy, Croatia, and elsewhere.

Turin: Museo di Antichità. Illustrated in Mariacher 1959, pl. 14.

Aquileia 1968, 57–58, no. 142, pl. B.5. A bottle said to date to the second century A.D.

Morin-Jean 1913, 66–71, figs. 59–62. Examples from Gaul, mostly ascribed to the second or early third century A.D.

Limburg 1971, 11, no. 16, fig. 1.

Trier 1977, 183, nos. 1141–1142, pl. 24.

Fishbourne 1971, 357, no. 87, fig. 142.

Isings 1957, 100, Form 84. With discussion and parallels listed.

Samothrace 1967, 43–44, no. 29, fig. 29. A bottle similar but not identical to most examples found in western Mediterranean contexts. From a tomb with gold-foil impressions of coins minted in A.D. 134 but containing multiple burials.

Prismatic and Mercury Bottles

525. C70.663 Pl. 33
VI-D, E, 0

Fragment of rim and neck of large prismatic bottle or flask; natural bluish-green. Iridescence and pitting. Horizontal rim, folded outward, then inward, and sloping obliquely to the outside; vestiges of a cylindrical neck, tapering downward. Blown (body possibly mold-blown). D ca. 12 cm; A Th neck .5 cm.

526. CG 586 Pl. 33
Forum East Corner, NE Passage, 0-I

Fragment of rim and neck of large prismatic bottle or flask natural bluish green. Iridescence and pitting. Outsplayed rim, folded outward, then inward, and sloping diagonally to outside; vestiges of a cylindrical neck, tapering downward. Blown (body possibly mold-blown). D ca. 9 cm; A Th neck .3 cm.

527. C70.680 Pl. 33
S.H. Street M, 0-I

Rim and fragment of neck and handle attachment (incomplete) of large prismatic bottle; natural bluish-green. Iridescence and pitting. Vertical collar-shaped rim, folded downward, then abruptly upward; cylindrical neck, tapering upward; handle attachment below rim. Blown, with drawn glass handle (body probably mold-blown). D 8.3 cm; A Th neck .5 cm.

528. CD 307 Pl. 33
Bas. N 7, I
Fragment of rim and neck of large prismatic bottle or flask; natural bluish-green. Iridescence and pitting. Nearly horizontal rim, folded outward, then inward, and sloping obliquely to the inside; vestiges of a cylindrical neck, tapering downward. Blown (body possibly mold-blown). D ca. 7.2 cm; A Th neck .5 cm.

529. CG 587 Pl. 33
Forum East Corner, NE Passage, 0-I
Fragment of rim and neck of large prismatic bottle or flask; natural bluish-green. Iridescence and pitting. Horizontal rim, folded outward, then inward; cylindrical neck. Blown (body possibly mold-blown). D ca. 7 cm; A Th neck .4 cm.

530. CD 1179 Pl. 33
Bas. ENW, I
Rim and fragment of neck of large bottle or flask; natural light green. Iridescence and pitting. Vertical, collar-shaped rim, folded downward, then abruptly upward; vestiges of a cylindrical neck, tapering upward. Blown (body possibly mold-blown). D 7 cm; A Th neck .3 cm.

531. C70.615 Pl. 33
S.H. Street M, 0-I
Five joining fragments of rim, neck, side, and handle of large prismatic bottle; natural bluish-green. Iridescence and pitting. Outsplayed rim, folded outward, then inward, and sloping obliquely to the outside; cylindrical neck; horizontal shoulder dips down at junction with neck; nearly vertical side; nine-ribbed hip handle attached to neck and at junction of shoulder and side. Mold-blown in four-sided body-mold, with drawn and tooled glass handle. D 6 cm; W shoulder 9.1 cm; A Th side .3 cm.

532. CD 1178 Pl. 33
Bas. ENW, I
Rim and fragment of neck and upper handle attachment (incomplete) of prismatic bottle; natural bluish-green. Iridescence and pitting. Horizontal rim, folded outward, then inward; cylindrical neck, tapering upward; tiny fragment of upper handle attachment on neck. Blown, with drawn glass handle (body probably mold-blown). D 5.5 cm; A Th neck .3 cm.

533. C70.661A Pl. 33
VI-D, E, 0
Rim and fragment of neck of large prismatic bottle or flask; natural light green iridescence and pitting. Outsplayed rim, folded outward, then inward, and sloping obliquely to outside; cylindrical neck, tapering upward. Scar of handle attachment on neck. Blown (body possibly mold-blown). D 5 cm; A Th neck .2 cm.

534. CE 2037 Pl. 33
A.P. Room 22, II Deposit VII
Decade prior to ca. A.D. 40/45

Rim and fragment of large bottle or flask; natural light green. Iridescence and pitting. Solid, horizontal rim with rounded edge; cylindrical neck, possibly from a prismatic bottle. Blown (body possibly mold-blown). D 4.7; A Th neck .2 cm.
Bibliography: Grose 1975, 45, no. 35.

535. CE 2158 Pl. 33
A.P. Room 22, II Deposit VII
Decade prior to ca. A.D. 40/45

Fragment of rim and neck of bottle or flask; natural light green. Iridescence and pitting. Outsplayed rim, folded outward, then inward, and sloping diagonally to outside; vestiges of cylindrical neck. Blown or mold-blown. D 4.3 cm; A Th neck .2 cm.
Bibliography: Grose 1975, 46, no. 37.

536. CF 2501 Pl. 33
A.P. Room 25A, III Deposit IX
Late first through early third century A.D.

Fragment of shoulder and side, with vestiges of neck of prismatic bottle; colorless. Iridescence, milky weathering, and pitting. Cylindrical neck; short horizontal shoulder dips down at junction with neck; straight side, tapering downward. Probably part of a Mercury bottle. Mold-blown in four-sided body-mold. Estimated W shoulder ca. 4.2 cm; A Th side .25 cm.

537. CE 2050 Pl. 33
A.P. Room 22, II Deposit VII
Decade prior to ca. A.D. 40/45

Lower half of hip handle of prismatic bottle; natural bluish-green. Iridescence and pitting. Five-ribbed handle broken at point where handle turns inward toward neck; shoulder attachment preserved. Drawn and tooled. Preserved H 5 cm; W 3.4 cm; A Th .5 cm.
Bibliography: Grose 1975, 46, no. 38.

538. CE 2160A and B Pl. 33
A.P. Room 22, II Deposit VII
Decade prior to ca. A.D. 40/45

Two non-joining fragments of hip handle of prismatic bottle; natural bluish-green. Iridescence and pitting. Larger fragment broken at a point where handle turns toward neck, pattern of ten narrow ribs preserved. Drawn and tooled. H 4.7 cm; W 2.5 cm; A Th .35 cm. Only uppermost fragment illustrated on pl. 33.
Bibliography: Grose 1975, 46, no. 38A.

Bottoms of Prismatic Bottles

539. C70.308 Pl. 34
S.H. Street M, 0

Three joining fragments of side and bottom of prismatic bottle; natural light green. Iridescence and pitting. Vertical side; slightly concave bottom. On underside, raised in molded relief, a Latin inscription in three lines: [e]X OF[ficina] / TITIEN[i] / YACIN[thi]. Mold-blown in four-sided body-mold. W side at bottom 6.5 cm; A Th side .2 cm.

540. CD 795 Pl. 34
Bas. ENW 2, III
Two joining fragments of side and bottom of prismatic bottle; natural light green. Vertical side; slightly concave bottom. On underside, raised in scarcely perceptible relief, the lower two lines of a four-line (?) Latin inscription: ALEX/ AD-- (r)I. Mold-blown in four-sided body-mold. W side at bottom 9 cm; A Th side .1 cm.

541. CF 2302 Pl. 34
A.P. Room 22 Exterior, IV–V
Eighteen joining fragments of side and bottom of prismatic bottle; natural bluish-green. Iridescence and pitting. Vertical side; slightly concave bottom. On underside, raised in molded relief, a pattern of three concentric ridges inscribing a six-pointed star (restored) at center of bottom. In each corner, a semicircular ridge with a raised dot at center. Mold-blown in four-sided body-mold. W side at bottom ca. 9 cm; A Th side .2 cm. Bottom restored symmetrically on pl. 34.

542. CG 387 Pl. 34
Forum East Corner, NE Passage, I
Four joining fragments of side and bottom of prismatic bottle; natural bluish-green. Iridescence and pitting. Vertical side; slightly concave bottom. On underside, raised in molded relief, a pattern of two concentric ridges with indications of a third. Mold-blown in four-sided body-mold. W side at bottom ca. 9 cm; A Th side .6 cm. Bottom restored symmetrically on pl. 34.

543. CE 2026 Pl. 34
A.P. Room 22, I
Fragment of bottom of prismatic bottle; natural bluish-green. Iridescence and pitting. Slightly concave bottom. On underside, raised in molded relief, a pattern of two, possibly three, concentric circular ridges. In each corner, a right-angled ridge. Mold-blown in four-sided body-mold. W side at bottom ca. 9 cm; A Th bottom .35 cm. Bottom restored symmetrically on pl. 34.

544. CE 2210 Pl. 34
A.P. Cistern 12-a-5, I
Fragment of side and bottom of prismatic bottle; natural bluish-green. Iridescence and pitting. Nearly vertical side; slightly concave bottom. On underside, raised in molded relief, raised ridges in an uncertain pattern. Mold-blown in four-sided body-mold. W side at bottom ca. 9 cm; A Th side .2 cm.

545. CE 2159C Pl. 34
A.P. Room 22, II Deposit VII
Decade prior to ca. A.D. 40/45
Fragment of side and bottom of small prismatic bottle; natural bluish-green. Iridescence and pitting. Vertical side; slightly concave bottom. On underside, raised in molded relief, a dot at center. Mold-blown in four-sided body-mold. W side at bottom 5 cm; A Th side .1 cm.
Bibliography: Grose 1975, 46, no. 39.

546. CE 2159A Pl. 34
A.P. Room 22, II Deposit VII
Decade prior to ca. A.D. 40/45
Fragment of side and bottom of small prismatic bottle or other storage vessel; natural bluish-green. Iridescence and pitting. Vertical side; slightly concave bottom. Undecorated. Mold-blown in four-sided body-mold. W side at bottom ca. 4–5 cm; A Th bottom .3 cm.
Bibliography: Grose 1975, 46, no. 39A.

547. CE 2159B
A.P. Room 22, II Deposit VII
Decade prior to ca. A.D. 40/45
Fragment of side and bottom of small prismatic bottle or other storage vessel; natural bluish-green. Iridescence and pitting. Shape and technique of manufacturing as in **No. 546**. W side at bottom ca. 4–5 cm; A Th bottom .4 cm.
Bibliography: Grose 1975, 46, no. 39B.

548. CF 2321 Pl. 34
A.P. Room 22, Exterior, III
Small fragment of bottom of prismatic bottle; natural bluish-green. Iridescence and pitting. Slightly concave bottom. On underside, raised in molded relief, a pattern of at least two concentric circular ridges surrounding a raised cross at center. Mold-blown. A Th bottom .85 cm.

*Small Unguent Bottles (**Nos. 549–574**)*

This final section devoted to storage wares from Cosa comprises a group of small naturally colored and colorless, thin-walled unguent bottles. Although their initial development in the first half of the first century A.D. paralleled that of their counterparts in colored fabrics (see pp. 143–144 above), naturally colored examples continued in production after the fashion for color disappeared. In fact, following the mid-first-century A.D., glassmakers throughout the empire fabricated quantities of these little containers to accommodate, one assumes, the vast market for perfumes, cosmetics, ointments, and medicines. The spherical and globular shapes associated with the Augustan and Julio-Claudian periods, however, progressively give way to more elongated shapes in the Flavian and subsequent periods. Bodies can be oval, conical, or tubular and support ever-taller necks that possess simple folded, tubular, or fire-polished rims. Some of these shapes appear to have persisted for many decades, even centuries, because the forms are basic to the glassblowers' craft and require, given their functions, little elaboration and no decoration.

PARALELLS AND DATED EXAMPLES OF SMALL UNGUENT BOTTLES

Cosa: **Nos. 549**, **555**, **565**, Deposit V, ca. 25/15 B.C. to ca. A.D. 40/45; **Nos. 550–551**, **558–559a**, **562–564**, Deposit VII, decades prior to ca. A.D. 40/45; **Nos. 557** and **561**, Deposit VIII, prior to ca. A.D. 50/55.

Bologna Coll. 1979, 98–139, nos. 96–192. Many examples that parallel the shapes of bottles found at Cosa.

Pavia Coll. 1983, 110, nos. 77–151. Numerous examples, many from first-century A.D. contexts. They also illustrate the same transformations in shapes during the first century that occur at Cosa.

Vatican Coll. 1975, 52–56, nos. 399–487, pls. 15–18. Numerous parallels to the pieces at Cosa.
Magdalensberg 1979, 103–111, nos. 793–851, pl. 6. Globular bottles said to be from early Augustan contexts, with comparanda cited; also others of Isings Form 8 on pp. 111–115, nos. 852–914, which resemble **Nos. 558–559** at Cosa. At Magdalensberg bottles of this latter shape are recovered from Augustan through Claudian levels and are thus consistent with the dating information provided by Deposit VII at Cosa.
Vindonissa 1960, 74–78, nos. 183–198. Discussion with additional references.
Samothrace 1967, 41–42, esp. nos. 19–20, 22–23, and 25; figs. 19–20, 22–23, and 25. These examples belong mostly to the first half of the first century A.D.
Cyprus 1956, 162, fig. 50. Most of the shapes found at Cosa are illustrated here, but these examples are all from Cypriot or eastern Mediterranean sources.
Isings 1957, 22–23, Form 6; 24, Form 8; 40–41, Form 26; 41, Form 27; 41–42, Forms 28 a–b; 97–100, Forms 82 a–d, illustrate the changing shapes of unguent bottles from the early first century A.D. into the third century A.D. Numerous references.

Evidence for dating the various forms of small unguent bottles found at Cosa is abundant but lacks precision due to the fact that many shapes lasted for a half century or longer. Although many examples from Cosa are from dated deposits, they furnish little information that would alter the basic chronological framework already established for these wares by Isings and others (see list above). Moreover, the pieces from Cosa span only two or two and a half centuries of production, and the bottles themselves continued to be made until the end of the Roman era. Although some of the forms known from Cosa may eventually prove to be predominantly Italian or western, examples of similar shapes from the Aegean, Cyprus, and the Syro-Palestinian region are much the same and reflect the fairly consistent international character of the group.

Small Unguent Bottles

549. CE 2192 — Pl. 35
A.P. Room 16, I — Deposit V
ca. 25/15 B.C. to ca. A.D. 40/45
Fragment of rim, neck, and side of small unguent bottle; natural bluish-green. Iridescence and pitting. Outsplayed rim with rounded edge; funnel-shaped neck, bulging slightly at center; side curves outward; body probably ovoid in shape. Blown. D 2.2 cm; A Th side .1 cm.

550. CE 2052 — Pl. 35
A.P. Room 22, II — Deposit VII
Decade prior to ca. A.D. 40/45
Rim (edge missing), neck, and shoulder of small unguent bottle; natural bluish-green. Iridescence and pitting. Outsplayed rim; cylindrical neck; flaring shoulder. Blown. D ca. 2.3 cm; A Th shoulder .05 cm.
Bibliography: Grose 1975, 48, no. 41.

551. CE 2038
A.P. Room 22, II — Deposit VII
Decade prior to ca. A.D. 40/45
Two joining fragments of neck and shoulder of small unguent bottle; natural bluish-green. Iridescence

and pitting. Cylindrical neck expands outward at shoulder. Shape probably similar to **No. 550**. Blown. D neck 1.5 cm; A Th neck .08 cm.

552. C69.421 Pl. 35
Forum SW, I
Fragment of rim, neck, and shoulder of small unguent bottle; natural bluish-green. Iridescence and pitting. Off-set rim with unfinished edge; funnel-shaped neck, bulging slightly at center; flaring shoulder. Blown. D 1.6 cm; A Th shoulder .07 cm.

553. C65.802 Pl. 35
V-D, Cistern 2
Rim, neck, and fragment of shoulder of small unguent bottle; natural bluish-green. Iridescence and pitting. Slightly inturned rim with unfinished edge; cylindrical neck, tapering downward; flaring shoulder. Blown. D 1.3 cm; A Th shoulder .1 cm.

554. C68.282 Pl. 35
S.H. Garden, 0
Body and bottom of small unguent bottle; natural bluish-green. Silvery iridescence and pitting. Ovoid body; slightly concave bottom. Blown. Max D body 3.7 cm; A Th side .1 cm.

555. CE 2011 Pl. 35
A.P. Room 16, I Deposit V
ca. 25/15 B.C. to ca. A.D. 40/45
Rim (edge missing), neck, and body of small unguent bottle (profile mostly restorable); natural bluish-green. Silvery iridescence and pitting. Outsplayed rim; funnel-shaped neck, bulging slightly at center; ovoid body; slightly concave bottom. Blown. Restored H ca. 6 cm; Max D body 3.1 cm; A Th side .1 cm.

556. C70.601 Pl. 35
S.H. Street M, 0-I
Neck and body of small unguent bottle; natural bluish-green. Iridescence and pitting. Funnel-shaped neck; almost spherical body; slightly concave bottom. Blown. Max D body 2.9 cm; A Th side .2 cm.

557. CD 948 Pl. 35
Bas. ENE 1, III Deposit VIII
Prior to ca. A.D. 50/55
Rim (incomplete), neck, and body of small unguent bottle (profile restorable); natural light green. Iridescence and pitting. Slightly outsplayed rim with rounded edge, folded inward; short funnel-shaped neck; ovoid body; slightly concave bottom. Blown. H 4.5 cm; D 1.6 cm; A Th side .09 cm.

558. CE 1968 Pl. 35
A.P. Room 22, II Deposit VII
Decade prior to ca. A.D. 40/45

Neck and body of small unguent bottle; natural bluish-green. Iridescence and pitting. Short funnel-shaped neck; long conical body; slightly concave bottom. Blown. Restored H ca. 9 cm; Max D body 3 cm; A Th side .1 cm.
Bibliography: Grose 1975, 46, no. 40.

559. CE 1967 Pl. 35
A.P. Room 22, II Deposit VII
Decade prior to ca. A.D. 40/45
Four joining fragments of body and bottom of small unguent bottle; natural bluish-green. Iridescence and pitting. Long conical body similar to **No. 558**; slightly concave bottom. Blown. H body 8 cm; Max D body 2.9 cm; A Th side .1 cm.
Bibliography: Grose 1975, 46, no. 40A.

a. CE 2032, A.P. Room 22, II, Deposit VII, decade prior to ca. A.D. 40/45. Side and bottom of small unguent bottle probably identical to **Nos. 558–559**; natural bluish-green. Max D 3 cm; A Th side .1 cm.
Bibliography: Grose 1975, 46, no. 40B.

560. C71.59 Pl. 35
V-E, 0
Body and bottom of small unguent bottle; natural bluish-green. Iridescence and pitting. Conical body; slightly concave bottom. Blown. Max D body 2.8 cm; A Th side .2 cm.

561. CD 1152A Pl. 35
Bas. ENE 3, III Deposit VIII
Prior to ca. A.D. 50/55
Fragment of neck and body of small unguent bottle; natural bluish-green. Iridescence and pitting. Cylindrical neck; piriform body; slightly concave bottom. Blown. Max D body 2.5 cm; A Th side .3 cm.

562. CE 2157A Pl. 35
A.P. Room 22, II Deposit VII
Decade prior to ca. A.D. 40/45
Three joining fragments of body and bottom of small unguent bottle; natural bluish-green. Iridescence and pitting. Ovoid body; slightly concave bottom. Blown. Max D body 2.3 cm; A Th side .08 cm.
Bibliography: Grose 1975, 48, no. 42.

563. CE 2157B Pl. 35
A.P. Room 22, II Deposit VII
Decade prior to ca. A.D. 40/45
Two joining fragments of body and bottom of small unguent bottle; natural bluish-green. Iridescence and pitting. Ovoid body; slightly concave bottom. Blown. Max D body 2.2 cm; A Th side .08 cm.
Bibliography: Grose 1975, 48, no. 42A.

564. CE 2157C Pl. 35
A.P. Room 22, II Deposit VII
Decade prior to ca. A.D. 40/45

Fragment of body and bottom of small unguent bottle; natural bluish-green. Slight iridescence and pitting. Ovoid body; slightly concave bottom. Blown. Max D body 2.5 cm; A Th side .1 cm.
Bibliography: Grose 1975, 48, no. 42B.

565. CE 2192A Pl. 35
A.P. Room 16, I Deposit V
ca. 25/15 B.C. to ca. A.D. 40/45

Fragment of body and bottom of small unguent bottle; bluish-green. Iridescence and pitting. Ovoid body; slightly concave bottom. Blown. Max D body 2.5 cm; A Th side .1 cm.

566. C69.308 Pl. 35
Forum SW, I

Neck and body of small unguent bottle (profile mostly restorable); natural bluish-green. Iridescence and pitting. Long cylindrical neck, bulging slightly at center; tooled constriction at junction with body; ovoid body; slightly concave bottom. Blown. Restored H ca. 7.6 cm; Max D body 2.1 cm; A Th side .1 cm.

567. C70.624 Pl. 35
S.H. Street 5, 0

Fragment of rim and neck, and fragment of body and bottom of small unguent bottle (profile mostly restorable); natural light green. Iridescence and pitting. Outsplayed rim with unfinished edge; long cylindrical neck; short conical body; slightly concave bottom. Blown. Restored H ca. 11 cm; Max D body 3.4 cm; A Th side .1 cm.

568. C70.603 Pl. 35
S.H. Street M, 0-I

Fragment of rim and neck of small unguent bottle; natural light green. Iridescence and pitting. Outsplayed rim with rounded edge, folded inward; long cylindrical neck. Blown. D 2.9 cm; A Th neck .1 cm.

569. C70.602 Pl. 35
S.H. Street 5, 0-I

Fragment of neck and body of small unguent bottle; natural light green. Iridescence and pitting. Long cylindrical neck with constriction at junction with body; short conical body; slightly concave bottom. Blown. Max D body 2.2 cm; A Th side .13 cm.

570. CG 373 Pl. 35
Comitium, SE Room, I–II

Body and bottom of small unguent bottle; natural bluish-green. Iridescence and pitting. Short conical body; slightly concave bottom. Blown. Max D body 3.2 cm; A Th side .2 cm.

571. CA 680 Pl. 35
Capitolium, Area 5b, 0

Lower part of body and bottom of small unguent bottle or other vessel; natural light green. Iridescence and pitting. Long cylindrical body, bulging slightly at bottom; convex bottom. Pontil scar visible on bottom. Blown. Max D body 1.5 cm; A Th side .2 cm.

572. CD 1193 Pl. 35
Bas. ENE 1, I
Fragment of side and bottom of small unguent bottle; natural bluish-green. Iridescence and pitting. Cylindrical body, bulging at bottom; convex bottom. Blown. Max D body 1.7 cm; A Th side .1 cm.

573. C70.600 Pl. 35
S.H. Street M, 0-I
Fragment of body and bottom of small unguent bottle; natural bluish-green. Iridescence and pitting. Extremely narrow, long cylindrical body, bulging at bottom. Blown. Max D body 1 cm; A Th side .1 cm.

574. CG 439 Pl. 35
Forum East Corner, NE Passage, I
Fragment of lower part of body and bottom of small unguent bottle; natural bluish-green. Iridescence and pitting. Long, narrow cylindrical body, tapering upward; convex bottom. Blown. Max D body .9 cm; A Th side .08 cm.

Funnels, Lamps, Inkwells, and Unidentified Vessels (***Nos. 575–580***)

Among the fragments of blown glass at Cosa are half a dozen pieces of uncertain forms, but probably once part of funnels, lamps, and glass inkwells. **Nos. 575–576**, for example, probably represent the rims and upper portions of funnels, such as those described in Isings 1957, 92, Form 74, and also known from Pompeii (unpublished, and now in the Museo Nazionale Archeologico, Naples). Alternatively, they could be the rims of Imperial flasks or bottles of an otherwise unidentified sort. More securely identified are **Nos. 577–578**, which must once have belonged to lamps dating to the late antique period, possibly later. **No. 578**, with its preserved solid base knob, must have been intended for use in a metal polycandelon such as the one illustrated in Vatican Coll. 1975, pl. 60. **No. 577** is more problematic since the body and bottom appear to have been blown into a mold. No close parallel is known, but a lamp in the Vatican Collection seems to share the same general form (Vatican Coll. 1975, 96, no. 875, pl. 60). If so, the examples at Cosa, like it, must be of medieval date. For a general discussion of late antique lamps, consult Sardis 1980, 38–53.

The final two pieces (**Nos. 579–580**) both possess a circular aperture at the top of the vessel. Since these apertures are relatively small, the shape is probably that of an inkwell rather than a pyxis. Blown inkwells of various forms and styles are discussed by Isings 1957, 93, Form 77, and in Berlin Coll. 1976, 77, no. 153, both with references.

Funnels, Lamps, Inkwells, and Unidentified Vessels

575. CF 2405 Pl. 36
A.P. Room 25 Exterior NW, I
Fragment of rim and neck (or side) of vessel; colorless with greenish tinge. Iridescence, pitting, and

discoloration. Outsplayed rim with solid thickened edge, folded outward; uneven, funnel-shaped neck or side. Shape uncertain, but probably similar to **No. 576**. Blown. D ca. 8 cm; A Th side .15 cm.

576. CF 2417 Pl. 36
A.P. Room 25 Exterior, I
Fragment of rim and neck (or side) of vessel; natural light green. Iridescence and pitting. Outsplayed rim with solid, thickened edge; uneven funnel-shaped neck or side. Shape uncertain, but probably similar to **No. 575**. Blown. D ca. 8 cm; A Th .1 cm.

577. CD 410 Pl. 36
Bas. ENE 3, I
Side and bottom of lamp or other vessel; pale yellow. Slight iridescence and pitting. Almost cylindrical body with upper section curving outward, then upward to rim; slightly concave bottom with raised circular disk at center; upper side convex. Blown, with lower part of body probably blown into a cylindrical mold. D bottom 4.8 cm; Min Th side .12 cm. Presumably ancient, but possibly medieval.

578. C72.47 Pl. 36
IV-B, Trench 2, 0
Basal knob and indications of body of lamp or other vessel; natural light green. Iridescence and pitting. Solid, cylindrical knob with convex end; slightly concave bottom of vessel. Blown. H knob 3 cm; D knob 2 cm. Presumably ancient, but possibly medieval.

579. C66.545 Pl. 36
S.H. Street M, 0-I
Rim and upper portion of body of vessel or object; light green. Iridescence and pitting. Inturned rim with carefully polished edge; circular aperture at top; side curves downward and inward. Shape uncertain. Blown. D 3.5 cm; D aperture 1.2 cm. Presumably ancient.

580. CE 1998 Pl. 36
A.P. Cistern 12-a-5, I
Two joining fragments of rim and side of inkwell, lamp, or other vessel; colorless. Silvery iridescence and pitting. Horizontal rim with solid, thickened edge; inturned side, slopes diagonally outward; circular aperture at top. Shape uncertain but possibly an inkwell or lamp. Blown. D top of vessel 2.5 cm; D aperture 1.7 cm; A Th side .08 cm.

ROMAN GLASS OBJECTS

Surprisingly, few glass objects other than vessels were recovered from Cosa. Most that were found are either disk-shaped gaming pieces or window panes and quarrels. In addition, there are four bezels of finger rings, two fragmentary bracelets, nineteen beads, three stirring rods or applicators, and twenty-eight mosaic tesserae. Jewelry and objects of personal adornment are not included in this catalogue but may appear in a volume devoted to minor objects from the site.

With the exception of two bezels, the glass objects at Cosa all date to the Imperial epoch. Such articles were produced in quantity only after the establishment of the Romano-Italian industry and the spread of glasshouses throughout the empire. This occurred at the end of the first century B.C. and in the first half of the first century A.D., especially following the proliferation of cast glass and the introduction of the blowing technique.

Window Glass (**Nos. 581–586**)

Fragments of window glass at Cosa were uncovered among the ruins of several buildings in the forum, in the soundings of the bath building, and from the excavated blocks of housing. This indicated that glazing was in general use at the site during the Imperial period. The relatively small number of examples (no more than fifty-eight panes), however, suggests that window glass was only sporadically employed and was probably reserved for specialized functions. This, at least, is the impression gained from a survey of glazing during the first century A.D. at Pompeii and Herculaneum. In both communities only a few of the major houses enjoyed glass windows, and among public buildings only bath complexes utilized the material. For examples, in the House of the Faun and the House of Cornelius Rufus at Pompeii certain narrow, slit windows opening into small cubicula were sealed with rectangular panes set in plaster,[54] and outside the Herculaneum gate the tepidarium of the bath of the Villa of Diomedes was outfitted with four square panes, each about 27 cm on a side, and framed in wood (Mau-Kelsey 1899, 351); cf. Nissen 1877, 596, where the window is said to measure 1.25 × 1.15 m).

In the public baths of the two towns glazing was a normal feature of the architecture. The circular windows in the semidomes of the vaulted ceilings of the *caldaria* in the Forum Baths of Pompeii and Herculaneum are both fitted with circular panes. At Herculaneum the pane is set in plaster; at Pompeii it is placed in a bronze frame with two pivots allowing the window to swivel open (Mau-Kelsey 1899, 198). More impressive is the use of glass in the Suburban Baths at Herculaneum, where several large rectangular windows facing the sea were fitted with multi-paned windows utilizing wooden frames and casements. Moreover, each window bay was provided with two such windows installed a few centimeters apart, in effect creating double glazing.[55] The reason for using windows with glass panes in bath buildings is obvious; they simultaneously retained heat while admitting light into otherwise dim interiors. This very point is made in the sixties A.D. by Seneca (*Ep*. 86.11), who remarked how the large glazed windows (*specularia*) of the baths of his day contrasted with the unlit rooms of such establishments in the Republican period.

Glass panes, when well preserved, are rarely large. Most appear to be square or rectangular, with sides averaging between 10 and 25 cm. An exceptionally large example, said to be from Italy and now in the British Museum, measures 30.5 × 53.5 cm, and it was trimmed from a larger sheet (Harden 1961, 49). Otherwise, the majority of panes found on archaeological sites tend to be small and were either plastered directly into the window space or mounted in wooden or metal frames or *transennae* (see discussion in Daremberg-Saglio 1877, s.v. *Fenestra*, 1039; also Harden 1961 and Haevernick and Hahn-Weinheimer 1955). The popularity of triangular panes or quarrels, such as **No. 585** at Cosa, also testifies to the existence of *transennae* with crossbars placed at various angles.

All the window glass at Cosa, as at most other sites, is of the matt/glossy variety. These panes were cast as sheets by ladle or crucible pouring on a form of wetted wood or possibly a form lined

[54] Information furnished by Dr. Anne Laidlaw, Hollins College.

[55] Unpublished but seen by the author.

with sand (consult Boon 1966).[56] The undersides of the panes are flat but have a characteristic "matt" appearance owing to contact with the casting surface. The upper surfaces are shiny but uneven. The outer edges, unless cut or glazed, are beveled and somewhat thicker than the rest of the sheet. Corners often exhibit tooling marks made by the craftsman as he pulled or pushed the hot glass into a roughly square shape (see, for examples, **Nos. 581**, **583–584**). On many examples one or more of the edges is cut with a knife or other tool during manufacture. For instance, at Cosa **Nos. 582** and **586** are cut at right angles to the beveled edge to achieve rectangular panes, and **No. 585** is cut on the diagonal to make a triangular quarrel.

For the most part Roman window glass is naturally colored, either bluish-green or light green, with a lesser amount in colorless fabrics. Rarely if ever is it truly transparent, and the customary thicknesses (.3–.6 cm), combined with numerous bubbles, further contribute to panes that are translucent but not clear. From this fact alone it follows that the Romans were not seriously interested in glazing windows in the modern manner. Instead, they chose to enclose only those windows where there was a practical need to retain heat (or exclude cold) while admitting daylight.

The date for the introduction of window glass coincides with the establishment of the Romano-Italian casting industry late in the first century B.C. and with the initial development of flat, sheet glass for a variety of purposes. Aside from window panes, such sheets (often in colored glass) were used in *opus sectile* paneling and veneers, in household screens and furnishings, and as the primary source for mosaic tesserae and other wall decoration.[57] According to Sear's 1977 study of Roman parietal mosaic, for example, glass tesserae, disks, rods, and other insets first occur in Italy late in the first century B.C. and become exceedingly popular in the first quarter and half of the next century (cf. pp 28–29 above). Actual finds of dated glass sheets and window panes from Magdalensberg (and elsewhere) confirm this chronological pattern. At Magdalensberg, for example, window glass has been discovered in levels attributed to early Augustan times (before A.D. 1) and to later Augustan, Tiberian, and Claudian levels (Magdalensberg 1979, 223–224, with a list of other sites also yielding panes from strata of the first half of the first century A.D.).

Another early, if unusual, pane was recovered at Rome during the excavations at the Regia in the Forum Romanum. There, a fragmentary pane was recovered from a drain at a point where it was crushed by the marble blocks used in the Calvinian reconstruction of the building in 37/36 B.C. (Grose 1977, 17–18, no. 12, figs. 1–2). The debris that accumulated in the drain primarily belonged to the last quarter of the first century B.C., with the latest material dating to the first decade of the next century. What is extraordinary about the pane is the fact that it is purple with opaque white streaks. Whether it originally was used in a window or as a panel in a grille or screen matters little because it still establishes that cast sheet glass (in all respects comparable to typical window glass) was in production by or about the turn of the millennium.

Parallels and Dated Examples of Window Glass

Magdalensberg 1979, 218–225, with a discussion and list of sites yielding dated examples of window glass, mostly from the Augustan through Julio-Claudian periods.

Luni II 1977, 289, CM 3396, pl. 157.14; CM 11226, pl. 157.15; CM 11268, pl. 157.16; CM 11245, pl. 157.17, with numerous unillustrated pieces.

[56] An alternative manufacturing method has long been advocated by Harden, who believes that most Roman window glass was made by cylinder-blowing; see Harden 1959, 8–16 and Harden 1961, esp. 9–49.

[57] For examples of colored sheet glass used as paneling or as inlays of various sorts, see Corning Coll. 1979, 224–254 and 263–264. Numerous comparable examples, including a column veneering in marbled mosaic glass (acc. no. 80.658), are found in the Toledo Museum of Art. Unpublished.

Ostia I 1968, 80, fig. 253, pl. 11. Colorless to bluish-green examples.
Harden 1961, 39–63. Full discussion of the topic.
Boon 1966, 41–47. Examples of panes from Roman sites in Wales and discussion of various possible techniques of manufacture.

Window Glass

581. CF 2502 Pl. 36
A.P. Room 25 Exterior, I
Five joining fragments of window pane with corner preserved; colorless with greenish tinge. Iridescence and pitting. Thick rounded edges on two sides of fragment; curved corner preserved; other sides broken. Upper surface uneven, underside flat; tooling mark at corner. Matt/glossy pane. Cast and tooled. Max L fragment 12.5 cm; A Th .3 cm.

582. CF 2503 Pl. 36
A.P. Room 25 Exterior, I
Fragment of window pane with corner preserved; colorless with greenish-yellow tinge. Iridescence and pitting. Thick rounded edge on long side of fragment; shortest side cut at a right angle with a tool; other sides broken. Upper side uneven, underside flat. Matt/glossy pane. Cast, tooled, and cut. Max L fragment 9 cm; A Th .3 cm.

583. C74.6 Pl. 36
VIII-F Sounding, 0
Fragment of window pane with corner preserved; colorless with greenish tinge. Iridescence, pitting, and decoloration. Thick, rounded edges on two sides of fragment; curved corner preserved; other sides broken. Upper side uneven, underside flat; three tooling marks at corner. Matt/glossy pane. Cast and tooled. Max L fragment 6.5 cm; A Th .4 cm.

584. CD 1194 Pl. 36
Bas. ENE, I
Fragment of window pane with corner preserved; colorless with greenish-yellow tinge. Iridescence and pitting. Thick rounded edges on two sides of fragment; curved corner preserved; other sides broken. Upper side uneven, underside flat; two tooling marks on upper side at corner. Matt/glossy pane. Cast and tooled. Max L fragment 6 cm; A Th .4 cm.

585. CG 588 Pl. 36
Forum East Corner, NE Passage I
Fragment of window pane with corner preserved; colorless. Iridescence and pitting. Thick rounded edge on longest side; shortest side broken; third side cut diagonally with a tool to form a triangular pane. Upper side uneven, underside flat. Matt/glossy pane. Cast, tooled, and cut. Max L fragment 6 cm; A Th .5 cm.

586. CD 1195 Pl. 36
Bas. ENE 1, I
Fragment of window pane with corner preserved; colorless. Iridescence and pitting. Thick rounded edge on left side of drawing; lower edge cut at a right angle with a tool; other sides broken. Upper side uneven, underside flat. Matt/glossy pane. Cast, tooled, and cut. Max L fragment 4 cm; A Th .55 cm.

Gaming Pieces (**Nos. 587–591**)

587. CF 432 Pl. 36
A.P. Room 22 Exterior, IIIA
Circular, disk-shaped gaming piece; opaque white. Milky weathering and pitting. Upper side convex, underside slightly concave. D 2.6 cm; Max Th .6 cm.

588. CF 986 Pl. 36
A.P. Room 25C, III Deposit IX
Late first through early third century A.D.
Circular, disk-shaped gaming piece; opaque white. Milky weathering and pitting. Upper side convex, underside flat. D 1.8 cm; Max Th .6 cm.

589. CD 899 Pl. 36
Bas. Nave, I
Circular, disk-shaped gaming piece; dark purple appearing opaque black. Iridescence and pitting. Upper side convex, underside flat. D 2.2 cm; Max Th .6 cm.

590. CB 1396
Capitolium Exterior North, 0
Circular, disk-shaped gaming piece; translucent blue. Iridescence and pitting. Upper side convex, underside flat. D 1.6 cm; Max Th .6 cm.

591. C68.571 Pl. 36
S.H. Room 6, I
Circular, disk-shaped gaming piece; golden-yellow. Iridescence and pitting. Upper side convex, underside flat. D 1.4 cm; Max Th .6 cm.

Rods and Tubes (**Nos. 592–594**)

592. CE 1969 Pl. 36
A.P. Room 11, I Deposit V
ca. 25/15 B.C. to ca. A.D. 40/45
Ring-shaped end of stirring rod, round in cross section; blue with spiraling thread of colorless glass. Iridescence and pitting. Vertical rod bent at one end to form circular ring. Drawn and tooled. Max D ring 2.6 cm; A Th rod .4 cm.

593. CE 1993 Pl. 36
A.P. Room 6, I
Small fragment of shaft of rod, round in cross section; pale light green. Iridescence and pitting. Rod tapers downward. Drawn rod, twisted clockwise. L fragment 3.3 cm; A Th .4 cm.

594. CF 2361
A.P. Room 25A, III Deposit IX
Late first through early third century A.D.

Small fragment of thin cylindrical tube, round in cross section and hollow; blue. Iridescence and pitting. L fragment 3.3 cm; D .3 cm.

Tesserae (***Nos. 595–596***)

595. CE 2066 Pl. 36
A.P. Room 14, I Deposit V
ca. 25/15 B.C. to ca. A.D. 40/45

Tessera, roughly square in cross section; dark blue. Iridescence and pitting. Dimensions .8 × .8 × .8 cm.

596. CF 2327 Pl. 36
A.P. Room 25C, III Deposit IX
Late first through early third century A.D.

Tessera, roughly rectangular in cross section; opaque turquoise blue. Iridescence and pitting. Dimensions .7 × .5 × .5 cm.

MEDIEVAL GLASS VESSELS

Cosa enjoyed sporadic medieval occupation. As stated in the first chapter, that occupation came to an abrupt end in August 1329 when a military expedition from Siena attacked and seized the site. Ironically, despite decades of medieval inhabitation, the excavation of the arx unearthed no recognizable medieval glass—with the possible exception of a lamp (**No. 577**) discussed earlier. Instead, it is in and around the ancient forum at Cosa, chiefly among the ruins of the Curia and neighboring jail, that medieval coins, ceramics, and glassware have been discovered, mostly as ground scatter. Without question, the glass belongs to the second medieval occupation by virtue of the types involved and the associated numismatic evidence (for the latter, consult Buttrey 1980, 56, nos. 343–344, 373, and 375). It would appear that the residents utilized the standing walls of the ancient structures, as well as the cisterns beneath them, as temporary or permanent shelters.

The glass consists of a mere twelve pieces, all small fragments, of which seven belong to four or more prunted beakers (**Nos. 597–603**), three to ornate chalices with twisted stems (**Nos. 604–606**), and the last two (**Nos. 607–608**) to beakers or bowls of uncertain shapes, but with kicked bottoms. Of these, the prunted beakers are now known to have been one of the most common and widespread types of drinking vessels in the high medieval period (Whitehouse 1983, 115–116). Although they were once considered to be fairly unusual, examples in fact have been recovered from virtually every inhabited community of this period. Precise shapes vary, but most examples are simple, thinly blown cylindrical beakers with upright or outsplayed rims and deeply kicked bottoms. In most instances the glass is bubbly and lightly colored, usually a pale yellow, as is true of the pieces at Cosa. Decoration takes the form of prunts of the same fabric, sometimes applied randomly but unusually in a checkerboard layout (**see No. 599**). Quite often, a horizontal thread is also wound around the bottom, affording a base-ring of sorts (again, **No. 599**), as well as around the junction of the side and rim (**Nos. 598** and **600**). Many examples also have a ring of projecting prunts at the bottom forming a "toed" base-ring (see Whitehouse above).

Among the earliest dated examples of prunted beakers are those of the first half of the twelfth century discovered at Corinth, where the factories that made them have also been excavated (Corinth

1952, esp. 87–88 and 114). Since Corinth was sacked in 1147 by the Normans and the artisans of the city forcibly removed to Sicily, G. D. Weinberg ingeniously has argued (based on the literary evidence) that the style was thereby transmitted to Italy from the Near East. Although this proposition has yet to be confirmed, it is clear from the other twelfth/thirteenth-century sites in the Aegean, such as Pergamon (see list below), that the style was popular throughout the Byzantine lands of the eastern Mediterranean before it became the standard fare of glassmakers in Italy and Sicily, as well as in Dalmatia and the Balkans, from the thirteenth to the fifteenth centuries (evidence summarized in Whitehouse 1983, 116). Thereafter, it passed into the repertoire of the transalpine glasshouses and inspired the creation of the typical prunted roemer of the fifteenth and sixteenth centuries.

Parallels and Dated Examples of Prunted Beakers

Cosa: **Nos. 597–600**, dating before the termination of the medieval occupation in August 1329.

Arezzo: Museo Civico, inv. 309. Rim fragment, presumably from the town or neighboring region. Unpublished.

Paestum: Museo Archeologico. Several fragments from the sanctuary of Bona Dea, University of Michigan–Perugia University excavations. Unpublished.

Aidone (Sicily): Antiquarium. An almost intact prunted beaker discovered during the course of a building project in the village with other fragmentary glass vessels and ceramics of the thirteenth century.

In addition to prunted beakers, stemmed chalices also appear to have been a relatively common tableware of the high medieval period. Stems comparable to **Nos. 604–606** at Cosa have also been recovered at Luni and at Rome. At Luni a fragmentary chalice was found in an area of the ancient city where a house of the "high medieval period" once stood. Although the vessel is catalogued with the ancient glass, it definitely belongs to the medieval period (Luni II 1977, 556, K 781, pl. 286.15). Just as the Luni fragment preserved part of the lower body of the vessel, so does the chalice in the Vatican Collection (Vatican Coll. 1975, 107, no. 971, pl. 63). Although it lacks a find-spot, it must have come from Rome or its environs. Like the examples from Luni, it too is decorated with an elongated zigzag pattern of opaque white threads encircling the lowest portion of the bowl of the vessel.

The final two medieval fragments (**Nos. 607–609**) represent the bottoms of beakers (or possibly bowls) with high kicks so typical of the age. Both are undecorated and made in the pale, naturally colored fabrics preferred in this period. Although neither can be assigned to complete forms, like the rest of the medieval glass at Cosa, they must predate the Sienese expedition of A.D. 1329.

*Prunted Beakers (**Nos. 597–603**)*

597. CG 389 Pl. 37
Curia/Comitium, 0

Two non-joining fragments of side of prunted beaker; colorless (?) with greenish-yellow tinge. Iridescence and pitting. Almost vertical, straight side, tapering downward. On exterior, nipple-shaped prunts in the same fabric as the body. Blown with applied decoration. Max D fragment 7.5 cm; A Th side .1 cm.

598. CG 59 Pl. 37
Curia/Comitium, I

Two joining fragments of side, probably near rim, of prunted beaker; pale yellow. Iridescence and pitting. Almost vertical, straight side, tapering downward. On exterior, small prunts and a horizontal thread in the same fabric as the body. Blown with applied decoration. Max D fragment ca. 5.5 cm; A Th side .1 cm.

599. CG 58 Pl. 37
Curia/Comitium, Forecourt NW, 0
Fragment of side and bottom of prunted beaker; pale yellow. Iridescence and pitting. Almost vertical, straight side, tapering downward; indication of a concave, kicked bottom. On exterior, small prunts applied in a checkerboard pattern; also a horizontal thread at junction of side and bottom. Blown with applied decoration. D bottom ca. 6 cm; A Th side .1 cm.

600. CE 292 Pl. 37
VIII-D, Building constructed into A.P. II, Square 59, II
Two small, non-joining fragments of side of prunted beaker; pale yellow. Iridescence and pitting. Straight side, tapering diagonally downward. On exterior, nipple-shaped prunts and a horizontal thread in the same fabric as the body. Blown with applied decoration. A Th side .1 cm. Only one fragment illustrated on pl. 37.

601. CE 2065
VIII-D, Building constructed into A.P. II, Square 59, I
Small fragment of side of prunted beaker; pale yellow. Iridescence and pitting. Straight side, sloping diagonally downward. On exterior, a nipple-shaped prunt in the same fabric as the body. Blown with applied decoration. Possibly from the same vessel as **No. 600**. A Th side .1 cm.

602. C68.701A
Forum ENW, 0
Tiny fragment of side of prunted beaker; colorless (?) with greenish-yellow tinge. Iridescence and pitting. On exterior, a prunt in the same fabric as the body. Blown with applied decoration. A Th side .1 cm.

603. C68.701B
Forum ENW, 0
Tiny fragment of side of prunted beaker; pale yellow. Iridescence and pitting. Straight side, tapering downward. On exterior, a prunt in the same fabric as the body. Blown with applied decoration. A Th side .1 cm.

*Chalices (**Nos. 604–606**)*

604. CG 383 Pl. 37
Curia/Comitium, SE Room, I
Two joining fragments of stem and base of chalice; pale green, possibly incompletely decolorized. Iridescence and pitting. Tall twisted solid stem with three projecting horizontal rings applied on the upper end; an almost flat base with a solid, downturned edge flares outward from stem; underside of base concave. Blown and tooled. H stem 10.5 cm; D base 5 cm; A Th base .15 cm.

605. C67.545 Pl. 37
Forum Extension NW, 0
Fragment of upper end of stem of chalice; light green. Iridescence and pitting. Tall solid stem, partially preserved, with three projecting horizontal rings applied at the upper end. Blown and tooled. H fragment 2.9 cm.

606. The description for **No. 606** is missing.

*Bases (**Nos. 607–608**)*

607. CG 57 Pl. 37
Curia/Comitium, NW Room, 0
Three joining fragments of bottom of beaker or other vessel; natural light green. Iridescence and pitting. Deeply kicked bottom. Blown and tooled. D bottom ca. 6 cm; A Th side .1 cm.

608. CF 2504 Pl. 37
Temple B (Forum), Forecourt Exterior SW, I
Fragment of side and bottom of beaker or other vessel; colorless. Iridescence and pitting, some brown discoloration. Nearly vertical, straight side; deeply kicked bottom; pontil wad on underside at center. Blown. D bottom ca. 4 cm; A Th side .1 cm.

Appendix

SAGGED OVERSIZED COLORLESS BOWLS (ROMAN REPUBLICAN OR AUGUSTAN) (**NOS. 609–613**)

During the course of the excavations in 1970–1971, chiefly in the domestic quarters of the town, five highly unusual rim fragments were discovered. Four are colorless with a slight yellowish-green tinge; the fifth, found near the forum in an earlier season, is greenish-yellow. All five are cast, probably using a sagging technique, then rotary-polished on both surfaces of the body and on the top, but not necessarily on the underside of the rim. Usually, the undersurface of the rim is noticeably uneven, with numerous tooling marks and creases caused by the folding of the glass. In fact, the rounded, irregular edge of the rim has the same appearance as the edges of Roman window glass (here, **Nos. 581–586**).

In size and shape, the fragments belong to one of two basic forms: either a broad, hemispherical bowl with a wide, nearly horizontal rim (**Nos. 609–610**) or a deeper hemispherical bowl with an outsplayed rim (**Nos. 611–613**). Presumably, both shapes originally had convex or possibly flat bottoms rather than base-rings.[58] Diameters are impressively large; the shallow bowls average 40–42 cm, the others are around 30 cm. Most fragments are so heavily weathered or corroded that the polishing on the surface and the good quality of glass beneath are obscured.

Although the five pieces so closely resemble one another that they must form a distinct type or group of table wares, published comparanda are few, with the exception of forty-five examples found at Magdalensberg 1979, 91–93, nos. 683–727, pls. 4–5.[59] There, as at Cosa, the pieces are too fragmentary to determine complete shapes, but they also exhibit massive diameters and fall into the same two basic forms. Twenty-eight of the examples are colorless-yellowish, eleven yellow-green, and six blue-green. At Magdalensberg the fragments were located in early Augustan (until ca. A.D. 1), Augustan (until ca. A.D. 15), and early Tiberian (until ca. A.D. 25) contexts. Unfortunately, at Cosa no example comes from a sealed deposit, but four were recovered from an area of excavated houses destroyed ca. 72/70 B.C. (but in some cases reoccupied ca. 25/15 B.C.) or from a temple courtyard fronting the forum, an area actively used until ca. 72/70 B.C. but also disturbed in early Imperial times.

The only other published parallel is the fragment of a deep hemispherical bowl discovered in Campania at the villa site of Francolise (Cotton 1979, 78, no. 4, fig. 16). The bowl, which is greenish in color, appears to be similar to **Nos. 611–613** at Cosa, but its description states that it is blown (despite the presence of polishing marks). Other presumed comparanda, all unpublished but cited in Magdalensberg 1979, 92, are said to be from Pompeii, Rome, and Bolsena, but these have not been seen by this author.[60]

In overall appearance these vessels most closely resemble the glassware of the Hellenistic period rather than the brightly colored polychrome and monochrome Romano-Italian wares of

[58] Czurda-Ruth in Magdalensberg 1979, 91, likewise argues that such bowls lacked base-rings.

[59] Two additional deep hemispherical bowls are described in Conimbriga 1976, 170, nos. 76 and 76A, pl. 37. Alarcão, however, catalogues them with the blown glass and does not mention the presence of polishing marks. Consequently, it is unclear whether they are parallels to the cast vessels under discussion. Also, no. 76A was recovered from a Flavian level, in sharp contrast to the Republican or Augustan examples from Cosa and Magdalensberg.

[60] A search for the examples said to be located in the Forum Antiquarium at Rome and the Museo Nazionale in Naples proved fruitless.

the Augustan and Julio-Claudian epochs. The nearest Hellenistic parallel is a peculiar vessel with a square horizontal rim surrounding a circular, shallow hemispherical bowl (Haevernick 1981, 294–298). There are three other examples of such "square" bowls, one from Crete, the other two from Magna Graecia, where one was found with a hoard of glass of the Canosa type and presumably dates to the second century B.C. (Bartoccini 1935, 225–262, pl. 1; Grose 1981, 63; and Grose 1983, 44). Like the bowls described here, these "square" vessels have thick walls, colorless fabrics, and rims resembling the edges of Roman window glass. They also appear to have been sagged and tooled over a former mold. Nonetheless, in shape and overall size these bowls differ considerably from those under discussion, as do the alleged dates.

Given the present state of information, little more can be concluded than that the large vessels at Cosa and Magdalensberg seem to be a phenomenon of the late Republican or Augustan periods and are geographically restricted to Italy and adjacent areas. These facts in turn suggest that the bowls were the products of an Italian workshop, probably operating in the first century B.C. However, given their curious resemblance to at least one category of Hellenistic glass, it is difficult to determine whether the bowls in question were the products of the same Romano-Italian factories specializing in the colored and colorless table wares described elsewhere in this catalogue or may have been the products of a workshop that stood in a direct relationship to the earlier traditions of Hellenistic glassmaking. Whatever their source, their uniformity and commanding size indicate that they were used for a specialized function that is no longer readily apparent.

Sagged Oversized Colorless Bowls

609. C70.620A Pl. 37
VI D, E (Garden area behind SUNY House), 0
Large fragment of rim and side of broad, shallow bowl or colossal plate; colorless with slight yellowish-green tinge, especially visible along fractures. Iridescence, some discoloration, and much pitting. Circular pockets of corrosion cut through the entire thickness of glass. Broad horizontal rim, sloping slightly obliquely to the inside; rim distinctly thicker near edge. Upper surface of rim even; underside and edge of rim noticeably uneven. Along edge of rim, a discontinuous tooled line, possibly owing to use of a knife or other tool to cut out as even an edge as possible. Convex curving side; thickness tapering downward to bottom. Cast, rotary-polished on both surfaces of side, but especially on the interior; some rotary-polishing on the underside of the rim near its junction with the side; numerous tooling marks on the underside of the rim. D ca. 42; W rim 3.5 cm; Max Th edge rim .55 cm; Min Th rim .3 cm; Max Th side .3 cm; Min Th side .1 cm.

610. C70.620B
VI D, E (Garden area behind SUNY House), 0
Fragment of rim and side of broad, shallow bowl or colossal plate; colorless with slight yellowish-green tinge, especially visible along fractures. Iridescence, much pitting, and some corrosion. Broad horizontal rim, sloping inward, and thicker at edge. Upper surface of rim even; underside and edge of rim uneven; convex curving side. Cast, rotary-polished on both surfaces of sides; tooling marks on underside of rim. D uncertain, but presumably as large as **No. 609**; W rim 4.7 cm; Max Th edge of rim .5 cm; Min Th rim .3 cm; A Th side .35 cm.

611. C70.619 Pl. 37
S.H. Room 21, 0
Fragment of rim and side of large, thick-walled bowl; colorless with distinct yellowish-green tinge. Iridescence, much pitting, and discoloration. Some corrosion. Horizontal rim with rounded edge, sloping obliquely to the outside. Upper surface of rim even, underside creased and uneven, especially near edge. Nearly straight side, tapering downward; thickness of side also tapers downward. Cast, rotary-polished on both surfaces of sides; distinct tooling marks on underside of rim, especially near the edge. D ca. 30 cm; Max Th rim .55 cm; Max Th side .5 cm; Min Th side .3 cm.

612. CF 2505 Pl. 37
Temple B (Forum), Courtyard, 0
Two joining fragments of rim and side of large, thick-walled bowl; colorless with distinct yellowish-green tinge. Iridescence and pitting. Fairly broad, horizontal rim with rounded edge, sloping obliquely outward. Upper surface of rim even; underside slightly creased and uneven, especially near rim. Nearly straight side, tapering downward; thickness of side also tapers downward. Cast, rotary-polished on both surfaces on sides and on upper side of rim at junction with side; distinct tooling marks on underside of rim, especially near the edge. Along the edge of the rim, a continuous tooling mark, probably owing to use of a knife or other tool to form as even an edge as possible. D ca. 30 cm; Max Th rim .5 cm; Max Th side .5 cm; Min Th side .4 cm.

613. C71.234 Pl. 37
H.T. (House of Quintus Fulvius), Room 9, 0
Small fragment of rim and side of large, thick-walled bowl; greenish-yellow, possibly incompletely decolorized. Iridescence and pitting. Black scaly weathering. Horizontal rim with rounded edge, sloping obliquely to the outside. Upper side of rim even owing to rotary-polishing; underside slightly creased but lacks definite tooling marks. Nearly straight side, tapering downward. Cast, rotary-polished on sides and on the upper surface of rim; underside of rim rotary-polished near junction with side. D ca. 20–22 cm; A Th rim and side .5 cm.

Concordance 1

Catalogue Number/Inventory Number/Plate Number

Cat. No.	Inv. No.	Pl. No.
1	CD 963, CD 1019	1
2	CF 2374	1
3	C69.382	1
4	C66.801	1
5	CE 2016	1
6	CE 2182A	1
7	CE 2183A	
8	CE 2109	
9	CE 840	
10	CE2110	
11	C69.121	1
12	C68.519	1
13	CE 2013	1
14	CE 2012A	
14a	CE 2012B	
15	C71.195A and B	1
16	C71.201	1
17	C69.381A and B	1
18	C68.271	1
19	C70.570	1
20	C69.155	2
21	CE 2039	2
22	C68.505	2
23	CE 2111	2
24	C69.233	
25	C69.322	
26	C68.657	
27	C68.673	
28	C68.674	
29	C71.193	2
30	CE 2015	2
31	C69.403	2
32	CE 2014	2
33	CD 2079	2
34	CD 1078	2
35	C66.800	2
36	CD 257	
37	C66.294	
38	C72.100	
39	CD 345	2
40	C66.802	
41	C70.569	2
42	C71.200	
43	CE 2117	
43a	C73.71	
44	C70.571	
45	CD 782	3
46	C66.295	3
47	C72.89	3
48	C69.405	3
49	C68.678	3
50	C69.406	3
51	C68.676A and B	3
52	CF 2409	
53	C68.675	3
54	C70.574	3
55	CD 1112	
56	C71.176	3
57	CE 2070	3
58	C66.807	3
59	C70.578	
60	C73.72	4
61	C68.677	4
62	C68.456	4
63	C72.41	4
64	C66.804	4
65	CE 836	
66a	CF 2408	
66b	C69.404	
66c	C70.576	
66d	CE 2180	
66e	CE 2179	
66f	CG 376	
67	CD 1092	4
68	C70.579	4
69	C70.575	4
70	CE 2047	5
71	C72.126	5
72	CD 1091	5
73	CF 2410	
74	C66.805	5
75	C66.808	5
76	CD 1093	5
77	CD 1094	5
78	CE 2119	
79	CD 1113	
80	CD 1109	5
81	C72.96	5
82	CE 2006	5
83	CE 2113	
84	CE 2121	
85	CE 2155	5
86	CD 1110	
87	CF 2335	
88	CE 2120	
89	CD 1111	

Cat. No.	*Inv. No.*	*Pl. No.*
90a	C68.695	
90b	C69.696	
90c	CE 2118	
90d	C65.675A	
90e	CE 2177	
90f	CE 2182	
91	CE 2058	
92	C60.640	6
93	CF 2421	6
94	C71.198	6
95	C68.684	6
96	CD 1119	6
97	CD 1121	6
98	C69.412	6
99	C69.434	
100	CE 2135	6
101	CE 2137	6
102	C68.682	
103	CD 1123	
104	CE 2018	6
105	C71.206	
106	C70.538	7
107	CE 2042	7
108	C69.416	7
109	C69.410	7
110	C68.685	7
111	CE 2139	7
112	C66.817	7
113	CD 1125	7
114	C69.414	7
115	C69.417	7
116	C70.589	
117	C71.204	
118	CE 2020	7
119	CD 1148A	7
120a	C71.203	
120b	C70.587	
120c	C68.681	
120d	C69.413	
120e	C70.586	
120f	C69.411	

Cat. No.	*Inv. No.*	*Pl. No.*
120g	C72.127	
120h	C66.843	
120i	C69.437	
120j	C68.698	
120k	C69.436	
120l	C68.699	
120m	C70.588	
120n	C68.697	
120o	C70.629	
120p	C69.438	
121	CE 2138	8
122	CD 1120	8
123	CD 1122, CD 1146	8
124	C72.94	8
125	CE 2115	8
126	CE 1973	8
127	CD 1102	8
128	CE 2071	8
129	CD 1114	8
130	CD 1145	
131	CD 1115	
132	CF 2438	
133	C69.407	
134	CE 1964	8
135	C68.679, C68.683	8
136	CE 2140	8
137	CE 2001	8
138	CF 2411	8
139	CE 1963	9
140	CF 2303	9
141	C70.585	9
142	C70.583	9
143	CD 1116	
144	CD 1101	9
145	C68.638	9
146	CD 1108	9
147	C68.375	9
148	CD 1076	9
149	C70.228	9
150	C73.73	
151	C67.553	

Cat. No.	*Inv. No.*	*Pl. No.*
152	CD 1077	9
153	CD 1107	9
154	CE 1216	9
155	CF 2439	9
156	CE 1996	9
157	CD 251	9
158	C71.202	
159	C69.680	
160a	CG 449	
160b	CF 2439A	
160c	CF 2407	
161	CE 2182	10
162	C71.197	10
163	CD 1144	10
164	CD 1105	10
165	CF 2454	10
166	C73.74	
167	CF 2460	
168a	CE 2124	
168b	CE 2127	
168c	CF 2461	
168d	CF 2453	
168e	CG 579	
169	CF 2311	10
170	CF 2394	10
171	CF 2459	
172	CF 2393	10
173	CF 2455	
174	C74.4	
175	CE 2129	10
176	C74.3	10
177	CF 2440	10
178	CE 2125	
179	CF 2418	10
180	CD 1103	
181	CF 2416	10
182	CF 2310	11
183	CF 2324	11
184	CF 2462	
185	CE 2132	
186	CD 2441	

Cat. No.	Inv. No.	Pl. No.
187	CD 1104	11
188	C69.167	11
189	C69.429	11
190	C70.618	11
191	CF 2456	
192	CF 2457	
193	C74.2	11
194	CE 2126	11
195	C68.694	11
196	CF 2445	11
197	C70.628	
198	CF 2420	11
199	CE 2205	
200	CD 1163	11
201	C74.5	
202	CF 2458	
203	CE 2130	11
204	CE 2183	11
205	C70.342	12
206	C67.554	12
207	CE 2035	12
208	C68.664	12
209	CD 1084	12
210	C66.810	12
211	CF 2431	12
212	CF 2400	23
213	CF 2334	12
214	C66.803	12
215	C70.572	12
216	CE 2027	
217	CD 1083	12
218	CF 2430	
219a	C70.630	
219b	CE 2313	
219c	CE 1984	
219d	CE 2141	
220	CF 2450	12
221	CE 2076	13
222	CD 1080	13
223	C66.809	
224	CF 2401	

Cat. No.	Inv. No.	Pl. No.
225	C65.674	13
226	CD 1081	13
227	CD 1118	13
228	CD 1085	13
229	CD 1127	
230	CG 55	13
231	CF 2307	13
232	CD 1074	13
233	CD 1073	13
234	CD 1164	13
235	CF 2345	13
236	CG 234	14
237	CF 2351	13
238	CF 2434	14
239	CF 2318	14
240	CF 2452	13
241	CF 2375	14
242	CG 447	14
243	CD 833	14
244	CE 1994	14
245	C68.661	14
246	C73.69	14
247	CG 446	14
248	CF 2325	14
249	C70.302	15
250	CD 1095	15
251	CE 2116	
252	C71.196	
253	C70.627	
254	CD 1088	15
255	CD 1089	
256	C71.194	15
257	CE 2048	15
258	C70.577	
259	CE 2008	15
260	C70.573	15
261	C70.626	
262	CF 2446	15
263	C68.662	15
264	CE 2064	15
265a	CF 2444	15

Cat. No.	Inv. No.	Pl. No.
265b	CF 2331	
265c	CE 2172	
266	C72.90	
267	CE 291	16
268	CD 1099	16
269	CF 2363	15
270	CD 1090	15
271	CE 2104	15
272	CF 2388	15
273	CE 2024	15
274	CG 540	16
275a	CD 774	16
275b	CG 581	
275c	CG 56	
275d	C69.439	
276	CF 2396	16
277	CD 1130	16
278	CF 2437	16
279	C70.623	16
280	CF 2463	16
281	CF 2464	16
282	CE 2142	16
283	CF 2465	16
284	CF 2353	16
285	CF 2395	16
286	CF 2305	16
287	CF 2428	16
288	CF 2427	16
289	C72.95	17
290	CE 2144	17
291	CE 2206	
292	CF 2426	17
293	CF 2309	17
294	CD 1131	17
295	CF 2425	17
296	CF 2360	
297	CE 2105	17
298	CF 2364	
299	CF 2308	17
300	CF 2429	
301	CD 1075	17

Cat. No.	*Inv. No.*	*Pl. No.*
302	CD 1143	
303	CF 2442	
304	C69.426	17
305	CE 2025	17
306	C70.611	17
307	CF 2432A	18
307a	CF 2432B	18
308	CE 1954	18
308a	CE 2145	18
309	CE 2146, CE 1958	18
310	CE 1965	18
311a	CE 1991	
311b	CE 2147	
311c	CE 1955	
311d	CE 1956	
311e	CE 1957	
312	CD 1136	
313a	CD 1135	18
313b	CD 1154	
313c	C70.631	
313d	C68.710	
313e	C66.872	
313f	C69.459	
313g	CE 2171B	
314	CD 1155	18
314a	CE 2122	
314b	CE 2133	
315	CE 1992	18
316	CE 2045	18
317	CE 1959	18
318	CE 2036A and B	18
319a	CE 1987	
319b	CE 2173	
320	CE 1976	19
321	C66.870	19
322	C72.56	19
323	CE 2148	19
324a	C68.712	19
324b	CF 2449	
324c	C69.457	
324d	C69.458	
324e	C66.871	
325	CE 1971	19
326	CF 2412	19
327	C70.613	19
328	CE 1970	19
329	C66.815	19
330	C66.812	19
331	C69.408	19
332	C72.135	
333	CE 1988	19
334	C66.811	19
335a	CD 1168	
335b	C70.584	
335c	C69.452	
336	CE 1167	19
337	CE 2171	19
338	CD 1166	19
339	C70.660	
340	C70.612	20
341	C72.44	20
342	CE 2041	20
343	CD 1156	20
344	CD 1157	20
345	CE 2185	20
346	C69.415	20
347	CD 1158	20
348	CD 1159	20
349	C70.632	20
350	CE 2019	20
351	CE 2149	20
352	CE 2150	
353	CE 1961	21
354	CE 2043	21
355	CE 1972	21
356	CE 2107	21
357	CD 1161	
358	CE 1966	21
359	CG 375	21
360	CE 2063	21
361	C69.409	21
362	C70.303	22
363	CE 2108	22
364	C68.600	22
365	CD 72	22
366	C71.207	22
367	CE 2189	22
368	CE 2188	
369	CE 2187	
370	CE 1139	22
371	CE 2201	
372	CE 2017	22
373	CE 1962	21
374	CE 2154	22
374a	CE 2151B	22
375	CE 2153C	22
375a	CE 2049	22
376	CE 2153A	
376a	CE 2151A	
377	CE 2153B	
378	CE 2151C	
379	CE 2033	22
380a	CE 2153D	
380b	CE 2153E	22
380c	CE 2153F	
380d	CE 2053	
380e	CE 2153G	22
380f	CE 2151E	22
380g	CE 2153I	22
380h	CE 2151J	
380i	CE 2151D	23
380j	CE 2151H	
380k	CE 2151F	23
380l	CE 2153H	23
380m	CE 2151G	23
380n	CE 2151I	
381	CE 2034	22
382	CE 2195A-I	
383	CD 1141	23
384	CD 1153A and B	23
385	CE 2003A	23
386	CE 2003B	23
387	C70.607	23

Cat. No.	Inv. No.	Pl. No.
388	C66.822	23
389	C66.881	23
390	C70.662	23
391	CF 2351A	23
392	CD 1182	24
393	CE 289	24
394	C66.878	24
395	CF 2473	24
396	CD 1184	24
397	CD 1183	24
398	C66.876	24
399	C72.140	24
400	C66.837	24
401	C66.777	24
402	CD 1181	24
403	CF 2449B	25
404	C72.138	25
405	CF 2390	25
406	CF 2467	25
407	CE 2209	25
408	CD 1175	25
409	CF 2468A	25
410	CF 2468B	25
411	CF2468C	25
412	CF 2358	25
413	C72.137B	25
414	CD 1176	26
415	CF 2471	26
416	CD 1177	26
417	CF 2469	26
418	CF 2470	26
419	CF 2450A	26
419a	CF 2350	26
420	CF 2496	26
421	CF 2507	
422	CF 2508	
423	C72.93	23
423a	C66.879	
423b	C66.880	
424	CE 2062	27
425	CE 2004	27

Cat. No.	Inv. No.	Pl. No.
426	CE 2133	
427	C69.424	27
428	CD 1169	27
429	CE 837	27
430	CE 2136	27
431	CD 1174A	27
432	CE 2008B	27
433	C68.495	27
445	CE 2164	28
446	CE 1188	28
447	CE 2054	28
447a	CE 2161	
448	C68.688	28
449	CD 1170	28
450	CE 2163	28
451	C70.568	28
452	CE 2169	28
453	CF 2356	28
454	CE 2056	29
455	CD 1106	29
456	C72.42	29
457	C66.823	29
458	C65.803	28
459	CD 1173	29
460	C70.582	29
461	C68.687	29
462	C66.875	29
463	CE 1990	29
464	C70.605	29
465	C70.581	29
466	C70.580	29
467	C66.828	29
468	CD 1100	30
469	C68.659	30
470	CF 2497A	30
470a	CF 2497B	
470b	CF 2497C	
470c	CF 2497D	
471	C69.419	30
472	CF 2339	30
473	CF 2498	30

Cat. No.	Inv. No.	Pl. No.
474	CD 1128	30
475	C70.593	30
476	CD 1087	30
477	C69.454	30
478	CG 577	30
479	CE 1997	30
480	C69.400A	30
481	C69.400B	30
482	C72.43	30
483	CB 1149	
484	C69.402	30
485	CD 308	30
486	C69.400C	30
487	C68.657A	30
488	CG 386	30
489	CG 382	30
490	CD 1162A	31
491	CD 1162B	31
492	CD 1190	31
493	CD 1189	31
494	CF 2373	31
495	CF 204	31
496	CE 2002	31
497	CF 2369	31
498	C67.72	31
499	CF 2367	31
500	CD 1096	32
501	CD 1191	32
502	CF 2342	32
503	CF 2328	32
504	CF 2377	32
505	C69.432	32
506	CD 1151	32
507	CD 1150	32
508	C69.453B	32
509	CE 2046	32
510	CE 2156	
511	C68.736	32
512	CE 2194	32
513	C69.455	32
514	CE 2193	32

Cat. No.	*Inv. No.*	*Pl. No.*
515	C66.839	32
516	CD 1138	32
517	CE 2030	32
518	CD 1192	32
519	CD 1097	33
520	C70.592	33
521	CF 2403	33
522	CA 665	33
523	CF 2402	33
524	C70.621	33
525	C70.663	33
526	CG 586	33
527	C70.680	33
528	CD 307	33
529	CG 587	33
530	CD 1179	33
531	C70.615	33
532	CD 1178	33
533	C70.661A	33
534	CE 2037	33
535	CE 2158	33
536	CF 2501	33
537	CE 2050	33
538	CE 2160A and B	33
539	C70.308	34
540	CD 795	34
541	CF 2302	34
542	CG 387	34
543	CE 2026	34
544	CE 2210	34
545	CE 2159C	34
546	CE 2159A	34
547	CE 2159B	
548	CF 2321	34
549	CE 2192	35
550	CE 2052	35

Cat. No.	*Inv. No.*	*Pl. No.*
551	CE 2038	
552	C69.421	35
553	C65.802	35
554	C68.282	35
555	CE 2011	35
556	C70.601	35
557	CD 948	35
558	CE 1968	35
559	CE 1967	35
559a	CE 2032	
560	C71.59	35
561	CD 1152A	35
562	CE 2157A	35
563	CE 2157B	35
564	CE 2157C	35
565	CE 2191A	35
566	C69.308	35
567	C70.624	35
568	C70.603	35
569	C70.602	35
570	CG 373	35
571	CA 680	35
572	CD 1193	35
573	C70.600	35
574	CG 439	35
575	CF 2405	36
576	CF 2417	36
577	CD 410	36
578	C72.47	36
579	C66.545	36
580	CE 1998	36
581	CF 2502	36

Cat. No.	*Inv. No.*	*Pl. No.*
582	CF 2503	36
583	C74.6	36
584	CD 1194	36
585	CG 588	36
586	CD 1195	36
587	CF 432	36
588	CF 986	36
589	CD 899	36
590	CB 1396	
591	C68.571	36
592	CE 1969	36
593	CE 1993	36
594	CF 2361	
595	CE 2066	36
596	CF 2327	36
597	CG 389	37
598	CG 59	37
599	CG 58	37
600	CE 292	37
601	CE 2065	
602	C68.701A	
603	C68.701B	
604	CG 383	37
605	C67.545	37
606	CG 436	
607	CG 57	37
608	CF 2504	37
609	C70.620A	37
610	C70.620B	
611	C70.619	37
612	CF 2505	37
613	C71.234	37

Total number of catalogue numbers	613
Total number of catalogued pieces (and listed entries)	709
Total number of inventory numbers	713
(4 pieces have two inventory numbers)	

Concordance 2

Inventory Number/Catalogue Number/Plate Number

Inv. No.	*Cat. No.*	*Pl. No.*
CA 665	522	33
CA 680	571	35
CB 1149	483	
CB 1396	590	
CD 1073	233	13
CD 1074	232	13
CD 1075	301	17
CD 1076	148	9
CD 1077	152	9
CD 1078	34	2
CD 1080	222	13
CD 1081	226	13
CD 1083	217	12
CD 1084	209	12
CD 1085	228	13
CD 1087	476	30
CD 1088	254	15
CD 1089	255	
CD 1090	270	15
CD 1091	72	5
CD 1092	67	4
CD 1093	76	5
CD 1094	77	5
CD 1095	250	15
CD 1096	500	32
CD 1097	519	33
CD 1099	268	16
CD 1100	468	30
CD 1101	144	9
CD 1102	127	8
CD 1103	180	
CD 1104	187	11
CD 1105	164	10
CD 1106	455	29
CD 1107	153	9
CD 1108	146	9
CD 1109	80	5
CD 1110	86	
CD 1111	89	
CD 1112	55	
CD 1113	79	
CD 1114	129	8
CD 1115	131	
CD 1116	143	
CD 1118	227	13
CD 1119	96	6
CD 1120	122	8
CD 1121	97	6
CD 1122, CD 1146	123	8
CD 1123	103	
CD 1125	113	7
CD 1127	229	
CD 1128	474	30
CD 1130	277	16
CD 1131	294	17
CD 1135	313a	18
CD 1136	312	
CD 1138	516	32
CD 1141	383	23
CD 1143	302	
CD 1144	163	10
CD 1145	130	
CD 1148A	119	7
CD 1150	507	32
CD 1151	506	32
CD 1152A	561	35
CD 1153A and B	384	23
CD 1154	313b	
CD 1155	314	18
CD 1156	343	20
CD 1157	344	20
CD 1158	347	20
CD 1159	348	20
CD 1161	357	
CD 1162A	490	31
CD 1162B	491	31
CD 1163	200	11
CD 1164	234	13
CD 1166	338	19
CD 1168	335a	
CD 1169	428	27
CD 1170	449	28
CD 1171	435	27
CD 1172A	443	28
CD 1172B	441	28
CD 1173	459	29
CD 1174A	431	27
CD 1174B	439	28
CD 1175	408	25
CD 1176	414	26
CD 1177	416	26
CD 1178	532	33
CD 1179	530	33
CD 1181	402	24
CD 1182	392	24
CD 1183	397	24

Inv. No.	*Cat. No.*	*Pl. No.*
CD 1184	396	24
CD 1189	493	31
CD 1190	492	31
CD 1191	501	32
CD 1192	518	32
CD 1193	572	35
CD 1194	584	36
CD 1195	586	36
CD 2079	33	2
CD 2441	186	
CD 251	157	9
CD 257	36	
CD 307	528	33
CD 308	485	30
CD 345	39	2
CD 410	577	36
CD 72	365	22
CD 774	275a	16
CD 782	45	3
CD 795	540	34
CD 833	243	14
CD 899	589	36
CD 948	557	35
CD 963, CD 1019	1	1
CE 1139	370	22
CE 1167	336	19
CE 1188	446	28
CE 1216	154	9
CE 1954	308	18
CE 1955	311c	
CE 1956	311d	
CE 1957	311e	
CE 1959	317	18
CE 1961	353	21
CE 1962	373	21
CE 1963	139	9
CE 1964	134	8
CE 1965	310	18
CE 1966	358	21
CE 1967	559	35
CE 1968	558	35

Inv. No.	*Cat. No.*	*Pl. No.*
CE 1969	592	36
CE 1970	328	19
CE 1971	325	19
CE 1972	355	21
CE 1973	126	8
CE 1976	320	19
CE 1984	219c	
CE 1987	319a	
CE 1988	333	19
CE 1990	463	29
CE 1991	311a	
CE 1992	315	18
CE 1993	593	36
CE 1994	244	14
CE 1996	156	9
CE 1997	479	30
CE 1998	580	36
CE 2001	137	8
CE 2002	496	31
CE 2003A	385	23
CE 2003B	386	23
CE 2004	425	27
CE 2006	82	5
CE 2008	259	15
CE 2008B	432	27
CE 2011	555	35
CE 2012A	14	
CE 2012B	14a	
CE 2013	13	1
CE 2014	32	2
CE 2015	30	2
CE 2016	5	1
CE 2017	372	22
CE 2018	104	6
CE 2019	350	20
CE 2020	118	7
CE 2024	273	15
CE 2025	305	17
CE 2026	543	34
CE 2027	216	
CE 2028	440	28

Inv. No.	*Cat. No.*	*Pl. No.*
CE 2030	517	32
CE 2031	444	28
CE 2032	559a	
CE 2033	379	22
CE 2034	381	22
CE 2035	207	12
CE 2036A and B	318	18
CE 2037	534	33
CE 2038	551	
CE 2039	21	2
CE 2041	342	20
CE 2042	107	7
CE 2043	354	21
CE 2045	316	18
CE 2046	509	32
CE 2047	70	5
CE 2048	257	15
CE 2049	375a	22
CE 2050	537	33
CE 2052	550	35
CE 2053	380d	
CE 2054	447	28
CE 2056	454	29
CE 2058	91	
CE 2062	424	27
CE 2063	360	21
CE 2064	264	15
CE 2065	601	
CE 2066	595	36
CE 2070	57	3
CE 2071	128	8
CE 2076	221	13
CE 2104	271	15
CE 2105	297	17
CE 2107	356	21
CE 2108	363	22
CE 2109	8	
CE 2111	23	2
CE 2113	83	
CE 2115	125	8
CE 2116	251	

Inv. No.	*Cat. No.*	*Pl. No.*
CE 2117	43	
CE 2118	90c	
CE 2119	78	
CE 2120	88	
CE 2121	84	
CE 2122	314a	
CE 2124	168a	
CE 2125	178	
CE 2126	194	11
CE 2127	168b	
CE 2129	175	10
CE 2130	203	11
CE 2132	185	
CE 2133	314b	
CE 2133	426	
CE 2135	100	6
CE 2136	430	27
CE 2137	101	6
CE 2138	121	8
CE 2139	111	7
CE 2140	136	8
CE 2141	219d	
CE 2142	282	16
CE 2144	290	17
CE 2145	308a	18
CE 2146, CE 1958	309	18
CE 2147	311b	
CE 2148	323	19
CE 2149	351	20
CE 2150	352	
CE 2151A	376a	
CE 2151B	374a	22
CE 2151C	378	
CE 2151D	380i	23
CE 2151E	380f	22
CE 2151F	380k	23
CE 2151G	380m	23
CE 2151H	380j	
CE 2151I	380n	
CE 2151J	380h	
CE 2153A	376	
CE 2153B	377	
CE 2153C	375	22
CE 2153D	380a	
CE 2153E	380b	22
CE 2153F	380c	
CE 2153G	380e	22
CE 2153H	380l	23
CE 2153I	380g	22
CE 2154	374	22
CE 2155	85	5
CE 2156	510	
CE 2157A	562	35
CE 2157B	563	35
CE 2157C	564	35
CE 2158	535	33
CE 2159A	546	34
CE 2159B	547	
CE 2159C	545	34
CE 2160A and B	538	33
CE 2161	447a	
CE 2163	450	28
CE 2164	445	28
CE 2169	452	28
CE 2171	337	19
CE 2171B	313g	
CE 2172	265c	
CE 2173	319b	
CE 2177	90e	
CE 2179	66e	
CE 2180	66d	
CE 2182	90f	
CE 2182	161	10
CE 2182A	6	1
CE 2183	204	11
CE 2183A	7	
CE 2185	345	20
CE 2187	369	
CE 2188	368	
CE 2189	367	22
CE 2191A	565	35
CE 2192	549	35
CE 2193	514	32
CE 2194	512	32
CE 2195A-I	382	
CE 2201	371	
CE 2205	199	
CE 2206	291	
CE 2207	436	27
CE 2209	407	25
CE 2210	544	34
CE 2313	219b	
CE 289	393	24
CE 291	267	16
CE 292	600	37
CE 836	65	
CE 837	429	27
CE 840	9	
CE2110	10	
CF 204	495	31
CF 2302	541	34
CF 2303	140	9
CF 2305	286	16
CF 2307	231	13
CF 2308	299	17
CF 2309	293	17
CF 2310	182	11
CF 2311	169	10
CF 2318	239	14
CF 2321	548	34
CF 2324	183	11
CF 2325	248	14
CF 2327	596	36
CF 2328	503	32
CF 2331	265b	
CF 2334	213	12
CF 2335	87	
CF 2339	472	30
CF 2342	502	32
CF 2345	235	13
CF 2350	419a	26
CF 2351	237	13
CF 2351A	391	23

Inv. No.	*Cat. No.*	*Pl. No.*
CF 2353	284	16
CF 2356	453	28
CF 2358	412	25
CF 2360	296	
CF 2361	594	
CF 2363	269	15
CF 2364	298	
CF 2367	499	31
CF 2369	497	31
CF 2373	494	31
CF 2374	2	1
CF 2375	241	14
CF 2377	504	32
CF 2388	272	15
CF 2390	405	25
CF 2393	172	10
CF 2394	170	10
CF 2395	285	16
CF 2396	276	16
CF 2400	212	23
CF 2401	224	
CF 2402	523	33
CF 2403	521	33
CF 2405	575	36
CF 2407	160c	
CF 2408	66a	
CF 2409	52	
CF 2410	73	
CF 2411	138	8
CF 2412	326	19
CF 2416	181	10
CF 2417	576	36
CF 2418	179	10
CF 2420	198	11
CF 2421	93	6
CF 2425	295	17
CF 2426	292	17
CF 2427	288	16
CF 2428	287	16
CF 2429	300	
CF 2430	218	
CF 2431	211	12
CF 2432A	307	18
CF 2432B	307a	18
CF 2434	238	14
CF 2437	278	16
CF 2438	132	
CF 2439	155	9
CF 2439A	160b	
CF 2440	177	10
CF 2442	303	
CF 2444	265a	15
CF 2445	196	11
CF 2446	262	15
CF 2449	324b	
CF 2449B	403	25
CF 2450	220	12
CF 2450A	419	26
CF 2452	240	13
CF 2453	168d	
CF 2454	165	10
CF 2455	173	
CF 2456	191	
CF 2457	192	
CF 2458	202	
CF 2459	171	
CF 2460	167	
CF 2461	168c	
CF 2462	184	
CF 2463	280	16
CF 2464	281	16
CF 2465	283	16
CF 2466	437	27
CF 2467	406	25
CF 2468A	409	25
CF 2468B	410	25
CF 2469	417	26
CF 2470	418	26
CF 2471	415	26
CF 2473	395	24
CF 2496	420	26
CF 2497A	470	30
CF 2497B	470a	
CF 2497C	470b	
CF 2497D	470c	
CF 2498	473	30
CF 2501	536	33
CF 2502	581	36
CF 2503	582	36
CF 2504	608	37
CF 2505	612	37
CF 2507	421	
CF 2508	422	
CF 432	587	36
CF 986	588	36
CF2468C	411	25
CG 234	236	14
CG 373	570	35
CG 375	359	21
CG 376	66f	
CG 382	489	30
CG 383	604	37
CG 386	488	30
CG 387	542	34
CG 389	597	37
CG 436	606	
CG 439	574	35
CG 446	247	14
CG 447	242	14
CG 449	160a	
CG 540	274	16
CG 55	230	13
CG 56	275c	
CG 57	607	37
CG 577	478	30
CG 579	168e	
CG 58	599	37
CG 581	275b	
CG 586	526	33
CG 587	529	33
CG 588	585	36
CG 59	598	37
C60.640	92	6

Inv. No.	*Cat. No.*	*Pl. No.*
C65.674	225	13
C65.675A	90d	
C65.802	553	35
C65.803	458	28
C66.294	37	
C66.295	46	3
C66.545	579	36
C66.777	401	24
C66.800	35	2
C66.801	4	1
C66.802	40	
C66.803	214	12
C66.804	64	4
C66.805	74	5
C66.807	58	3
C66.808	75	5
C66.809	223	
C66.810	210	12
C66.811	334	19
C66.812	330	19
C66.815	329	19
C66.817	112	7
C66.822	388	23
C66.823	457	29
C66.825	434	27
C66.828	467	29
C66.837	400	24
C66.839	515	32
C66.843	120h	
C66.870	321	19
C66.871	324e	
C66.872	313e	
C66.875	462	29
C66.876	398	24
C66.878	394	24
C66.879	423a	
C66.880	423b	
C66.881	389	23
C67.545	605	37
C67.553	151	
C67.554	206	12
C67.72	498	31
C68.271	18	1
C68.282	554	35
C68.375	147	9
C68.456	62	4
C68.495	433	27
C68.505	22	2
C68.519	12	1
C68.571	591	36
C68.600	364	22
C68.638	145	9
C68.657	26	
C68.657A	487	30
C68.659	469	30
C68.661	245	14
C68.662	263	15
C68.664	208	12
C68.673	27	
C68.674	28	
C68.675	53	3
C68.676A and B	51	3
C68.677	61	4
C68.678	49	3
C68.679, C68.683	135	8
C68.681	120c	
C68.682	102	
C68.684	95	6
C68.685	110	7
C68.687	461	29
C68.688	448	28
C68.694	195	11
C68.695	90a	
C68.697	120n	
C68.698	120j	
C68.699	120l	
C68.701A	602	
C68.701B	603	
C68.710	313d	
C68.712	324a	19
C68.736	511	32
C69.121	11	1
C69.155	20	2
C69.167	188	11
C69.233	24	
C69.308	566	35
C69.322	25	
C69.381A and B	17	1
C69.382	3	1
C69.400A	480	30
C69.400B	481	30
C69.400C	486	30
C69.402	484	30
C69.403	31	2
C69.404	66b	
C69.405	48	3
C69.406	50	3
C69.407	133	
C69.408	331	19
C69.409	361	21
C69.410	109	7
C69.411	120f	
C69.412	98	6
C69.413	120d	
C69.414	114	7
C69.415	346	20
C69.416	108	7
C69.417	115	7
C69.419	471	30
C69.421	552	35
C69.423	438	27
C69.424	427	27
C69.426	304	17
C69.429	189	11
C69.432	505	32
C69.434	99	
C69.436	120k	
C69.437	120i	
C69.438	120p	
C69.439	275d	
C69.452	335c	
C69.453B	508	32
C69.454	477	30

Inv. No.	*Cat. No.*	*Pl. No.*
C69.455	513	32
C69.457	324c	
C69.458	324d	
C69.459	313f	
C69.680	159	
C69.696	90b	
C70.228	149	9
C70.302	249	15
C70.303	362	22
C70.308	539	34
C70.342	205	12
C70.538	106	7
C70.568	451	28
C70.569	41	2
C70.570	19	1
C70.571	44	
C70.572	215	12
C70.573	260	15
C70.574	54	3
C70.575	69	4
C70.576	66c	
C70.577	258	
C70.578	59	
C70.579	68	4
C70.580	466	29
C70.581	465	29
C70.582	460	29
C70.583	142	9
C70.584	335b	
C70.585	141	9
C70.586	120e	
C70.587	120b	
C70.588	120m	
C70.589	116	
C70.592	520	33
C70.593	475	30
C70.600	573	35
C70.601	556	35
C70.602	569	35
C70.603	568	35
C70.605	464	29
C70.607	387	23
C70.611	306	17
C70.612	340	20
C70.613	327	19
C70.615	531	33
C70.618	190	11
C70.619	611	37
C70.620A	609	37
C70.620B	610	
C70.621	524	33
C70.623	279	16
C70.624	567	35
C70.626	261	
C70.627	253	
C70.628	197	
C70.629	120o	
C70.630	219a	
C70.631	313c	
C70.632	349	20
C70.660	339	
C70.661A	533	33
C70.662	390	23
C70.663	525	33
C70.680	527	33
C71.176	56	3
C71.193	29	2
C71.194	256	15
C71.195A and B	15	1
C71.196	252	
C71.197	162	10
C71.198	94	6
C71.200	42	
C71.201	16	1
C71.202	158	
C71.203	120a	
C71.204	117	
C71.206	105	
C71.207	366	22
C71.234	613	37
C71.59	560	35
C72.100	38	
C72.126	71	5
C72.127	120g	
C72.135	332	
C72.136	442	28
C72.137B	413	25
C72.138	404	25
C72.140	399	24
C72.41	63	4
C72.42	456	29
C72.43	482	30
C72.44	341	20
C72.47	578	36
C72.56	322	19
C72.89	47	3
C72.90	266	
C72.93	423	23
C72.94	124	8
C72.95	289	17
C72.96	81	5
C73.69	246	14
C73.71	43a	
C73.72	60	4
C73.73	150	
C73.74	166	
C74.2	193	11
C74.3	176	10
C74.4	174	
C74.5	201	
C74.6	583	36

Glossary

Technical Terms and Conventions

Additional terms relating to weathering, material, shape, decoration, and techniques of manufacturing are defined in the relevant sections of chapter 3. The glossary in S. Goldstein, *Pre-Roman and Early Roman Glass in the Corning Museum of Glass* (Corning, NY 1979) 12–14, contains other terms. Here, I have sought only to include vocabulary used frequently in this volume or terms misunderstood or occasionally misused in some excavation reports.

annealing The slow process of gradually cooling the finished object by placing it in a subsidiary part of the glass furnace. Since glass contracts as it cools, controlled annealing prevents strain cracking and crazing owing to the differential cooling of the surface and the interior of the glass. It also allows strains built up during the manufacturing process to be released.

batch The mixture of raw materials, together with cullet, colorants, and other materials, that when fused under high temperatures results in glass.

blow pipe A hollow metal tube, usually made of iron, used to blow glass. Our only knowledge of blow pipes from antiquity rests with thirteen fragmentary iron tubes, one at least 17 inches long, from a fourth-century A.D. Roman context in Mérida, Spain. Five of the tubes have wedge-shaped ends, as is typical of modern blow pipes, and all were formed by winding multiple strips of iron spirally around a mandrel to create the tube (Price 1976, 117). Nothing is known about the earliest blow pipes devised at the time that blowing was discovered. For further discussion, see Charleston 1980, 31–32, and Grose 1984, 32–34.

blowing spirals Spiraling patterns of tiny bubbles produced by the expansion and rotation of the vessel as it is being inflated on a blow pipe.

cullet Scrap glass or chunks of raw glass added to a batch to lower the melting temperature and shorten the melting time. In modern operations 15 to 30 percent of the batch is cullet. In the first century A.D. the Roman poet Martial (epigram 1.41) records the existence of hawkers at Rome who bartered broken glass, presumably for reuse as cullet, in exchange for *pallentia sulfurata*, usually considered primitive versions of sulfur matches.

cut-out The technique of folding the glass outward in order to form a tubular ledge under the rim or around the body of a vessel, or to form a tubular base-ring at the bottom. See **No. 238** for a closed cut-out side and **No. 354** for an open cut-out base.

devitrification The term is often used interchangeably, albeit inaccurately, with "weathering." More correctly, it refers to the formation of crystals in the glass as it cools, owing either to improper annealing or to the faulty composition of the original batch.

fabric Technically, the term refers to the chemical composition of the glass as determined by scientific analysis. In this catalogue it is used occasionally to mean "glass" *per se*, as in the statement: "The decoration is in the same 'fabric' as the vessel."

frit In modern factories the constituent raw materials of glass are preheated to form a granular, even friable material called frit. It has a sandy, pitted appearance. Once reheated, it melts more rapidly than the original raw materials. In excavation reports, glass or even metal slag is often but incorrectly identified as frit.

gather or **gob** A mass of molten glass picked up on the end of a blow pipe. The term is sometimes used interchangeably but incorrectly with "paraison."

glass As defined by the American Society for Testing Materials, "The inorganic product of fusion which has cooled to a rigid condition without crystallization." In antiquity, glass was the result of a fused melt of three primary ingredients: sand (silicon dioxide), which forms the basic structure; an alkali oxide such as soda ash or natron, which causes the batch to melt at lower temperatures; and lime (usually dolomitic or other limestones), which acts as a stabilizer and makes glass less soluble in water. Once these ingredients have formed a homogeneous fused melt (around 800°C), a random non-crystalline molecular structure results. Once cooled, this characteristic structure remains even after the glass has become rigid at room temperature. Glass is not a specific chemical compound, such as Egyptian blue (copper-calcium-tetrasilicate: CuCo CaO $_4$Si0$_2$). There are innumerable recipes for glass, and each recipe results in a glass with different properties.

kick A deep, usually conical concavity at the bottom or on the side of a vessel. It is made with a tool while the glass is hot. Many medieval vessels are characterized by deep kicks at the bottom. See **Nos. 607** and **608**.

knocked-off The technique of breaking a vessel from the blow pipe at a predetermined place, usually by drawing a wet tool along the desired line of breakage, then tapping the glass sharply. This method results in a rough edge, which later is usually polished smooth.

marver A flat surface, often marble, used to roll the gather during manufacture. This is done to make the body of the vessel smooth and even, to press applied decoration into the vessel's wall, and generally to assist in shaping the vessel. Core-formed bottles (**Nos. 1–2**) have trailed decoration marvered flush to the surface.

paraison A gather or gob of glass collected from the crucible, then initially inflated on a blow pipe.

pontil or **punty** A long, solid metal rod used to hold a vessel while working on it, once it has been broken off from the blow pipe. At the point where the pontil is attached, the vessel exhibits a "pontil scar," which may be removed by polishing. Because of the apparent absence of such scars on Roman blown glasses of the early first century A.D., it seems that there was an interval of several decades or more before the pontil was introduced as a standard piece of equipment (Charleston 1980, 32–33). On occasion, at the point of attachment, a blob of glass or "pontil wad" remains (see **No. 481**, pl. 30).

Terms to Be Avoided or Used with Caution

glass paste or **pâte du verre** In some publications the material from which glass objects such as gaming pieces (here, **Nos. 587–591**) or even core-formed bottles (**Nos. 1–2**) are made is described as being glass paste or *pâte du verre*. Both terms refer to the use of powdered glass placed in molds and sintered, or fused, to form an article. The term and technique *pâte du verre*, however, are best reserved for objects manufactured in that manner in and after the nineteenth century once they were commercially popularized by French glassmakers such as Henri Cros, Gallè, DeLatte, and Lalique. There is no direct evidence that such a procedure was ever employed by ancient craftsmen. As a term for describing ancient glasses, it is anachronistic and inappropriate.

metal An unnecessary term used to refer to the substance "glass."

millefiori glass The term, in Italian literally meaning "thousand flowers," was first used in the Renaissance to describe tapestries with rich floral backgrounds. In the early nineteenth century, if not before, the term was applied to ancient mosaic glass because certain vessels used one or more canes that had the appearance of "flowers"—that is, a cane with a polychrome center surrounded by colorful "petals." In the mid- to late nineteenth century the word passed into standard usage when the mosaic technique was revived by the Murano factories. As a general term to describe ancient mosaic vessels, however, it fails to suggest the true nature of the technique involved (see pp. 29–30 above) or to account for the many varieties of mosaic glass (*inter alia*, network, striped, short-strip, marbled, and composite wares), none of which bears close resemblance to flowers or floral designs. Accordingly, the term should be avoided and the more general phrase "mosaic glass" used in its place. The only exception is a vessel, such as **No. 22**, which employs a cane design that indeed resembles a flower. In such a case the vessel can be described as being manufactured in a mosaic technique but utilizing a "millefiori" pattern.

pillar-molded bowls A quaint term introduced in nineteenth-century England to describe cast ribbed bowls in general (here, **Nos. 32–90e**). The term was adopted from a glass-forming technique employed by English craftsmen of that era by which they strengthened vessels by blowing them into metal body-molds that imparted pronounced vertical ribs. In classifying ancient glass, this term, if used at all, should designate only those ribbed bowls with evenly spaced, pronounced ribs, such as **Nos. 45–47** in this catalogue.

sand-core technique Core-formed vessels such as **Nos. 1–2** were long thought to have been made around a core composed solely of sand. Analyses of core material scraped from such vessels, however, have shown that the cores were made of several materials, including sand but not sand alone. The term "sand-core" should be discarded and preference given to "core-formed" or "core-made."

waste This term appropriately describes the waste materials associated with a glass factory. On Roman sites it takes the form of numerous drippings, chunks of raw glass, partially fused pieces of glass (often attached to portions of melting pans or crucibles), wasters, cullet, and *moiles* (the upper sections of paraisons that wrapped around the ends of blow pipes but were discarded after the vessel had been knocked off the pipe). In archaeological reports amorphous, "glassy" chunks of burned ceramic, metal, or melted glass are occasionally but erroneously called slag and cited as evidence for glass-working on the site.

wasters The correct term for vessels ruined and discarded during manufacture at a factory. Such vessels would often be reused as cullet. On excavated sites malformed or partly melted vessels accidentally warped by exposure to high temperatures are sometimes called wasters and even cited as evidence for the presence of glassmaking at the site.

Cosa
A Select Bibliography

Abbreviations for standard reference works and periodicals conform to those recommended by the *American Journal of Archaeology*: http://www.ajaonline.org/submissions/abbreviations.

Andreae, B. "Archäologische Grabungen im Bereich der Soprintendenzen von Nord- und Mittelitalien, 1949–1959." *AA* 86 (1959) 164–166.

Babcock, Charles L. "An Inscription of Trajan Decius from Cosa." *AJP* 83 (1962) 147–158.

Bace, Edward J. "The Inscriptions of Cosa." *AJA* 84 (1980) 193.

———. "Cosa: Inscriptions on Stone and Brick-Stamps." University of Michigan doctoral dissertation, 1983.

Blanck, Horst. "Archäologische Funde und Grabungen in Mittelitalien, 1959–1966." *AA* 97(1970) 283–284.

Boethius, Axel. *Etruscan and Early Roman Architecture.* The Pelican History of Art. Harmondsworth, England, rev. ed., 1978, 39, 50, 54–55, 57–58, 63, 116, 121–128, 131–134, 139, 145–146, 149, 151–152, 154, 156, 158–159, 165, 179, 183, 190–192, 195, and 232.

——— and John Ward-Perkins. *Etruscan and Roman Architecture.* The Pelican History of Art. Harmondsworth, England, 1970, 33, 46–50, 55, 97–98, 100–101, 103, 105, 110–113, 117, 122–124, 127–130, 132, 134–135, 151–152.

Boitani, Francesca, et al. *Le città etrusche*. Milan, 1973. Trans. as *Etruscan Cities*, 1975, 120–128.

Brendel, Otto. "Ganymede Group from Cosa." *AJA* 73 (1969) 232.

Brown, Frank E. "Cosa." *AJA* 53 (1949) 382.

———. "Cosa: Exploration in Etruria." *Archaeology* 2 (1949) 2–10.

———. "Cosa." *AJA* 55 (1951) 184.

———. "*Cosa I, History and Topography* (*MAAR* 20). Rome, 1951.

———. "Cosa." *AJA* 56 (1952) 135–136.

———. "Cosa." *EAA* 2 (1959) 869–870, with bibliography to 1959.

———. "Scavi a Cosa Ansedonia (1965–1966)." *BdA* 52 (1967) 37–41.

———. "Excavations at Cosa, 1965–1968." *AJA* 73 (1969) 232.

———. "Cosa." *EAA, Suppl.* Rome, 1970, 263–264.

———. "Cosa." *StEtr* 41 (1973) 523–524.

———. "Cosa." In *The Princeton Encyclopedia of Classical Sites*, Princeton, 1976, 245–246.

———. *Cosa: The Making of a Roman Town.* Jerome Lectures 13. Ann Arbor, MI, 1980.

———, Emeline H. Richardson, and L. Richardson, jr. *Cosa II, The Temples of the Arx. MAAR* 26. Rome, 1960.

———. *Cosa III: The Buildings of the Forum. MAAR* 37. Ann Arbor, MI, 1993.

Bruno, Vincent. "Fragments of a Temple Decoration from the Arx of Cosa." *AJA* 73 (1969) 232.

———. "Antecedents of the Pompeian First Style." *AJA* 73 (1969) 305–317.

———. "A Town House at Cosa." *Archaeology* 23 (1970) 232–241.

——— and Russell T. Scott, *Cosa IV: The Houses. MAAR* 38. Ann Arbor, MI, 1993.

Buttrey, Theodore V. *Cosa: The Coins. MAAR* 34. Rome, 1980.

Carandini, Andrea, et al. *Schiavi e padroni nell'Etruria romana, La villa di Settefinestre dallo scavo alla mostra*. Bari, 1979. With extensive bibliography on Cosa and environs prior to the opening in 1948 of the American Academy's excavations; also citations for studies resulting from the Italian excavations at Settefinestre.

Castagnoli, F. "La centuriazione di Cosa." *MAAR* 24 (1956) 149–165.

Collins, Jacquelyn. "A Late Antique Shrine of Liber in the Forum at Cosa." *AJA* 73 (1969) 234.

———. "The Marble Sculptures from Cosa." Columbia University doctoral dissertation, 1970.

Collins-Clinton, Jacquelyn. *A Late Antique Shrine of Liber Pater at Cosa.* Études preliminaires aux religions orientales dans l'empire romain 64. Leiden, 1977.

D'Arms, John H. "Republican Senators' Involvement in Commerce in the Late Republic: Some Ciceronian Evidence." In *Roman Seaborne Trade* (*MAAR* 36), Rome, 1980, 77–89, esp. 81–84.

Dyson, Stephen L. *Cosa: The Utilitarian Pottery. MAAR* 33. Rome, 1976.

———. "Settlement Patterns in the *Ager Cosanus*: The Wesleyan University Survey, 1974–1976." *JFA* 5 (1978) 251–268.

———. "New Methods and Models in the Study of Roman Town-Country Systems." *The Ancient World* 2 (1980) 91–95.

———. "Some Reflections on the Archaeology of Southern Etruria." *JFA* 8 (1981) 79–83.

———. "Survey Archaeology: Reconstructing the Roman Countryside." *Archaeology* 34 (1981) 31–37.

Fentress, Elizabeth. *Cosa V: An Intermittent Town. MAAR Suppl* 2. Ann Arbor, MI, 2003.

Fitch, Clio R. "The Lamps from Cosa." *Scientific American* 247 (1982) 148–158.

Gazda, Elaine K. "The Use of Tufo in Roman Underwater Construction." *AJA* 79 (1975) 151. Abstract bound in separate booklet.

Grose, David F. "Roman Glass of the First Century A.D., A Dated Deposit of Glassware from Cosa, Italy." In *Annales du 6e congrès de l'Association internationale pour l'histoire du verre*, Liège, 1975, 31–52.

———. "The Glass from the Roman *colonia* of Cosa." *Bulletin de l'Association internationale pour l'histoire du verre* 7 (1973–1976, pub. 1977) 175–182.

———. "Early Blown Glass: The Western Evidence." *JGS* 19 (1977) 8–29, esp. 20, 25–27.

———. "The Formation of the Roman Glass Industry." *Archaeology* 35 (1982) 40–47.

Hanson, J. A. *Roman Theater-Temples*. Princeton, 1959, 37–39.

Harris, W. V. *Rome in Etruria and Umbria*. Oxford, 1971, 97, 106, 147–148, 155–159, 202–208.

Laidlaw, Anne. "The Painted Decoration of the Cosa Basilica." Yale University doctoral dissertation, 1963.

———. "Reconstructions of the First Style Decorations in the House of Sallust in Pompeii." In *In Memoriam to Otto J. Brendel, Essays in Archaeology and the Humanities*, Mainz, 1976, 105–113, esp. 108–109.

———. "A Reconstruction of the First Style Decoration in the Alexander Exedra of the House of the Faun." In *Neue Forschungen in Pompeii*, Berlin, 1975, 39–46, esp. 44–45.

Lewis, J. D. "Excavations of the Ancient Port of Cosa." *AJA* 74 (1970) 199.

———. "Cosa: An Early Roman Harbor." In *Marine Archaeology*, Colston Papers 23, London, 1973, 233–259.

Lugli, G. *La tecnica edilizia romana*. Rome, 1957, 111–115, 459–460.
McCann, Anna Marguerite. "Excavations of the Ancient Port of Cosa, 1968." *AJA* 73 (1969) 241–242.
———. "Excavations of the Ancient Port of Cosa, 1972." *AJA* 77 (1973) 220.
———. "The Harbor and Fishery Remains at Cosa, Italy." *JFA* 6 (1979) 391–411.
——— and J. D. Lewis. "The Ancient Port of Cosa." *Archaeology* 23 (1970) 201–208.
McKay, A. G. *Houses, Villas and Palaces in the Roman World*. Ithaca, NY, 1975, 60–63.
MacKendrick, Paul. *The Mute Stones Speak*. New York, 1960, esp. 101–116.
Manacorda, Daniele. "The Ager Cosanus and the Production of Amphorae of Sestius: New Evidence and a Reassessment." *JRS* 68 (1979) 122–131.
Moevs, Maria Teresa Marabini. "The Dating of the Italic 'Megarian' Bowls in the Light of the Finds from Cosa." *AJA* 67 (1963) 215.
———. "From Campanian Black to Sigillata Red and the Rise of Orange Gloss and Metallic Gloss." *AJA* 71 (1967) 192.
———. "The Iron Age Tradition in Roman Pottery of the Late Republican–Early Imperial Period." *AJA* 71 (1967) 192.
———. *The Roman Thin Walled Pottery from Cosa, 1948–1954. MAAR* 32. Rome, 1973.
———. *ACO in Northern Etruria: The Workshop of Cusonius at Cosa. MAAR* 34. Rome, 1980.
———. *Italo-Megarian Ware at Cosa. MAAR* 34. Rome, 1980.
———. "Decorated Late Italian Sigillata: New Evidence for an Absolute Chronology." *AJA* 84 (1980) 224.
———. "New Evidence for an Absolute Chronology of Decorated Late Italian Sigillata." *AJA* 84 (1980) 319–327.
———. "Le muse di Ambracia." *BdA* 95 (1981) 1–58.
Montuoro, Paola Zancani. "Il faro di Cosa in ex-voto a Vulci?" *RivistArch*, series 3.2 (1979) 5–29, with contribution by F. D. Brown, 21–23.
Morel, J-P. "L'atelier des petites estampilles." *MÉFR* 81 (1969) 94–99.
Nagle, D. B. "The Etruscan Journey of Tiberius Gracchus." *Historia* 25 (1976) 487–489.
Paqua, M. Lopez. "La vera origine di Cosa Vulcente." *Studi Etruschi* 22 (1952–1953) 411–420.
Quilici, L. "Ville dell'agro cosano con fronte a toretti." *RivistArch* 23 (1978) 11–64.
Richardson, Lawrence, jr. "Excavations at Cosa in Etruria 1948–1952." *Antiquity* 27 (1953) 102–103.
———. "Cosa." *AJA* 57 (1953) 215–216.
———. "Cosa." *AJA* 58 (1954) 138.
———. "Cosa." *AJA* 59 (1955) 307–308.
———. "Cosa." *FA* 6 (1951) no. 2491, 7 (1952) no. 1988, 8 (1953) no. 2173, 9 (1954) no. 2838.
———. "Cosa and Rome, Comitium and Curia." *Archaeology* 10 (1957) 49–55.
Salmon, E. T. *Roman Colonization under the Republic*. Ithaca, NY, 1970, 28–29, 62–63, 68.
Santangelo, Maria. *L'Antiquarium di Orbetello*. Rome, 1954.
Scott, Ann R. *Cosa: The Black Glaze Pottery Revisited. MAAR Suppl* 5. Ann Arbor, MI, 2008.
Scott, Russell T. "The Arx of Cosa, 1965–1968." *AJA* (1969) 245.
———. "Cosa 1969–1970, The Houses." *AJA* 75 (1971) 273.
———. "Pre-Arretine: New Evidence from Cosa." *AJA* 82 (1978) 249. Abstract bound in separate booklet.
———. "A New Inscription of the Emperor Maximinus at Cosa." *Chiron* 11 (1981).
Seaborne Commerce of Ancient Rome: Studies in Archaeology and History, ed. J. D'Arms and E. Kopff. *MAAR* 36. Rome, 1980.

Taylor, Doris M. *Cosa: Black Glaze Pottery. MAAR* 25. Rome, 1957, 67–193.

Taylor, Lily Ross. *Roman Voting Assemblies.* Jerome Lectures 8. Ann Arbor, MI, 1966, 21–23, 107–108.

Tongue, William. "The Brick-Stamps of Cosa." *AJA* 54 (1950) 263.

Von Gerkan, A. "Zur Datierung der Kolonie Cosa." In *Scritti in onore di G. Libertini*, Florence, 1958, 149–156.

Will, Elizabeth Lyding. "Les amphores de Sestius." *Revue archéologique de l'est et du centre-est* 7 (1956) 224–244.

———. "The Sestius Amphoras from Cosa." *AJA* 79 (1975) 151. Abstract bound in separate booklet.

———. "New Light on the Sestius Question." *AJA* 80 (1976) 194. Abstract bound in separate booklet.

———. "The Sestius Amphoras: A Reappraisal." *JFA* 6 (1979) 339–350.

Short Titles

Prepared by Katherine A. Larson

Abbreviations for standard reference works and periodicals conform to those recommended by the *American Journal of Archaeology*: http://www.ajaonline.org/submissions/abbreviations.

Alarcão 1965	Alarcão, Jorge, and Adilia de. 1965. *Vidros romanos de Conimbriga*. Coimbra: Museu monográfico de Conimbriga.
Alarcão 1968	Alarcão, Jorge de. 1968. "Vidros romanos de museus do Alentejo e Algarve." *Conimbriga* 7:7–39.
Albintimilium 1950	Lamboglia, Nino. 1950. *Gli scavi di Albintimilium e la cronologia della ceramica romana,* part 1: *Campagne di scavo 1938–1940*. Bordighera: Istituto internazionale di studi Liguri.
Alexandra 1972	Alexandra, Olga. 1972. "Κερατσινίου 54 καὶ Πλάτωνος (οἰκόπεδον Ζορμπᾶ)." *ArchDelt* 27 (B1):115–118.
Alexiou 1965	Alexiou, Stylianos. 1965. "Ἀρχαιότητες καὶ μνημεῖα κεντρικῆς καὶ ἀνατολ. Κρήτης." *ArchDelt* 20 (B3):549–557.
Aquileia 1968	[See Calvi 1968] Calvi, Maria Carina. 1968. *I vetri romani del museo di Aquileia*. Aquileia: Associazione nazionale per Aquileia.
Ashdod II–III 1971	Dothan, M. 1971. *Ashdod II–III: The Second and Third Seasons of Excavations, 1963, 1965, Soundings in 1967*. Jerusalem: Israel Exploration Society.
Avigad 1962	Avigad, Nahman. 1962. "The Expedition to the Judean Desert 1961." *IEJ* 12 (3):169–183.
Avigad 1972	Avigad, Nahman. 1972. "Excavations in the Jewish Quarter of the Old City of Jerusalem, 1971." *IEJ* 22:193–200.
Barag 1962	Barag, Dan. 1962. "The Expedition to the Judean Desert 1961: Glass Vessels from the Cave of Horror." *IEJ* 12 (4):208–214.
Barag 1969	Barag, Dan. 1969. "Flower and Bird and Snake-Thread Glass Vessels." In *Annales du Congrès de l'Association Internationale pour l'Histoire du Verre* 4:55–66. Liège.
Barag 1978	Barag, Dan. 1978. "Hanita tomb XV: A Tomb of the Third and Early Fourth Century CE." *Atiqot* 13.
Barovier 1974	Barovier, Rosa. 1974. "Roman Glassware in the Museum of Murano and the Muranese Revival of the Nineteenth Century." *JGS* 16:111–119.
Bartoccini 1935	Bartoccini, R. 1935. "La tomba degli ori di Canosa." *Iapigia* 6:225–262.

Berger 1960	Berger, Ludwig. 1960. *Römische Gläser aus Vindonissa.* Veröffentlichungen der Gesellschaft Pro Vindonissa. Basel: Birkhäuser.
Berger 1977	Berger, Ludwig. 1977 (1978). "Neufund eines Gladiatorenbechers aus Vindonissa." In *Jahresbericht der Gesellschaft Pro Vindonissa 1977,* 62–68.
Berger 1980	Berger, Ludwig, and Michel Jouve. 1980. "Un fragment de verre à ruban d'or découvert à Béthisy-Saint-Martin (Oise)." *Revue archéologique de l'Oise* 18:9–13.
Berlin Coll. 1976	Platz-Horster, Gertrud. 1976. *Antike Gläser* (Exhib. Cat. Antikenmuseum Berlin, Staatliche Museen Preussischer Kulturbesitz). Berlin: Antikenmuseum.
Bibracte 1899	Bulliot, Jacques Gabriel. 1899. *Fouilles du Mont Beuvray (ancienne Bibracte) de 1867 à 1895.* Autun: Dejussieu.
Biella 1950	Carducci, Carlo. 1950–1951. "La necropoli romana di Biella." *Bollettino della Società piemontese di archeologia e di belle arti* 4–5:23–39.
BMFA Coll. 1968	Saldern, Axel von. 1968. *Ancient Glass in the Museum of Fine Arts, Boston.* Boston: Museum of Fine Arts.
Bologna Cat. 1964	Mansuelli, Guido Achille, and Maria Vergnani. 1964. *Arte e civiltà romana nell'Italia settentrionale dalla repubblica alla tetrarchia* (Exhib. Cat. Bologna, Palazzo dell'Archiginnasio). Bologna: Edizioni Alfa.
Bologna Coll. 1979	Meconcelli Notarianni, Gioia. 1979. *Vetri antichi nelle collezioni del Museo Civico Archeologico di Bologna.* Bologna: Istituto per la storia di Bologna.
Bomford Coll. 1976	Thomas, Nicholas. 1976. *Ancient Glass: The Bomford Collection of Pre-Roman & Roman Glass on Loan to the City of Bristol Museum & Art Gallery.* Bristol: City of Bristol Museum and Art Gallery.
Boon 1966	Boon, George C. 1966. "Roman Window Glass from Wales." *JGS* 8:41–45.
British Museum Coll. 1981	Harden, Donald B. 1981. *Catalogue of Greek and Roman Glass in the British Museum,* vol. 1. London: Trustees of the British Museum.
Brock and Mackworth-Young 1949	Brock, J. K., and G. Mackworth-Young. 1949. "Excavations in Siphnos." *Annual of the British School at Athens* 44:1–92, pls. 1–37.
Brown 1951	Brown, Frank Edward. 1951. *Cosa I: History and Topography. MAAR* 20.
Brown 1960	Brown, Frank Edward, Emeline Hill Richardson, and L. Richardson, jr. 1960. *Cosa II: The Temples of the Arx. MAAR* 26.
Brown 1980	Brown, Frank Edward. 1980. *Cosa: The Making of a Roman Town.* Ann Arbor: University of Michigan Press.
Brown 1993	Brown, Frank Edward, Emeline Hill Richardson, and L. Richardson, jr. 1993. *Cosa III: The Buildings of the Forum: Colony, Municipium, and Village. MAAR* 37.
Bruno and Scott 1993	Bruno, Vincent J., and Russell T. Scott. 1993. *Cosa IV: The Houses. MAAR* 38.
Buechner 1960	Buechner, Thomas S. 1960. "The Glass from Tarrha." *Hesperia* 29 (1):109–117.

Buttrey 1980	Buttrey, T. V. 1980. *Cosa: The Coins*. *MAAR* 34.
Calvi 1965	Calvi, Maria Carina. 1965. "La coppa vitrea di Aristeas nella collezione Strada." *JGS* 7:9–16.
Calvi 1968	Calvi, Maria Carina. 1968. *I vetri romani del museo di Aquileia*. Aquileia: Associazione nazionale per Aquileia.
Calvi 1973	Calvi, Maria Carina. 1973. "Vetri preromani in collezioni ticinesi." In *Quaderni ticinesi di numismatica e antichità classiche* 2:53–61. Lugano.
Calvi 1975	Calvi, Maria Carina. 1975. "I vetri di Brescia romana." In *Atti del convegno internazionale per il XIX centenario della dedicazione del 'Capitolium' e per il 150° anniversario della sua scoperta*, 205–216. Brescia: Ateneo.
Camulodunum 1947	Harden, Donald B. 1947. "Glass." In *Camulodunum. First Report on the Excavations at Colchester, 1930–1939*, ed. C. F. C. Hawkes and M. R. Hull, 287–307. Oxford: Oxford University Press.
Canivet 1969	Canivet, Maria Teresa Fortuna. 1969. "I vetri romani di Cornus conservati al museo di Cagliari." *JGS* 11:19–26.
Carmona 1931	Bonsor, Jorge Eduardo. 1931. *An Archaeological Sketchbook of the Roman Necropolis at Carmona*. Hispanic Notes & Monographs. New York: Hispanic Society of America.
Charleston 1980	Charleston, Robert J. 1980. "Our Forefathers in Glass." *Glass Technology* 21:27–36.
Charlesworth 1959	Charlesworth, Dorothy. 1959. "Roman Glass in Northern Britain." *Archaeologia Aeliana* 37:33–58.
Charlesworth 1966	Charlesworth, Dorothy. 1966. "Roman Square Bottles." *JGS* 8:26–40.
Conimbriga 1965	Alarcão, Jorge de. 1965. *Vidros romanos de Conimbriga*. Coimbra: Ministério da educação nacional.
Conimbriga 1976	Alarcão, Jorge [de], and Robert Etienne (eds). 1976. *Fouilles de Conimbriga*, VI: *Ceramiques diverses et verres*. Paris: E. de Boccard.
Constable-Maxwell Coll. 1979	*Constable-Maxwell Collection of Ancient Glass*. 1979. London: Sotheby's.
Corinth 1952	Davidson, Gladys R. 1952. *Corinth*, XII: *The Minor Objects*. Princeton: American School of Classical Studies at Athens.
Corning Coll. 1979	Goldstein, Sidney M. 1979. *Pre-Roman and Early Roman Glass in the Corning Museum of Glass*. Corning, New York: Corning Museum of Glass.
Corning Coll. 1980	Charleston, Robert J. 1980. *Masterpieces of Glass from the Corning Museum of Glass*. Corning Museum of Glass Monograph. New York: H. N. Abrams.
Cotton 1979	Cotton, M. Aylwin. 1979. *The Late Republican Villa at Posto, Francolise*. Supplementary Publication of the British School at Rome. London: British School at Rome.
Crivelli 1979	Crivelli, Aldo. 1979. "I vetri romani di Locarno." *Verbanus* 1:21–39. (Alberti Editore).

Cumae 1913 — Gàbrici, E. 1913. Cuma. *Monumenti antichi* 22:449–871.

Cumae 1925 — Levi, Alda. 1925. "Cuma (necropolis), tomba 'a schiena' del periodo greco-sannitico." *NSc* 1925:85–90.

Cummings 1980 — Cummings, Keith. 1980. *The Technique of Glass Forming*. London: Batsford.

Cyprus 1956 — [See Vessberg 1956] Vessberg, Olof. 1956. "Glass." In *The Swedish Cyprus Expedition*, vol. IV, part 3: *The Hellenistic and Roman Periods in Cyprus*, ed. O. Vessberg and A. Westholm, 128–174, 193–219. Stockholm.

Damascus Coll. 1969 — Abû-l-Faraj Al-'Ush, M., 'Adnān Joundī, and Bashīr Zouhdī. 1969. *Catalogue du musée national de Damas.* Damascus: Direction générale des antiquités et des musées.

Daremberg-Saglio 1877 — Daremberg, Charles, and Edm. Saglio. 1877. *Dictionnaire des antiquités grecques et romains d'après les textes et les monuments,* 5 vols. Paris: Hachette.

de Franciscis 1963 — de Franciscis, Alfonso. 1963. "Vetri antichi scoperti ad Ercolano." *JGS* 5:137–139.

de Solla Price 1974 — de Solla Price, Derek John. 1974. *Gears from the Greeks: The Antikythera Mechanism: A Calendar Computer from ca. 80 B.C.* Transactions of the American Philosophical Society 64.7. New York: Science History Publications.

Déchelette II 1904 — Déchelette, Joseph. 1904. *Les vases céramiques ornés de la Gaule romaine (Narbonnaise, Aquitaine et Lyonnaise)*, 2 vols. Paris: A. Picard & fils.

Delos 1970 — Bruneau, Philippe. 1970. *L'ilôt de la maison des comédiens*. Exploration archéologique de Délos 27. Athens: École française d'Athènes.

Dillon 1907 — Dillon, Edward. 1907. *Glass*. New York: G. P. Putnam.

Doppelfeld 1966 — Doppelfeld, Otto. 1966. *Römisches und fränkisches Glas in Köln*. Schriftenreihe der archäologischen Gesellschaft Köln. Köln: Greven.

Dura-Europos 1963 — Clairmont, Christoph W. 1963. *The Excavations at Dura-Europos Conducted by Yale University and the French Academy of Inscriptions and Letters,* Final Report IV, part 5: *The Glass Vessels*. New Haven: Yale University Press.

Dyson 1976 — Dyson, Stephen L. 1976. *Cosa: The Utilitarian Pottery*. *MAAR* 33.

Eggers 1951 — Eggers, Hans Jürgen. 1951. *Der römische Import im freien Germanien*, 2 vols. Atlas der Urgeschichte 1. Hamburg: Hamburgisches Museum für Völkerkunde und Vorgeschichte.

Eisen 1927 — Eisen, Gustavus A., and Fahim Kouchakji. 1927. *Glass, Its Origin, History, Chronology, Technic and Classification to the Sixteenth Century*, 2 vols. New York: W. E. Rudge.

Ekholm 1936 — Ekholm, Gunnar. 1936. "Orientalische Glasgefässe in Skandinavien." *Eurasia septentrionalis antiqua* 10:61–72.

Fentress et al. 2003 — Fentress, Elizabeth, et al. 2003. *Cosa V: An Intermittent Town, Excavations 1991–1997*. *MAAR* Supplementary Volume 2.

Fishbourne 1971	Harden, Donald B., and Jennifer Price. 1971. "The Glass." In *Excavations at Fishbourne 1961–1969*, vol. II: *The Finds*, ed. B. W. Cunliffe, 317–368, pls. 25–28. London: Society of Antiquaries.
Fitzwilliam Coll. n.d.	*Glass at the Fitzwilliam Museum*. [1978]. Cambridge: Cambridge University Press.
Fortuna 1965	Fortuna, Maria Teresa. 1965. "I vetri soffiati della necropoli di Akko." *JGS* 7:17–25.
Fossing 1940	Fossing, Poul. 1940. *Glass Vessels before Glass-Blowing*. Copenhagen: E. Munksgaard.
Fremersdorf 1932	Fremersdorf, Fritz. 1932. "Alexandrinisches Buntglas aus einer Grabummauerung in Köln." *Germania* 16:278–286.
Fremersdorf 1938	Fremersdorf, Fritz. 1938. "Die Henkelkanne von Cortil-Noirmont im brüsseler Museum, eine kölnische Arbeit." *L'antiquité classique* 7:201–214.
Fremersdorf 1939	Fremersdorf, Fritz. 1939. "Erzeugnisse kölner Manufakturen in den Funden von Kastell Saalburg und Zugmantel." *Saalburg Jahrbuch* 9:6–22.
Fremersdorf 1955	Fremersdorf, Fritz. 1955. "Gutachten über die römischen Gläser im freien Germanien." *Jahrbuch des Römisch-Germanischen Zentralmuseums Mainz* 2:243–245.
Fremersdorf 1958	Fremersdorf, Fritz. 1958. *Römisches Buntglas in Köln*. Denkmäler des *römischen Köln* 3. Köln: Verlag der Löwe.
Fremersdorf 1959	Fremersdorf, Fritz. 1959. *Römische Gläser mit Fadenauflage in Köln (Schlangenfadengläser und Verwandtes)*. Denkmäler des römischen Köln 5. Köln: Verlag der Löwe.
Fremersdorf 1967	Fremersdorf, Fritz. 1967. *Die römischen Gläser mit Schliff, Bemalung und Goldauflagen aus Köln*, 2 vols. Denkmäler des römischen Köln 8. Köln: Verlag der Löwe.
Frova 1976	Frova, Antonio. 1976. "Bocca di Magra." In *Archeologia in Liguria: Scavi e scoperte, 1967–1975*, 55–58. Genoa: Soprintendenza archeologica della Liguria.
Genty 1972	Genty, P. Y. 1972. "Un type de gobelet en verre bien représenté dans le centre de la Gaule." *Revue archéologique du centre de la France* 11 (1–2):69–75.
Graue 1974	Graue, Jörn. 1974. *Die Gräberfelder von Ornavasso: Eine Studie zur Chronologie der späten Latène- und frühen Kaiserzeit*. Hamburger Beiträge zur Archäologie. Hamburg: H. Buske.
Gréau Coll. 1903	Froehner, Wilhelm. 1903. *Collection Julien Gréau: Verrerie antique, émaillerie et poterie appartenant à M. John Pierpont Morgan*. Paris.
Greifenhagen 1962	Greifenhagen, Adolf. 1962. "Ancient Glass in the Berlin-West Museum." *JGS* 4:61–66.
Grose 1973	Grose, David F. 1973. "The Glass from the Roman *Colonia* at Cosa." *Bulletin de l'Association Internationale pour l'Histoire du Verre* 7:175–182. Liège.

Grose 1975	Grose, David F. 1975. "Roman Glass of the First Century A.D.: A Dated Deposit of Glassware from Cosa, Italy." In *Annales du Congrès de l'Association Internationale pour l'Histoire du Verre* 6:31–52. Liège.
Grose 1977	Grose, David F. 1977. "Early Blown Glass: The Western Evidence." *JGS* 19:9–29.
Grose 1978	Grose, David F. 1978. "Ancient Glass." *Museum News: The Toledo Museum of Art* 20 (3):67–90.
Grose 1979	Grose, David F. 1979. "The Syro-Palestinian Glass Industry in the Later Hellenistic Period." *Muse* 13:54–67.
Grose 1981	Grose, David F. 1981. "The Hellenistic Glass Industry Reconsidered." In *Annales du Congrès de l'Association Internationale pour l'Histoire du Verre* 8:61–72. Liège.
Grose 1982	Grose, David F. 1982. "The Hellenistic and Early Roman Glass from Morgantina (Serra Orlando), Sicily." *JGS* 24:20–29.
Grose 1983	Grose, David F. 1983. "The Formation of the Roman Glass Industry." *Archaeology* 36 (4):38–45.
Grose 1983b	Grose, David F. 1983. Review of *Ancient and Byzantine Glass from Sardis*, by Axel von Saldern. *AJA* 87 (2):284–285.
Grose 1984	Grose, David F. 1984. "Glass Forming Methods in Classical Antiquity: Some Considerations." *JGS* 26:25–34.
Grose 1989	Grose, David F. 1989. *The Toledo Museum of Art: Early Ancient Glass*. New York: Hudson Hills Press.
Grose 1991	Grose, David F. 1991. "Early Imperial Roman Cast Glass: The Translucent Coloured and Colourless Fine Wares." In *Roman Glass: Two Centuries of Art and Invention*, ed. M. Newby and K. Painter, 1–18. Occasional Paper 13. London: Society of Antiquaries.
Gudiol y Cunill 1925	Gudiol y Cunill, Joseph. 1925. *Catálech dels vidres que integren la collecció Amatller*. Barcelona: Giró.
Haevernick 1967	Haevernick, Thea Elisabeth. 1967. "Zur Verbreitung der 'zarten Rippenschalen'." *Jahrbuch des Römisch-Germanischen Zentralmuseums Mainz* 14:153–160.
Haevernick 1981	Haevernick, Thea Elisabeth. 1981. *Beiträge zur Glasforschung: Die wichtigsten Aufsätze von 1938 bis 1981*. Mainz am Rhein: Philipp von Zabern.
Haevernick and Hahn-Weinheimer 1955	Haevernick, Thea Elisabeth, and Paula Hahn-Weinheimer. 1955. "Untersuchungen römischer Fenstergläser." *Saalburg Jahrbuch* 14:65–73.
Haltern 1901	Ritterling, Emil. 1901. "Die Fundstücke." In *Haltern und die Altertumsforschung an der Lippe*. Münster: Aschendorffsche Buchhandlung.
Haltern 1909	Koepp, Friedrich. 1909. *Ausgrabungen bei Haltern*. Mitteilungen der Altertums-Kommission für Westfalen. München: Aschendorff.
Hamelin 1953	Hamelin, P. 1953. "Matériaux pour servir à l'étude des verreries de Bégram." *Cahiers de Byrsa* 3:121–128.

Hamelin 1954	Hamelin, P. 1954. "Matériaux pour servir à l'étude des verreries de Bégram." *Cahiers de Byrsa* 4:153–183.
Harden 1933	Harden, Donald B. 1933. Ancient Glass. *Antiquity* 7:419–428.
Harden 1934	Harden, Donald B. 1934. "Snake-Thread Glasses Found in the East." *JRS* 24:50–55.
Harden 1935	Harden, Donald B. 1935. "Romano-Syrian Glasses with Mould-Blown Inscriptions." *JRS* 25:163-86.
Harden 1936	Harden, Donald B. 1936. *Roman Glass from Karanis*. Ann Arbor: University of Michigan Press.
Harden 1940	Harden, Donald B. 1940. "Roman Mould-Blown Glasses." *The Connoisseur* 106:102–105.
Harden 1945	Harden, Donald B. 1944–1945. "Two Tomb Groups of the First Century A.D. from Yahmour, Syria, and a Supplement to the List of Romano-Syrian Glasses with Mould-Blown Inscriptions." *Syria* 24:81–95, 291–292.
Harden 1959	Harden, Donald B. 1959. "New Light on Roman and Early Medieval Window-Glass." *Glastechnische Berichte* 32 (8):8–16.
Harden 1961	Harden, Donald B. 1961. "Domestic Window-Glass: Roman, Saxon and Medieval." In *Studies in Building History*, ed. E. M. Jope, 39–63. London: Odhams Press.
Harden 1968	Harden, Donald B. 1968. "The Canosa Group of Hellenistic Glasses in the British Museum." *JGS* 10:21–47.
Harden 1969	Harden, Donald B. 1969. "Ancient Glass, II: Roman." *Archaeological Journal* 126:44–77.
Harris 1927	Harris, James Rendel. 1927. "Glass Chalices of the First Century." *Bulletin of the John Rylands Library, Manchester* 2:286–295.
Hayward 1962	Hayward, Jane. 1962. "Roman Mold-Blown Glass at Yale University." *JGS* 4:48–60.
Hoffmann 1965	Hoffmann, Herbert, and Patricia F. Davidson. 1965. *Greek Gold: Jewelry from the Age of Alexander*. Mainz am Rhein: Philipp von Zabern.
Honey 1946	Honey, William Bowyer. 1946. *Glass: A Handbook for the Study of Glass Vessels of All Periods and Countries & A Guide to the Museum Collection, Victoria and Albert Museum*. London: Ministry of Education.
Isings 1957	Isings, Clasina. 1957. *Roman Glass from Dated Finds*. Archaeologica Traiectina 2. Groningen: J. B. Wolters.
Isings 1964	Isings, Clasina. 1964. "Glass from Ulpia Noviomagus, Nijmegen." *Bulletin Antieke Beschaving* 39:174–179.
Israeli 1964	Israeli, Y. 1964. "Sidonian Mold-Blown Glass Vessels in the Museum Haaretz." *JGS* 6:34–41.
Israeli 1983	Israeli, Y. 1983. "Ennion in Jerusalem." *JGS* 25:65–69.
Karanis 1936	[See Harden 1936] Harden, Donald B. 1936. *Roman Glass from Karanis*. Ann Arbor: University of Michigan Press.

Kassel Coll. 1967 — Spartz, Edith. 1967. *Antike Gläser: Vollständiger Katalog*. Kataloge der staatlichen Kunstsammlungen Kassel. Kassel: Staatliche Kunstsammlungen.

Kisa 1899 — Kisa, Anton. 1899. "Antike Gläser mit Fadenverzierung." *Kunst und Kunsthandwerk* 2:137–147.

Kisa 1908 — Kisa, Anton. 1908. *Das Glas im Altertume*, 3 vols. Leipzig: K. W. Hiersemann.

Knossos 1971 — Hayes, John W. 1971. "Four Early Roman Groups from Knossos." *Annual of the British School at Athens* 66:249–275.

La Baume n.d. — La Baume, Peter. [1970]. *Glas der antiken Welt*, 1. Wissenschaftliche Kataloge des Römisch-Germanischen Museums Köln 1. Köln: Römisch-Germanisches Museum der Stadt Köln und Archäologische Gesellschaft Köln.

Laviosa 1958 — Laviosa, C. 1958. "Un nuovo bicchiere romano-siriaco con iscrizione." *Bollettino d'arte* 43:171–173.

Lazarides 1957 — Lazarides, Demetrios I. 1957. "Ἀνασκαφὴ τῆς νεκροπόλεως Ἀμφιπόλεως." *Praktika* 1957:70–72.

Leveau 1970 — Leveau, Philippe. 1970. "Trois tombeaux monumentaux à Cherchel." *Bulletin d'archéologie algérienne* 4:101–148.

Limburg 1971 — Isings, Clasina. 1971. *Roman Glass in Limburg*. Archaeologica Traiectina 9. Groningen: Wolters-Noordhoff.

Lombardy Cat. 1964 — Roberti, Mario Mirabella, and Anna Maria Tamassia. 1964. *Catalogo della mostra dei vetri romani in Lombardia*. (Exhib. Cat. Palazzo del'Arengario, Milan). Milan: Soprintendenza alle antichità della Lombardia.

Luni I 1973 — Frova, Antonio. 1973. *Scavi di Luni*, I: *Relazione preliminare delle campagne di scavo 1970–1971*. Rome: L'Erma di Bretschneider.

Luni II 1977 — Frova, Antonio. 1977. *Scavi di Luni*, II: *Relazione delle campagne di scavo 1972–1973–1974*. Rome: L'Erma di Bretschneider.

Magdalensberg 1979 — Czurda-Ruth, Barbara. 1979. *Die römischen Gläser vom Magdalensberg*. Kärntner Museumsschriften 65. Klagenfurt: Verlag des Landesmuseums für Kärnten.

Maioli 1974 — Maioli, Maria. 1974. "I vetri della necropoli romana de 'Le Palazzette' nel territorio di Classe." *Felix Ravenna* 107–108:15–36.

Marconi 1932 — Marconi, Pirro. 1932. "Vetri romani nel museo di Zara." *Bollettino d'arte* 26:33–41.

Marconi 1934 — Marconi, Pirro, and Luigi Serra. 1934. *Il Museo nazionale delle Marche in Ancona (100 illustrazioni)*. "Itinerari dei musei e monumenti d'Italia." Rome: Istituto poligrafico dello stato/Libreria dello stato.

Mariacher 1959 — Mariacher, Giovanni. 1959. *I vetri della raccolta Fortuny al museo di Murano*. Venice: Zinco.

Mariacher 1966 — Mariacher, Giovanni. 1966. *Vetri di Zara restaurati. Bollettino dei Musei civici veneziani* 11 (4).

Mariacher 1988 — Mariacher, Giovanni. 1988. *Glass: From Antiquity to the Renaissance*. Revised edition London: Cassell.

Mau-Kelsey 1899 — Mau, August, and Francis W. Kelsey. 1899. *Pompeii, Its Life and Art*. New York: Macmillan Company.

McCann et al. 1987 — McCann, Anna Marguerite, Joanne Bourgeois, Elaine K. Gazda, John Peter Oleson, and Elizabeth Lyding Will. 1987. *The Roman Port and Fishery of Cosa: A Center of Ancient Trade*. Princeton, N.J.: Princeton University Press.

McClellan 1983 — McClellan, Murray C. 1983. "Recent Finds from Greece of First-Century A.D. Mold-Blown Glass." *JGS* 25:71–78.

McClellan 1984 — McClellan, Murray C. 1984. "Core-Formed Glass from Dated Contexts." Ph.D. diss. University of Pennsylvania, Philadelphia.

Moevs 1973 — Moevs, Maria Teresa Marabini. 1973. *The Roman Thin Walled Pottery from Cosa (1948–1954). MAAR* 32.

Moevs 1980a — Moevs, Maria Teresa Marabini. 1980. "Italo-Megarian Ware at Cosa." *MAAR* 34:161–227.

Moevs 1980b — Moevs, Maria Teresa Marabini. 1980. "Aco in Northern Etruria: The Workshop of Cusonius at Cosa." *MAAR* 34:231–280.

Morgantina 1982 — [See Grose 1982] Grose, David F. 1982. "The Hellenistic and Early Roman Glass from Morgantina (Serra Orlando), Sicily." *JGS* 24:20–29.

Morin-Jean 1913 — Morin-Jean (Jean Alexis Joseph Morin). 1913. *La verrerie en Gaule sous l'empire romain: Essai de morphologie et de chronologie*. Paris: H. Laurens éditeur.

Mutz 1972 — Mutz, Alfred. 1972. *Die Kunst des Metalldrehens bei den Römern: Interpretationen antiker Arbeitsverfahren auf Grund von Werkspuren*. Basel: Birkhäuser.

Neuburg 1949 — Neuburg, Frederic. 1949. *Glass in Antiquity*. London: Theodore Brun.

Newark Coll. 1976 — Auth, Susan Handler. 1976. *Ancient Glass at the Newark Museum from the Eugene Schaefer Collection of Antiquities*. Newark: Newark Museum.

Nida-Heddernheim 1974 — Welker, Edith. 1974. *Die römischen Gläser von Nida-Heddernheim*. Schriften des frankfurter Museums für Vor- und Frühgeschichte 3. Frankfurt am Main: Waldemar Kramer.

Nissen 1877 — Nissen, Heinrich. 1877. *Pompeianische Studien zur Städtekunde des Altertums*. Leipzig: Breitkopf und Härtel.

Norling-Christensen 1968 — Norling-Christensen, H. 1968. "Hohe Glasbecher vom Pompeji-Typ mit einer Verzierung, die meistens aus eingeschliffenen, dichtgestellten Furchen oder Facetten besteht." In *Provincialia: Festschrift für Rudolf Laur-Belart*, ed. E. Schmid, L. Berger, and P. Bürgin, 410–427. Basel: Schwabe.

Novaesium 1904	Lehner, Hans. 1904. "Die Einzelfunde von Novaesium." *Bonner Jahrbücher* 111–112:313–319, 416–417.
Oliver 1967	Oliver, Andrew, Jr. 1967. "Late Hellenistic Glass in the Metropolitan Museum." *JGS* 9:13–33.
Oliver 1968	Oliver, Andrew, Jr. 1968. "Millefiori Glass in Classical Antiquity." *JGS* 10:48–70.
Oppenländer Coll. 1974	Saldern, Axel von, Birgit Nolte, P. La Baume, and Thea Elisabeth Haevernick. 1974. *Gläser der Antike: Sammlung Erwin Oppenländer*. Mainz am Rhein: Philipp von Zabern.
Ornavasso 1895	Bianchetti, Enrico, Paola Piana Agostinetti, Carlo Carducci, and Felice Gino Lo Porto. 1895. *I sepolcreti di Ornavasso: cento anni di studi*, 2 vols. Atti della Società di archeologia e belle arti per la provincia di Torino. Rome: L. Nardini.
Ostia I 1968	Carandini, Andrea. 1968. *Ostia,* I: *Le terme del nuotatore: Scavo del'ambiente IV.* Studi miscellanei. Rome: De Luca.
Ostia II 1970	Carandini, Andrea. 1970. *Ostia,* II: *Le terme del nuotatore: Scavo del'ambiente I.* Studi miscellanei. Rome: De Luca.
Ostia III 1973	Carandini, Andrea. 1973. *Ostia,* III: *Le terme del nuotatore: Scavo degli ambienti III, VI, VIII.* Studi miscellanei. Rome: De Luca.
Ostia IV 1977	Carandini, Andrea. 1977. *Ostia,* IV: *Le terme del nuotatore: Scavo dell'ambiente XVI e dell'area XXV.* Studi miscellanei. Rome: De Luca.
Oxé and Comfort 1968	Oxé, August, and Howard Comfort. 1968. *Corpus vasorum arretinorum. A Catalogue of the Signatures, Shapes and Chronology of Italian Sigillata.* Bonn: Rudolf Habelt.
Parlama 1973	Parlama, Liana. 1973. "Καλύβια Ἤλιδος." *ArchDelt* 28 (B1):204–206.
Parlasca 1976	Parlasca, Klaus. 1976. "Zur Verbreitung ptolemäischer Fayencekeramik ausserhalb Ägyptens." *Jahrbuch des Deutschen Archäologischen Instituts* 91:135–156.
Pavia Coll. 1983	Maccabruni, Claudia. 1983. *I vetri romani dei musei civici di Pavia: Lettura di una collezione*. Pavia: Ticinum Edizioni.
Pfeffer and Haevernick 1958	Pfeffer, Waldtraut von, and Thea Elisabeth Haevernick. 1958. "Zarte Rippenschalen." *Saalburg Jahrbuch* 17:76–88.
Pompeii 1978	Ward-Perkins, J. B., and Amanda Claridge. 1978. *Pompeii* A.D. *79.* (Exhib. Cat. Museum of Fine Arts, Boston, vol. 2). New York: Knopf.
Pompeii 1984	F. Zevi, ed. 1984. *Pompei 79.* Naples: Macchiaroli (reprint of 1979 edition).
Price 1974	Price, Jennifer. 1974. "Some Roman Glass from Spain." In *Annales du Congrès de l'Association Internationale pour l'Histoire du Verre* 6:65–84. Liège.
Price 1976	Price, Jennifer. 1976. "Glass." In *Roman Crafts*, ed. D. Strong and D. Brown, 111–126. New York: New York University Press.

Ritterling 1912 — Ritterling, Emil. 1912. *Das frührömische Lager bei Hofheim am Taunus.* Annalen des Vereins für nassauische Altertumskunde und Geschichtsforschung 40. Wiesbaden.

ROM Coll. 1975 — Hayes, John W. 1975. *Roman and Pre-Roman Glass in the Royal Ontario Museum: A Catalogue*. Toronto: Royal Ontario Museum.

Rowell 1958 — Rowell, Henry T. 1958. "The Gladiator Petraites and the Date of the Satyricon." *Transactions of the American Philological Association* 89:14–24.

Russi (Ravenna) 1977 — Mazzeo Saracino, L. 1977. "Russi (Ravenna). Campagna di scavo 1971." *NSc* 31:5–156.

Saldern 1975 — Saldern, Axel von. 1975. "Two Achaemenid Glass Bowls and a Hoard of Hellenistic Glass Vessels." *JGS* 17:40–46.

Salona 1975 — Clairmont, Christoph W. 1975. *Excavations at Salona, Yugoslavia, 1969–1972. Conducted for the Department of Classics, Douglass College, Rutgers, The State University of New Jersey*. Park Ridge, N.J.: Noyes Press.

Samaria-Sebaste 1957 — Crowfoot, G. M. 1957. "The Glass." In *Samaria Sebaste,* III: *The Objects,* ed. J. W. Crowfoot, G. M. Crowfoot, and K. Kenyon, 403–422. London: Palestine Exploration Fund.

Samothrace 1967 — Dusenbery, Elsbeth B. 1967. "Ancient Glass from the Cemeteries of Samothrace." *JGS* 9:34–49.

Sangiorgi Coll. 1914 — Sangiorgi, G. 1914. *Collezione di vetri antichi dalle origini al V sec. d.C.* Milan and Rome: Bestetti e Tumminelli.

Sardis 1980 — Saldern, Axel von. 1980. *Ancient and Byzantine Glass from Sardis*. Archaeological Exploration of Sardis 6. Cambridge, MA.: Harvard University Press.

Schuler 1959 — Schuler, Frederic. 1959. "Ancient Glassmaking Techniques: The Blowing Process." *Archaeology* 12 (2):116–122.

Sear 1977 — Sear, Frank B. 1977. *Roman Wall and Vault Mosaics*. Heidelberg: F. H. Kerle.

Simonett 1941 — Simonett, Christoph. 1941. *Tessiner Gräberfelder: Ausgrabungen des archäologischen Arbeitsdienstes in Solduno, Locarno-Muralto, Minusio und Stabio, 1936 und 1937*. Monographien zur Ur- und Frühgeschichte der Schweiz 3. Basel: Birkhäuser.

Simoni 1963 — Simoni, P. 1963. "Una necropoli romana a Salò (Brescia)." *Annali del Museo di Gavardo* 2:1–48.

Spinazzola 1928 — Spinazzola, Vittorio. 1928. *Le arti decorative in Pompei e nel museo nazionale di Napoli.* Milan: Bestetti e Tumminelli.

Staré Hradisko 1961 — Meduna, Jirí. 1961. *Staré hradisko: Katalog der Funde im Museum der Stadt Boskovice*. Fontes archaeologiae moravicae 2. Brno: Archeologický ústav Československé akademie.

Stern 1981 — Stern, E. Marianne. 1981. "Hellenistic Glass from Kush (Modern Sudan)." In *Annales du Congrès de l'Association Internationale pour l'Histoire du Verre* 8:35–50. Liège.

Sturgis 1894	Sturgis, Russell. 1894. "The Coleman Collection of Antique Glass." *The Century Magazine* 48:554–558.
Taylor 1957	Taylor, Doris Mae. 1957. "Cosa: Black-Glaze Pottery." *MAAR* 25:65–193.
Tel Anafa 1970	[See Weinberg 1970] Weinberg, Gladys Davidson. 1970. "Hellenistic Glass from Tel Anafa in Upper Galilee." *JGS* 12:17–27.
Thill 1975	Thill, Gérard. 1975. "Frühkaiserzeitliche Grabbeigaben von Hellingen (G. H. Luxemburg)." *Archäologisches Korrespondenzblatt* 5 (1):69–79.
Tipasa 1967	Lancel, Serge. 1967. *Verrerie antique de Tipasa*. Paris: E. de Boccard.
Trier 1977	Goethert-Polaschek, Karin. 1977. *Katalog der römischen Gläser des Rheinischen Landesmuseums Trier*. Trierer Grabungen und Forschungen 9. Mainz am Rhein: Philipp von Zabern.
Trowbridge 1930	Trowbridge, Mary Luella. 1930. *Philological Studies in Ancient Glass*. Studies in Language and Literature 13 (3–4). Urbana: University of Illinois.
Ulrich 1914	Ulrich, R. A. 1914. *Die Gräberfelder in der Umgebung von Bellinzona, Kt. Tessin*, 2 vols. Zürich: Verlag des Schweiz. Landesmuseums.
Vasa 1958	Harden, Donald B. 1958. "The Glass." In *Roman Tombs at 'Kambi', Vasa*, ed. Joan du Plat Taylor, 46–60. Nicosia: Department of Antiquities.
Vatican Coll. 1975	Fremersdorf, Fritz. 1975. *Antikes, islamisches und mittelalterliches Glas sowie kleinere Arbeiten aus Stein, Gagat und verwandten Stoffen in den vatikanischen Sammlungen Roms*. Catalogo del Museo Sacro della Biblioteca Apostolica Vaticana. Vatican City: Biblioteca Apostolica Vaticana.
Verulamium 1972	Charlesworth, Dorothy. 1972. "The Glass." In *Verulamium Excavations,* vol. 1, ed. S. S. Frere, 196–215. Reports of the Research Committee of the Society of Antiquaries of London 28. London: Society of Antiquaries.
Vessberg 1956	Vessberg, Olof. 1956. "Glass." In *The Swedish Cyprus Expedition*, vol. 4, part 3: *The Hellenistic and Roman Periods in Cyprus*, ed. O. Vessberg and A. Westholm, 128–174, 193–219. Stockholm.
Vetera 1917	Lehner, H. 1917. "Ausgrabungsberichte des Provinzialmuseums in Bonn. Vetera, Ausgrabungen in den Jahren 1913/14–1917." *Bönner Jahrbücher* 124:165–168.
Villa Potenza 1971	Mercando, L. 1971. "Villa Potenza: rinvenimento in proprietà AGIP." *NSc* 25:402–417.
Ville 1964	Ville, G. 1964. *Les coupes de Trimalcion figurant des gladiateurs et une série de verres sigillés gaulois*. In *Hommages à Jean Bayet*, ed. Marcel Renard and Robert Schilling, 722–733. Collection Latomus 70. Paris: Latomus.
Vindonissa 1960	[See Berger 1960] Berger, Ludwig. 1960. *Römische Gläser aus Vindonissa*. Veröffentlichungen der Gesellschaft Pro Vindonissa. Basel: Birkhäuser.
Volterra 1975	Cristofani, M. 1975. "Scavi nella necropoli del Portone (1971): Tomba ellenistica." *NSc* 29:5–35.

Waddesdon Manor Coll. 1977	Charleston, Robert J. 1977. "Glass." In *The James A. de Rothschild Collection at Waddesdon Manor*, vol. 8: *Glass and Enamels*. Fribourg: National Trust, Office du Livre.
Wadi ed-Daliyeh 1974	Weinberg, Gladys Davidson. 1974. "Discoveries in the Wâdi ed-Dâliyeh: Glass Vessels." *Annual of the American Schools of Oriental Research* 41:103–105.
Weinberg 1961	Weinberg, Gladys Davidson. 1961. "Hellenistic Glass Vessels from the Athenian Agora." *Hesperia* 30 (4):380–392.
Weinberg 1965	Weinberg, Gladys Davidson. 1965. "The Glass Vessels." In G. D. Weinberg, Virginia R. Grace, et al., "The Antikythera Shipwreck Reconsidered." *Transactions of the American Philosophical Society* 55 (3):30–39.
Weinberg 1970	Weinberg, Gladys Davidson. 1970. "Hellenistic Glass from Tel Anafa in Upper Galilee." *JGS* 12:17–27.
Weinberg 1972	Weinberg, Gladys Davidson. 1972. "Mold-Blown Beakers with Mythological Scenes." *JGS* 14:26–47.
Weinberg 1973	Weinberg, Gladys Davidson. 1973. "Notes on Glass from Upper Galilee." *JGS* 15:35–51.
Werner 1953	Werner, J. 1953. "Zu älterkaiserzeitlichen Glasbechern (Eggers Form 185–187)." *Germania* 31:61–64.
Wheaton Coll. 1971	Dusenbery, Elsbeth B. 1971. "Ancient Glass in the Collections of Wheaton College." *JGS* 13:9–33.
Whitehouse 1983	Whitehouse, David. 1983. "Medieval Glass in Italy: Some Recent Developments." *JGS* 25:115–120.
Wright 1980	Wright, Kathleen Slane. 1980. "A Tiberian Pottery Deposit from Corinth." *Hesperia* 49 (2):135–177.
Yadin 1963	Yadin, Yigael. 1963. *The Finds from the Bar Kokhba Period in the Cave of Letters*. Jerusalem: Israel Exploration Society.
Yale Coll. 1980	Matheson, Susan B. 1980. *Ancient Glass in the Yale University Art Gallery*. New Haven: Yale University Art Gallery.
Yalouris 1966	Yalouris, N. 1966. "Ἑλληνιστικὸν νεκροτάφειον Γιαλόβης Παλαιοναυαρίνου (Κορυφάσιον)." *Archaiologikon Deltion* 21 (B1):164–165.

Plates

Profile Drawings by John Huffstot

Scale 1:2 except where otherwise noted.

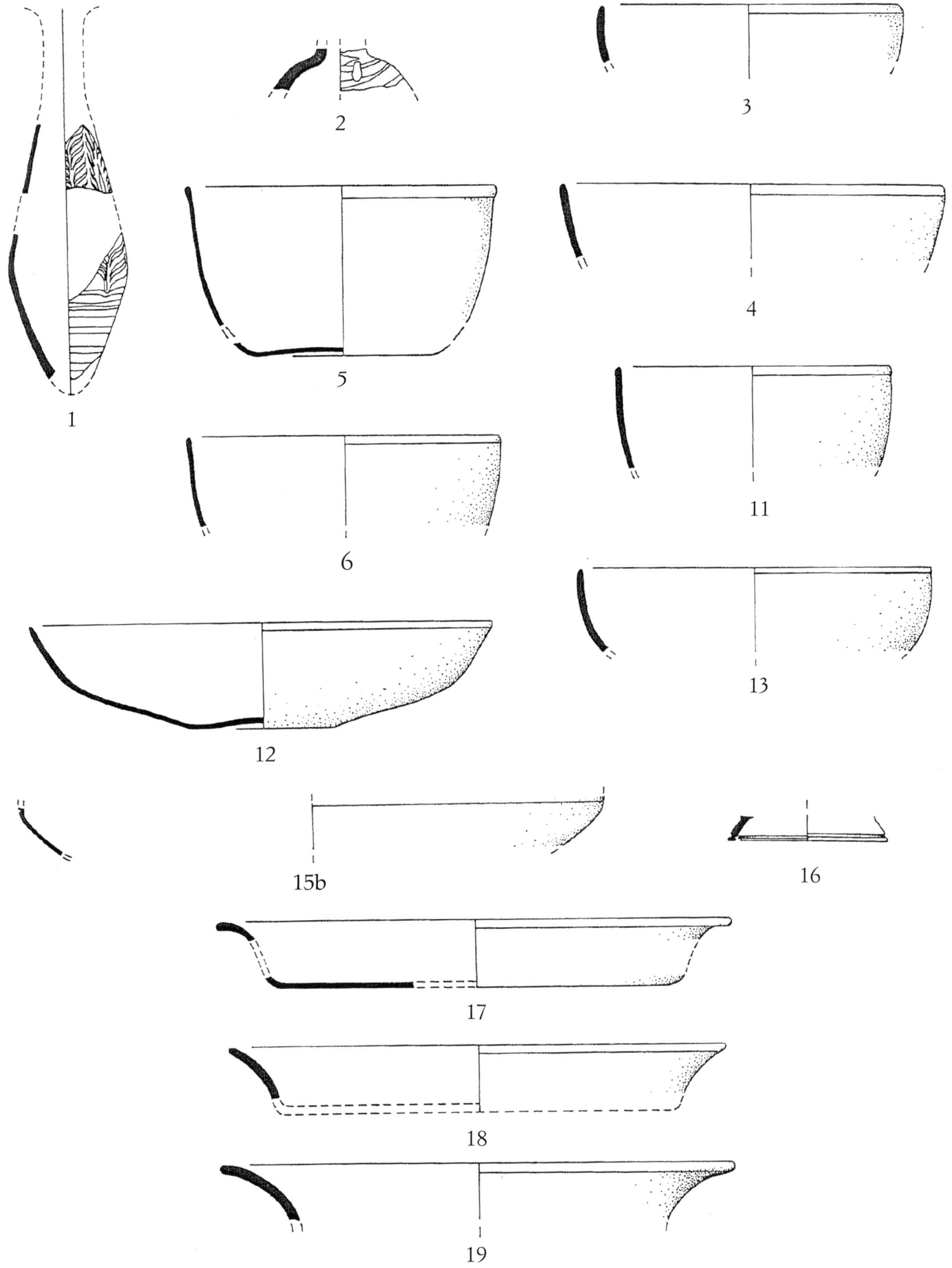

Plate 1. Core-made unguent bottles. Cast mosaic-glass vessels: network, striped, gold-band, and composite patterns.

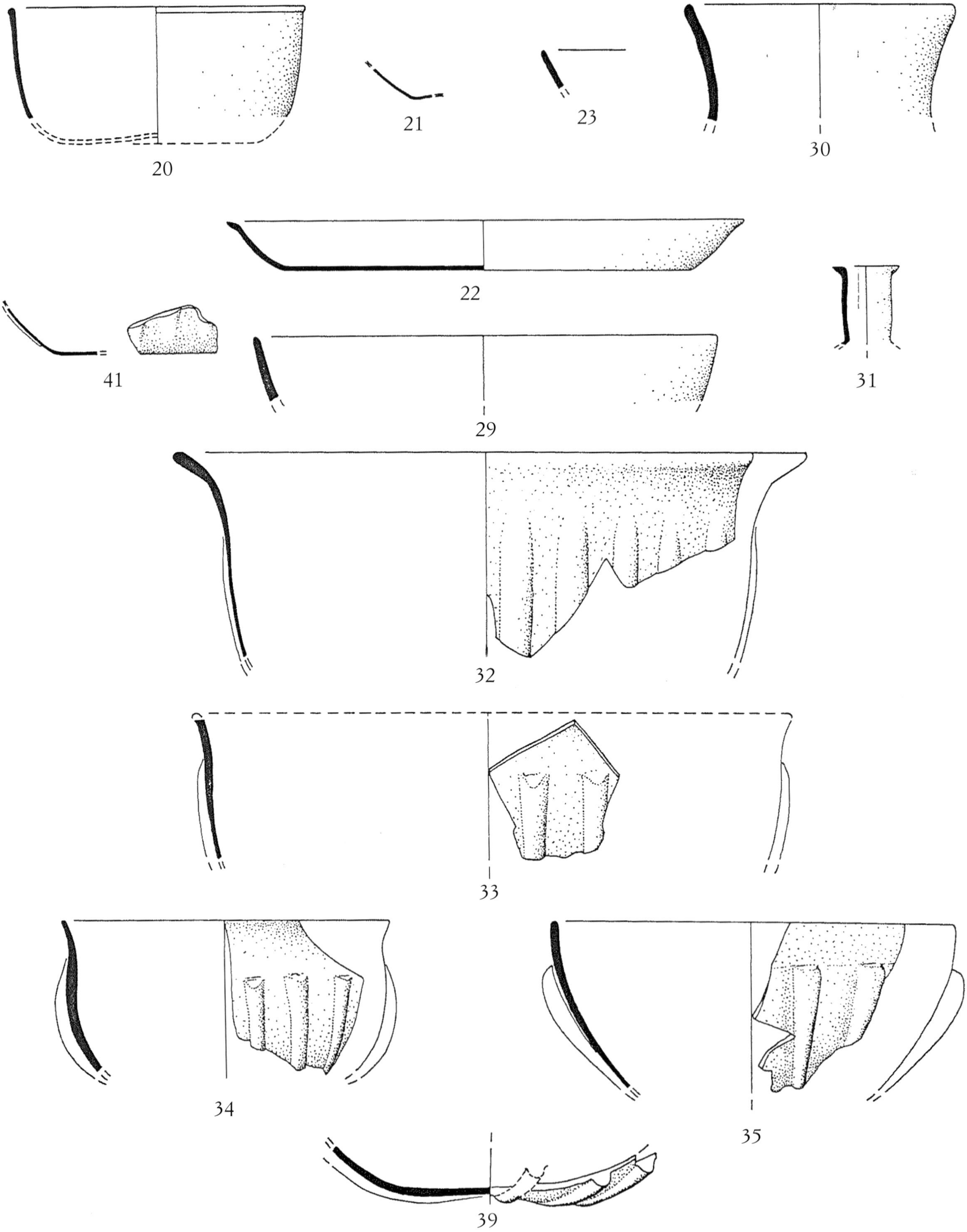

Plate 2. Cast mosaic-glass vessels: composite and marbled patterns. Cast marbled mosaic-glass ribbed bowls.

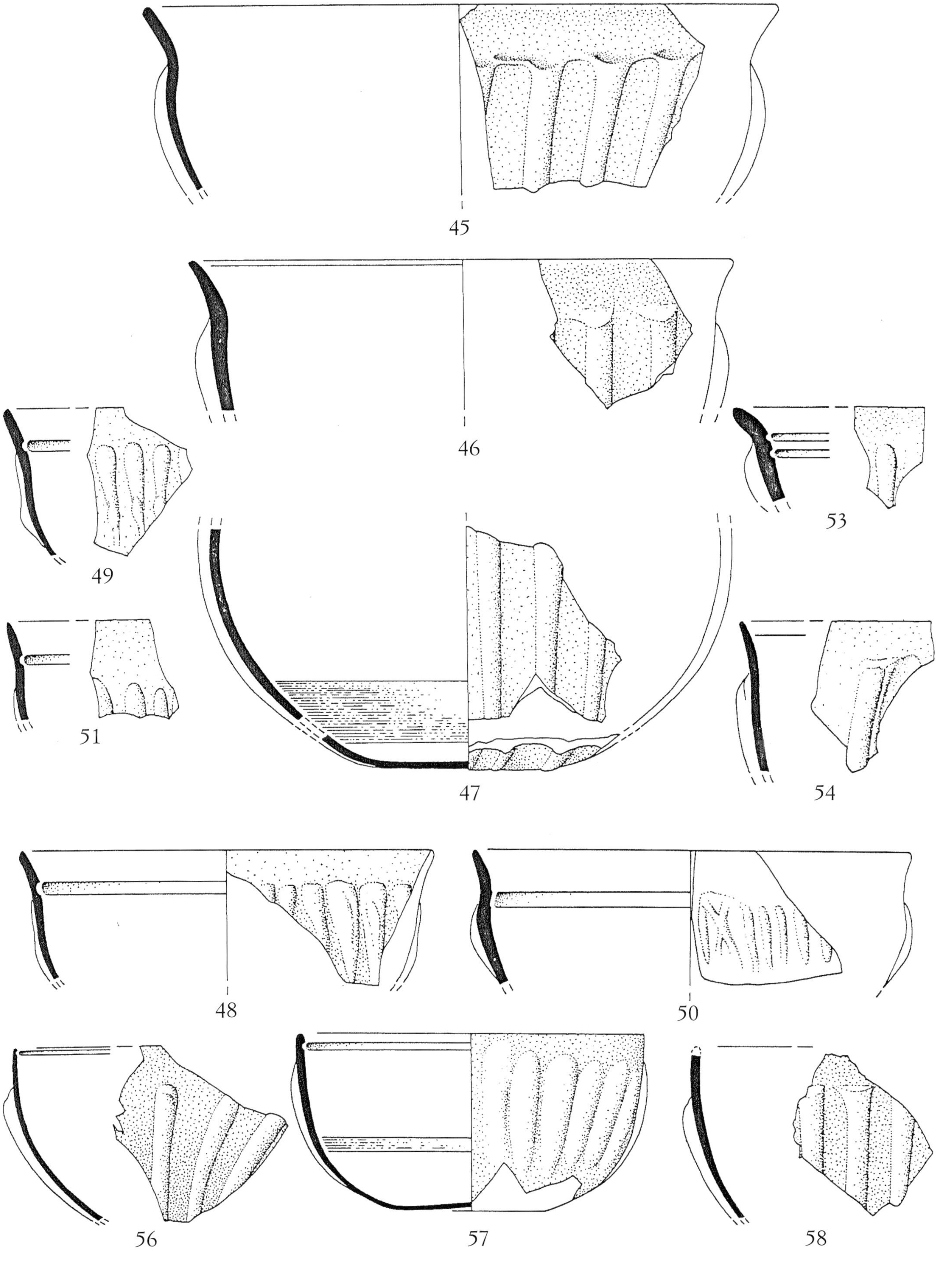

Plate 3. Cast monochrome colored ribbed bowls.

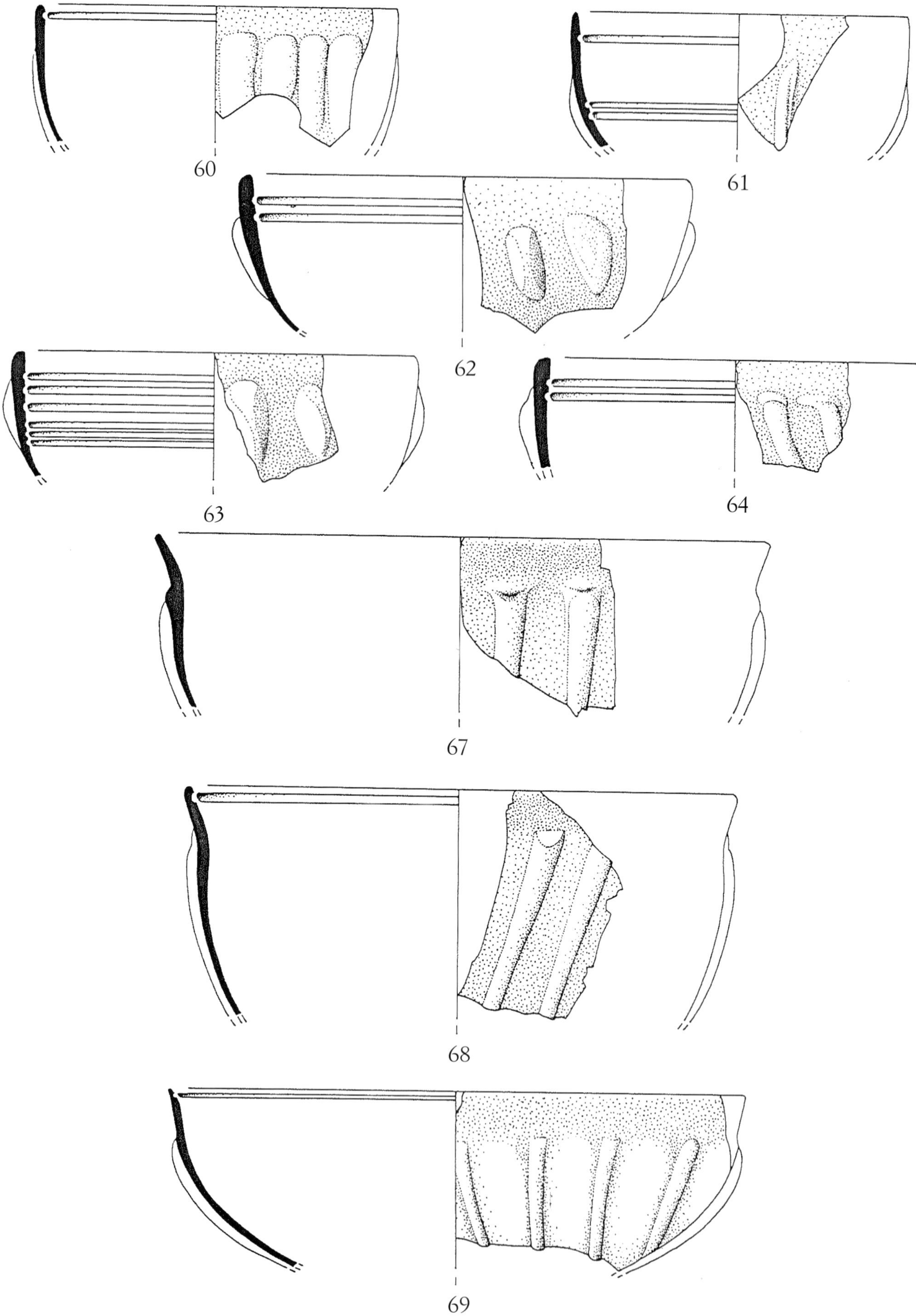

Plate 4. Cast monochrome colored ribbed bowls and monochrome naturally colored ribbed bowls.

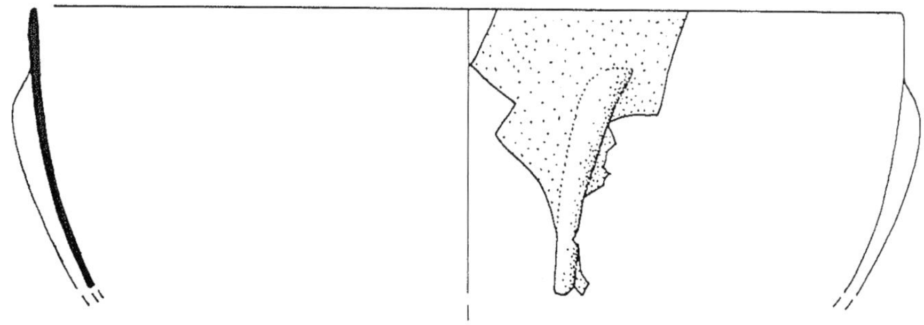

71

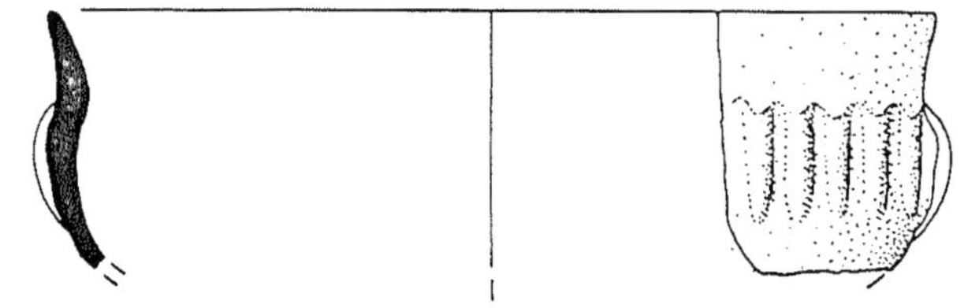

70

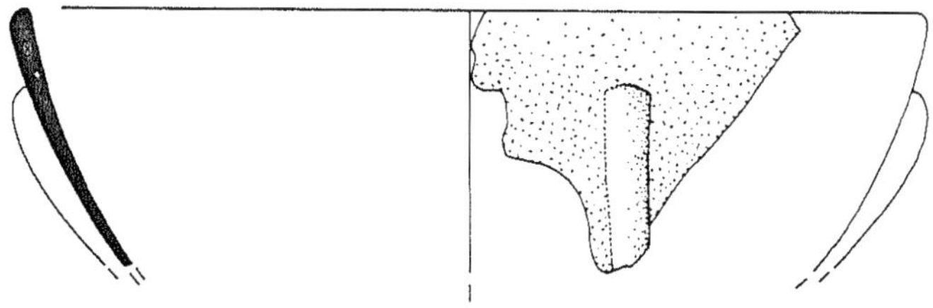

72

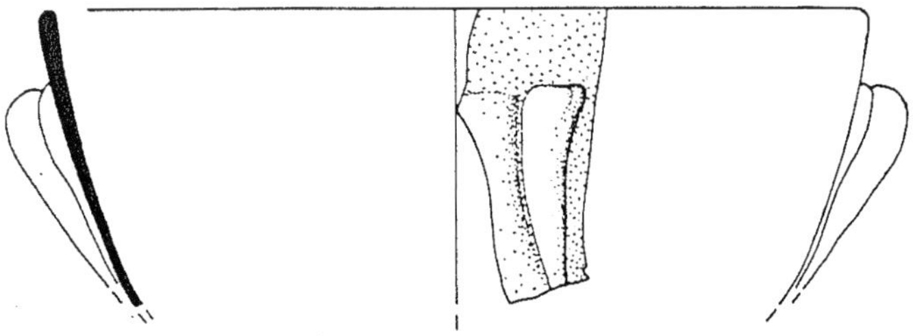

74

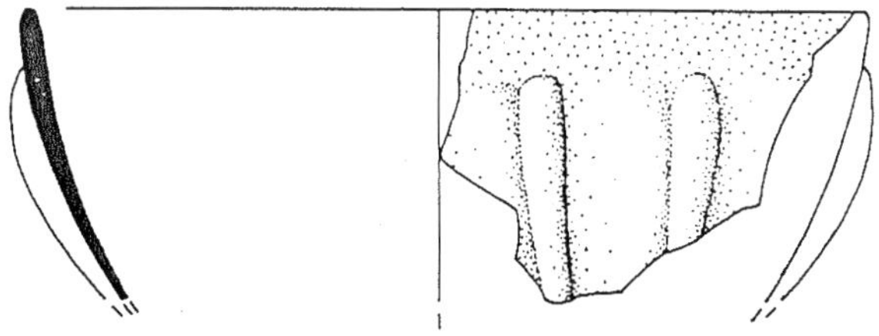

75

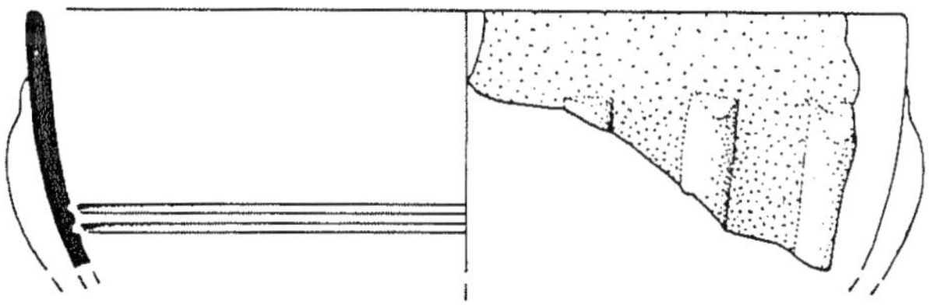

76

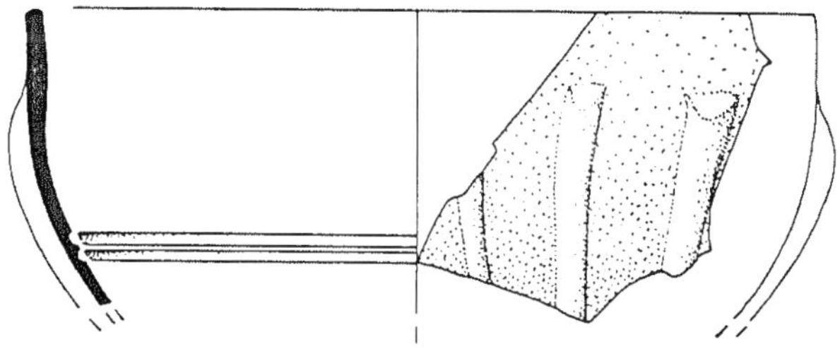

77

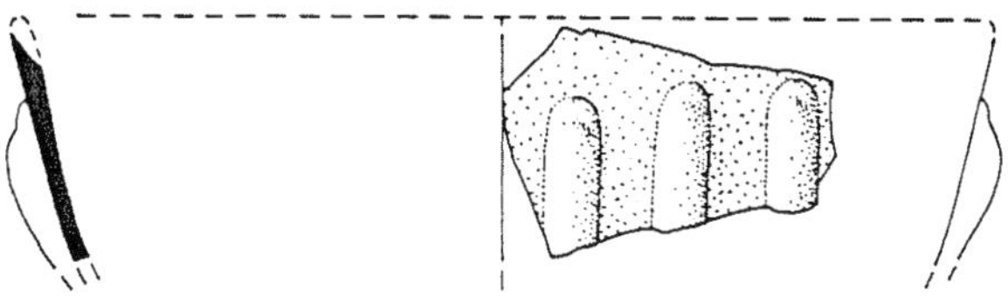

80

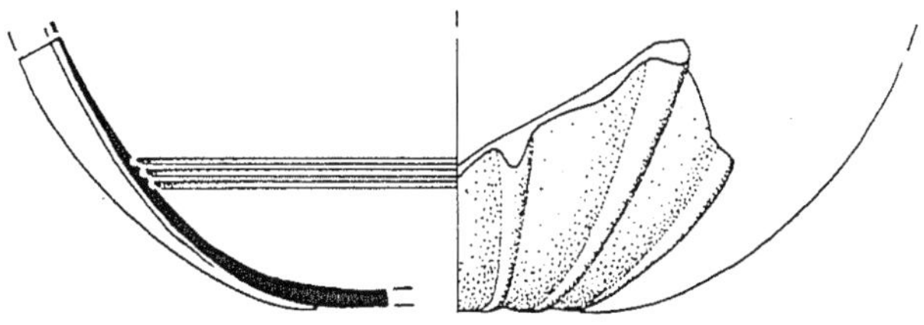

81

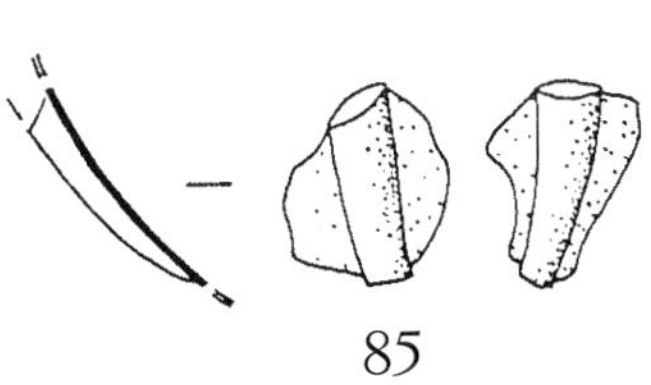

85

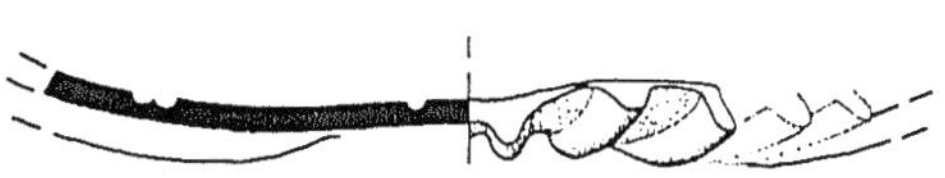

82

Plate 5. Cast monochrome naturally colored ribbed bowls.

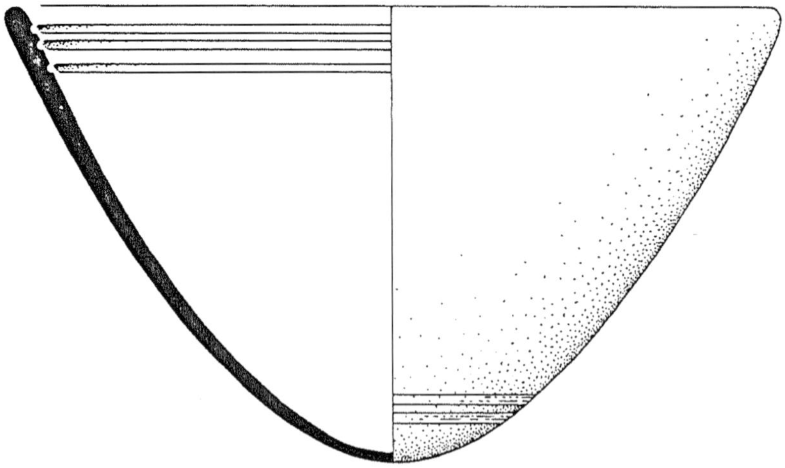

92

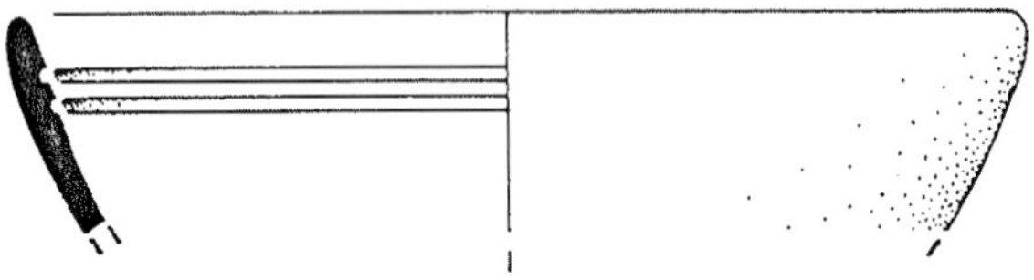

93

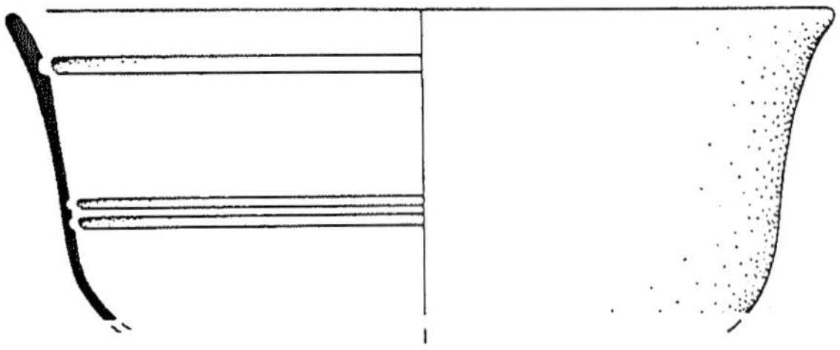

94

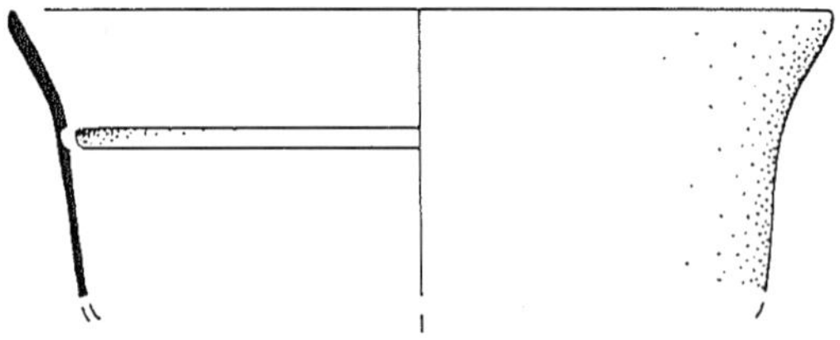

95

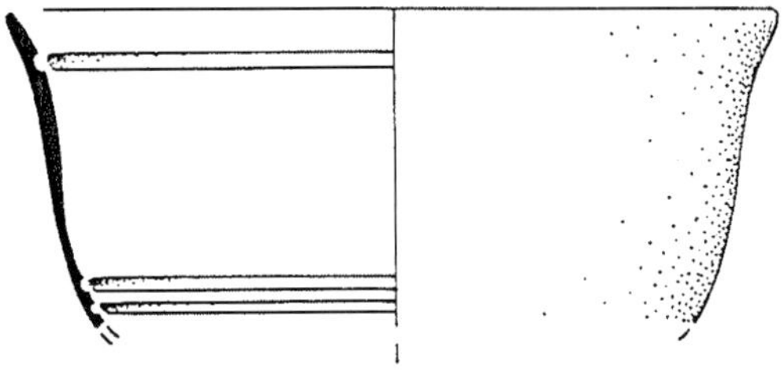

96

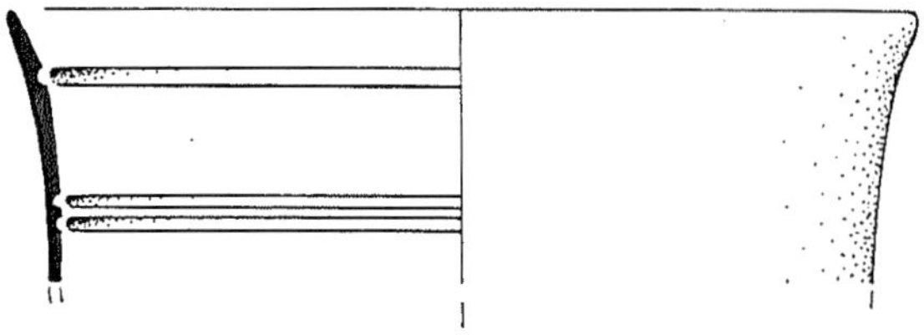

97

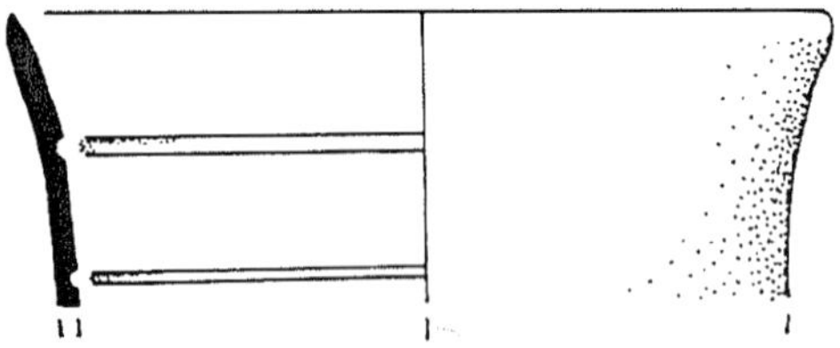

98

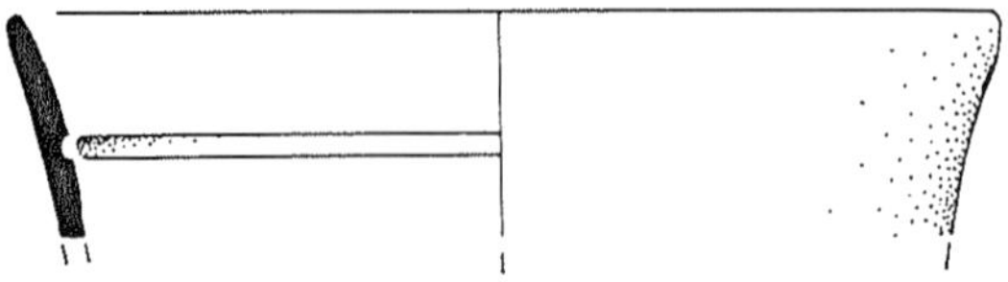

100

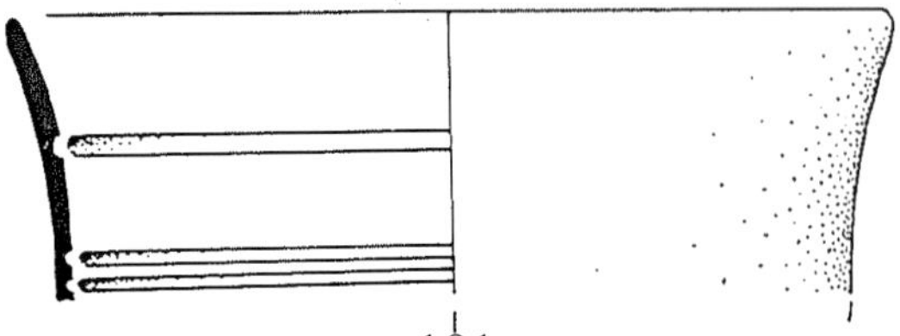

101

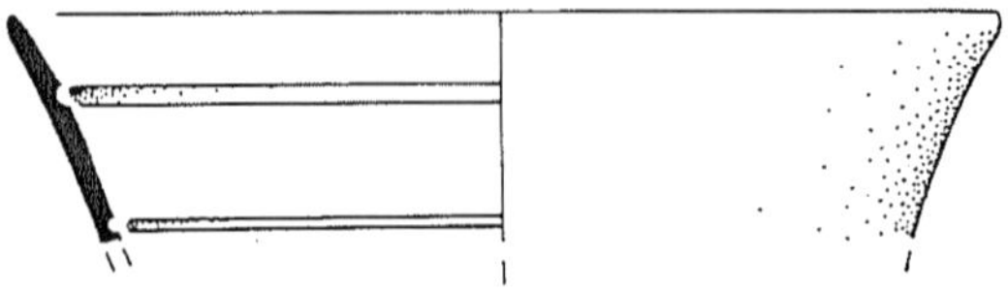

104

Plate 6. Cast grooved bowls. Cast linear-cut bowls, monochrome colored.

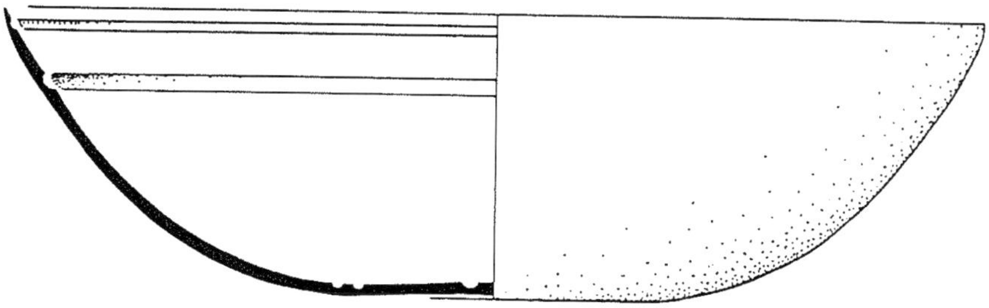
106

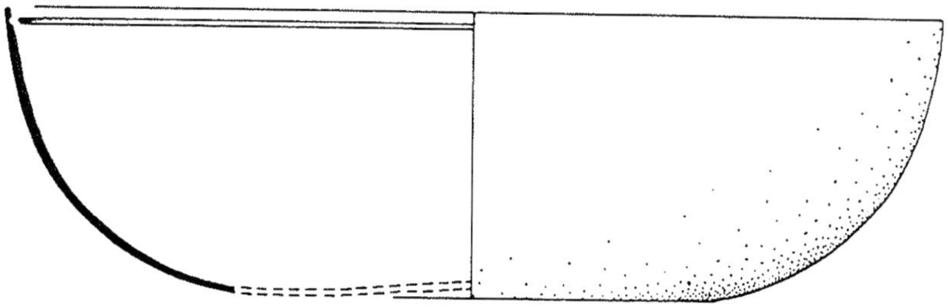
107

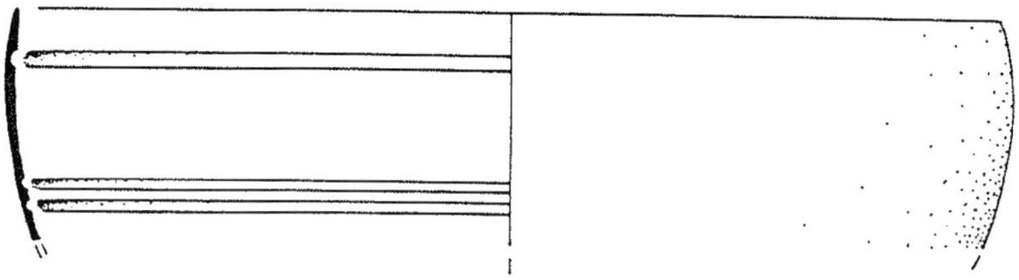
108

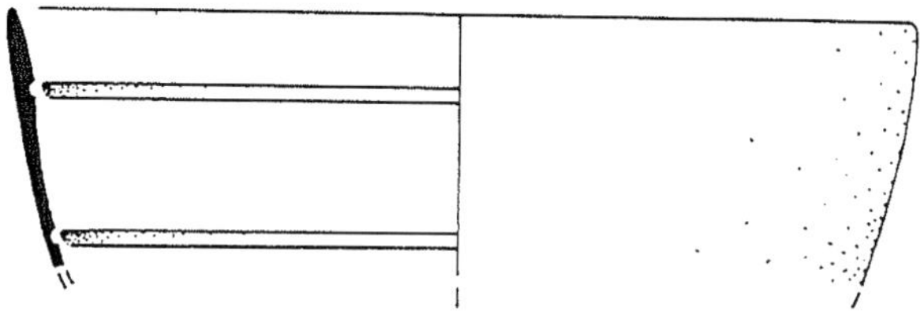
115

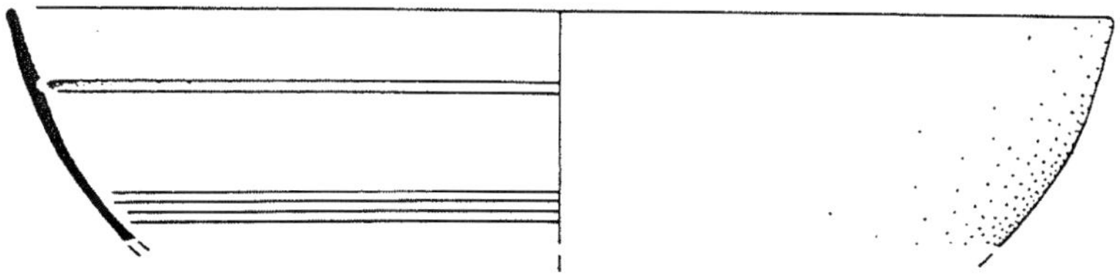
109

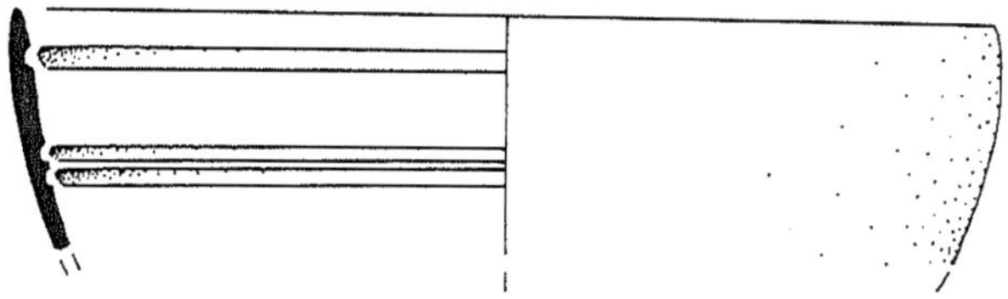
114

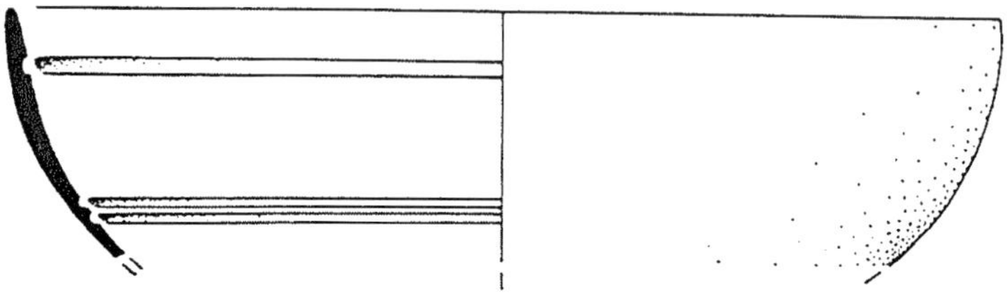
110

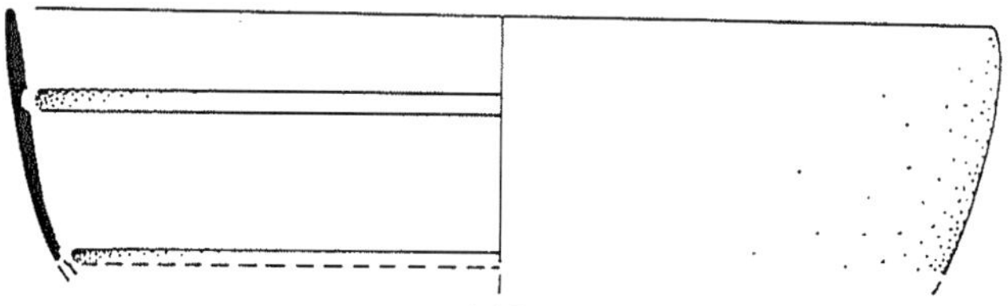
113

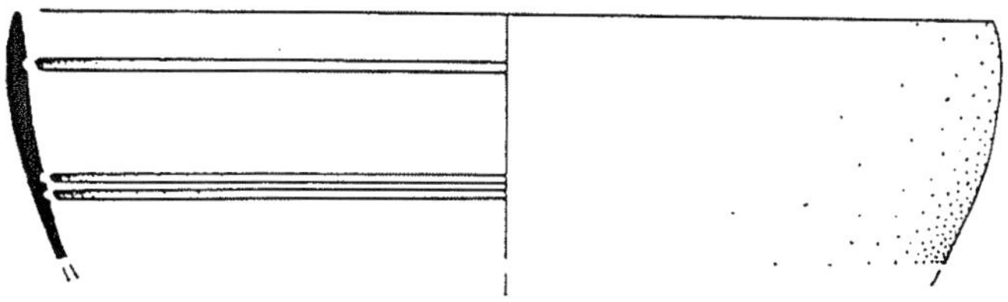
111

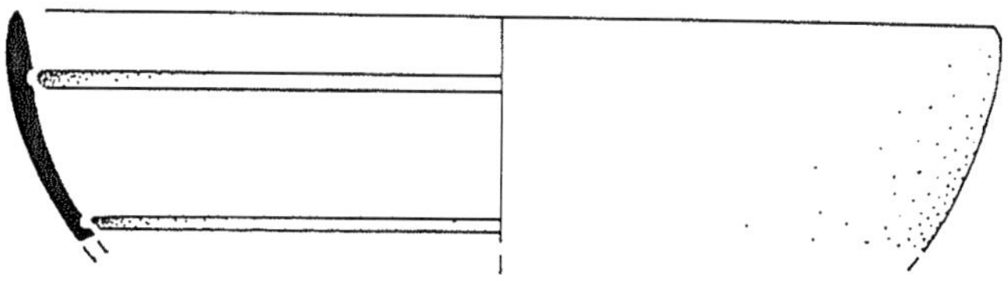
112

118

119

Plate 7. Cast linear-cut bowls, monochrome colored, naturally colored, and colorless.

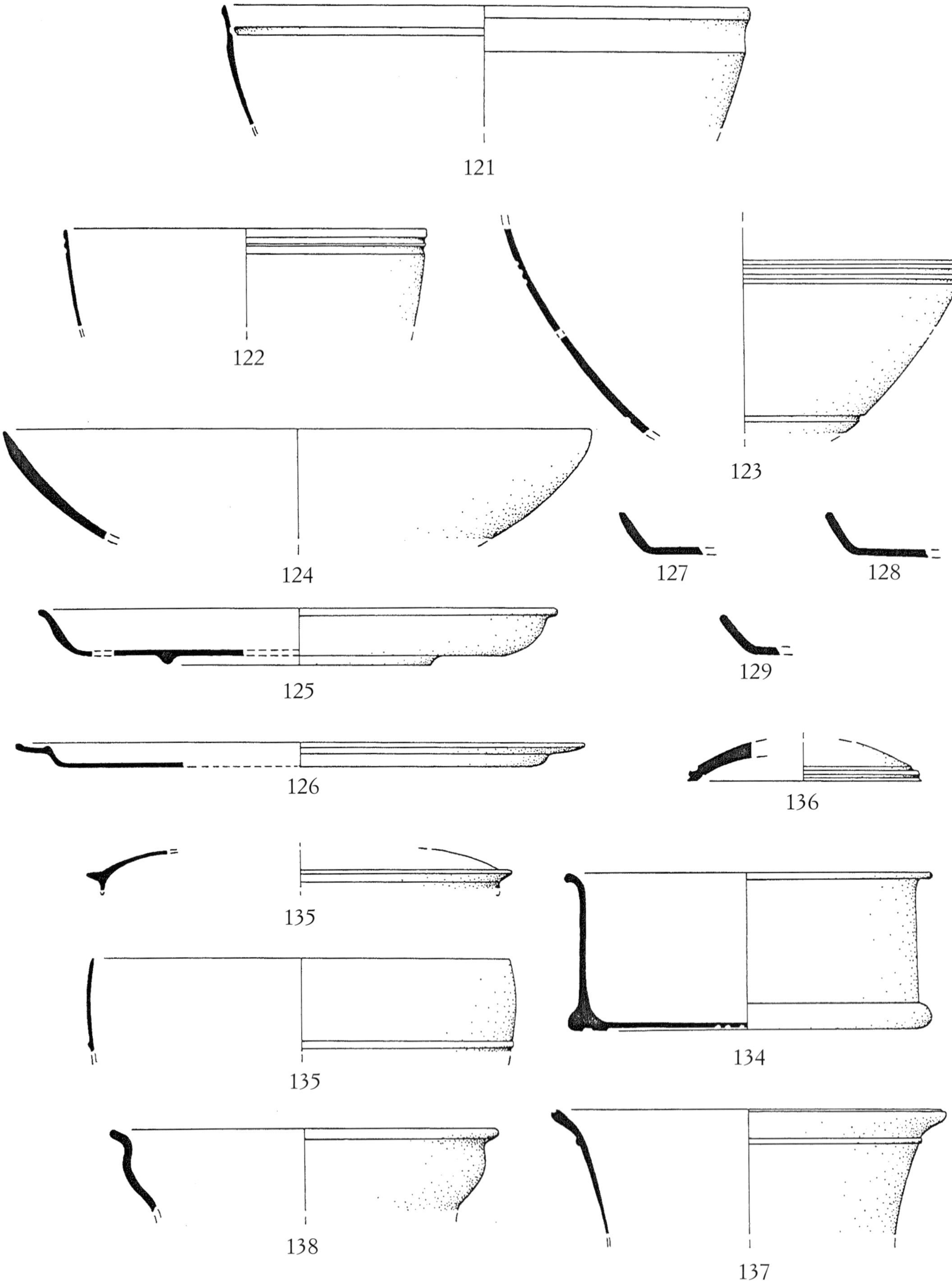

Plate 8. Cast linear-cut bowls: variants. Cast translucent colored fine wares.

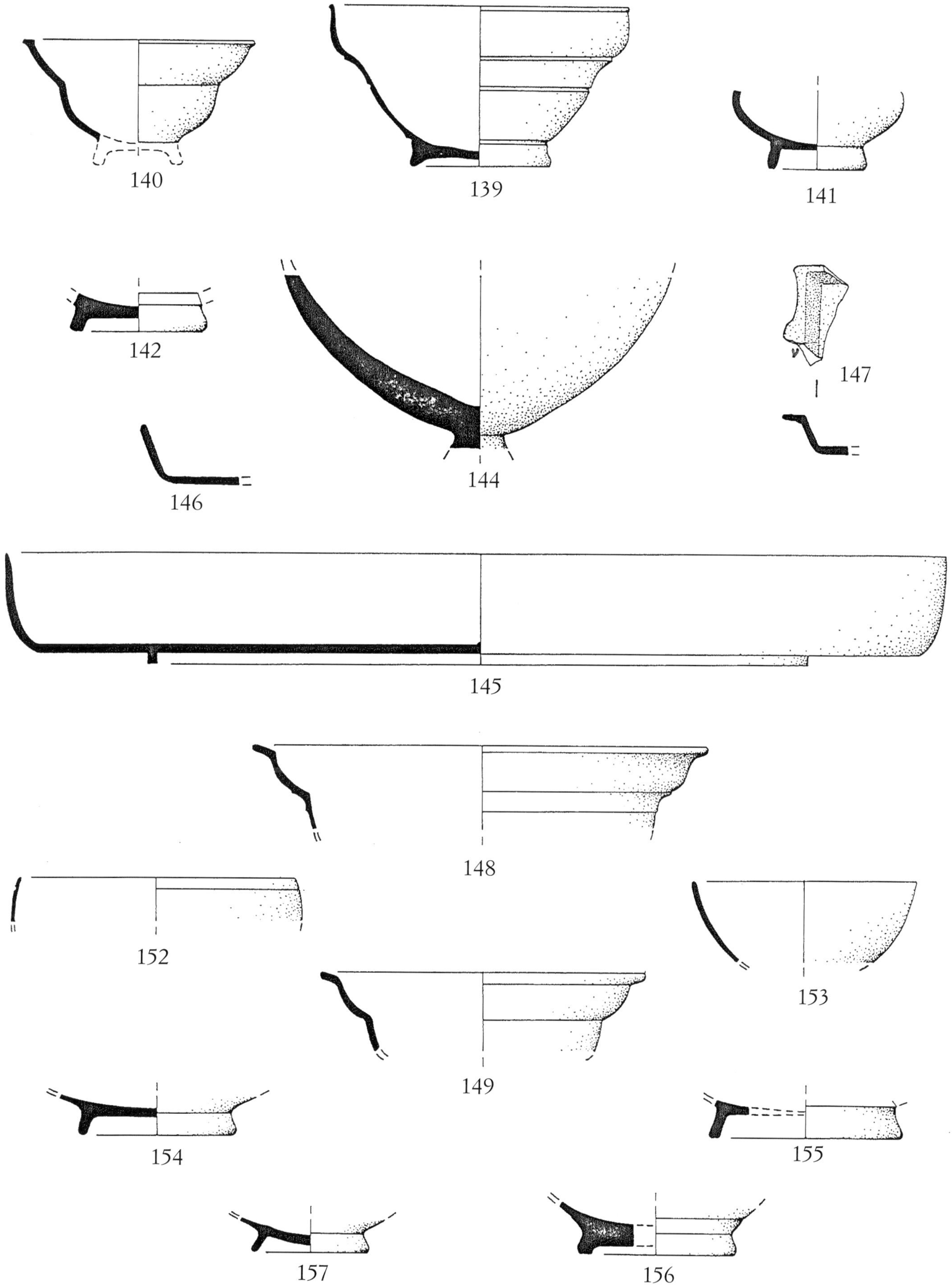

Plate 9. Cast translucent colored fine wares. Cast opaque colored fine wares.

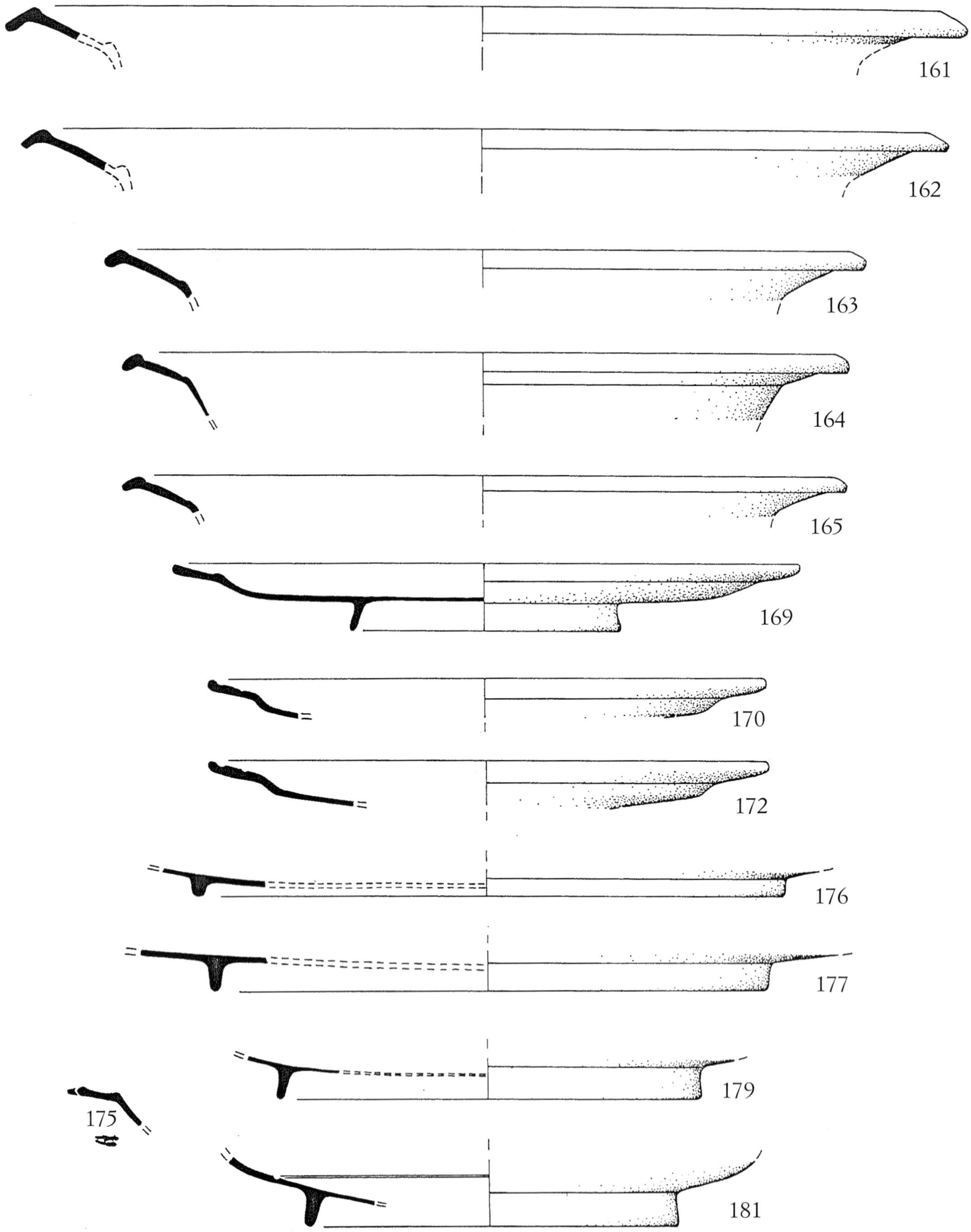

Plate 10. Cast colorless vessels with broad, overhanging rims.

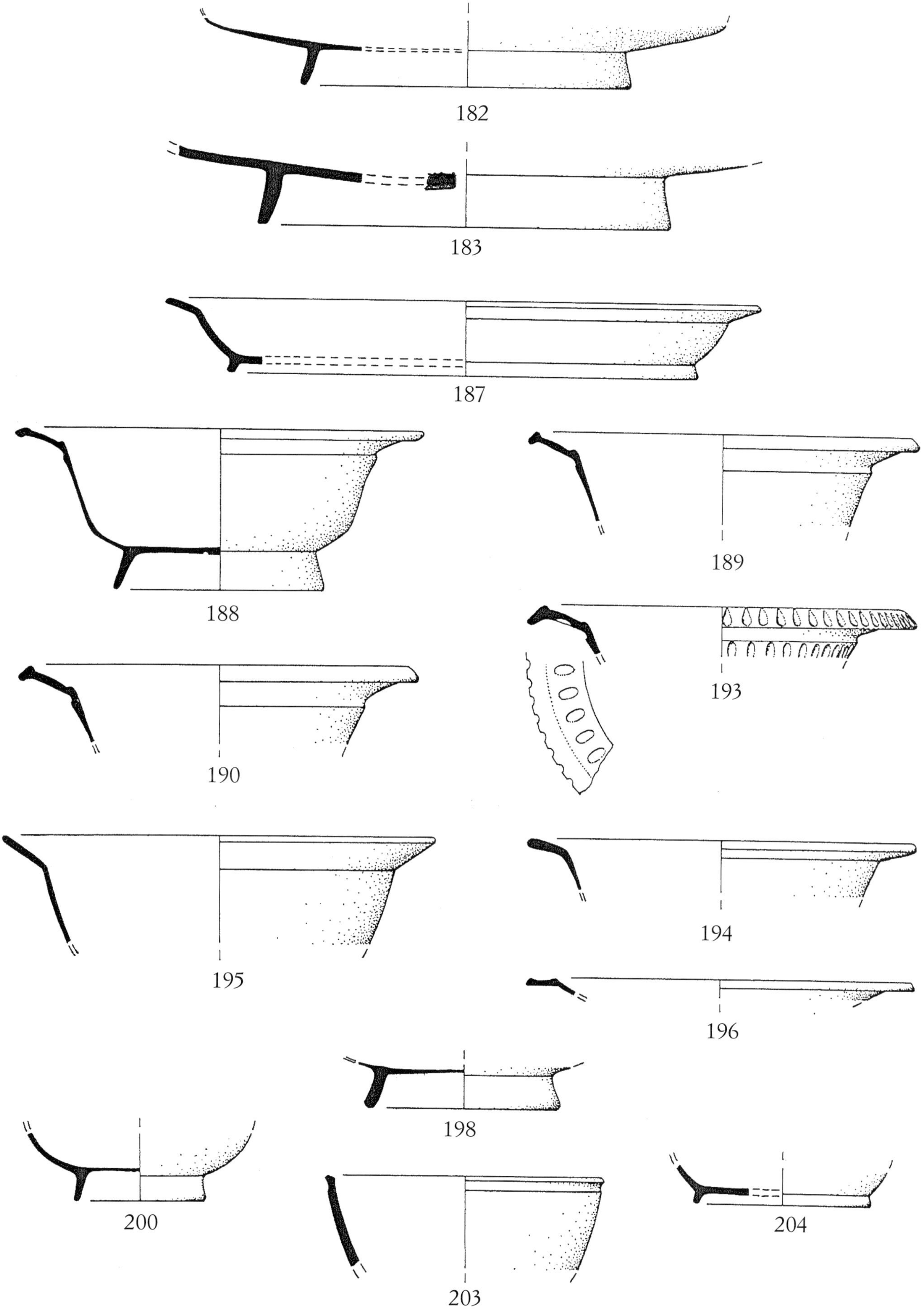

Plate 11. Cast colorless vessels with broad, overhanging rims. Cast colorless vessels miscellaneous.

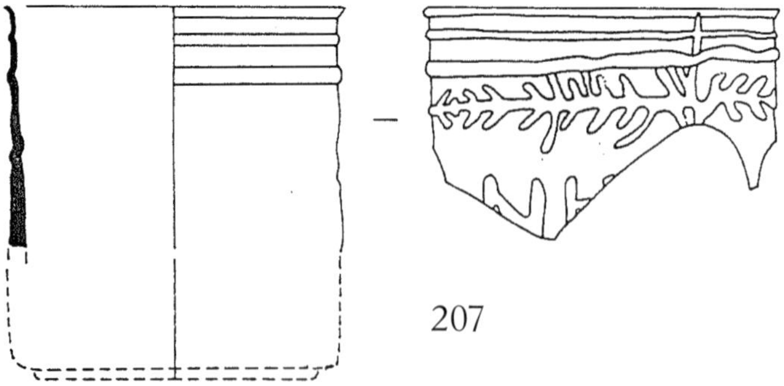

207

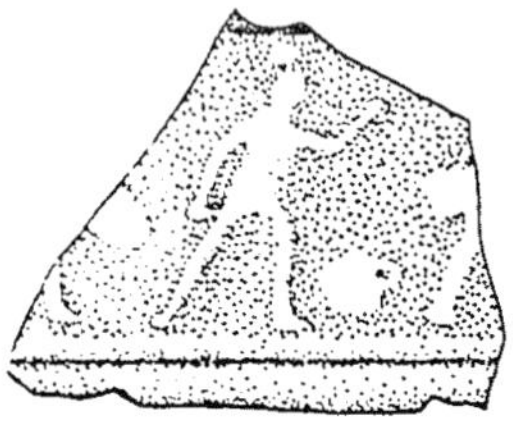

205

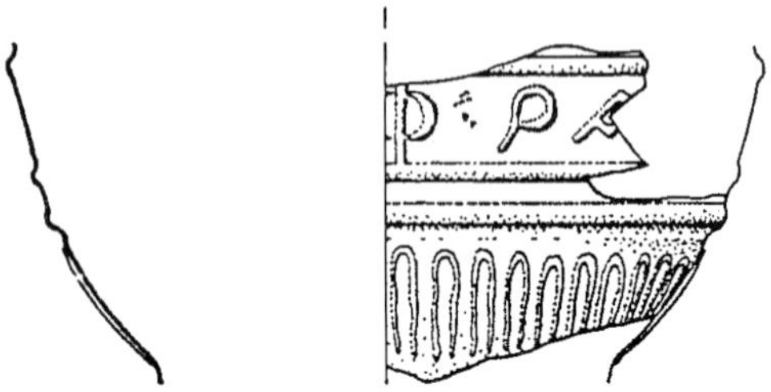

206

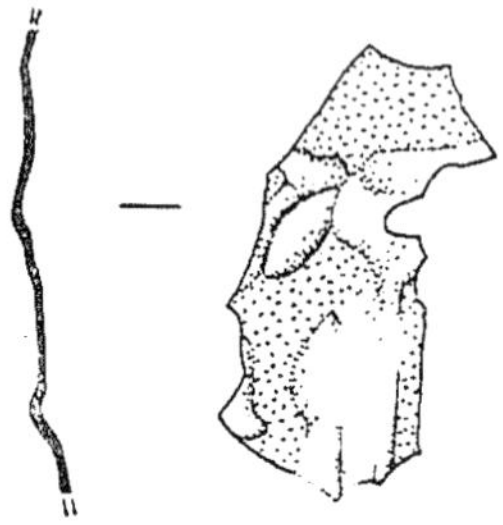

220

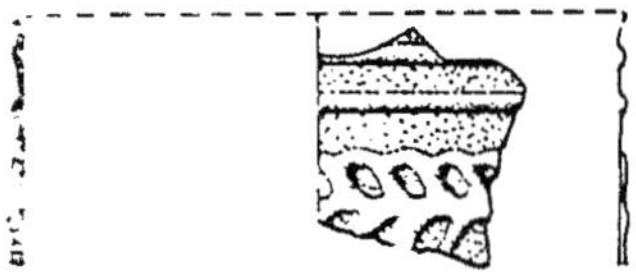

209

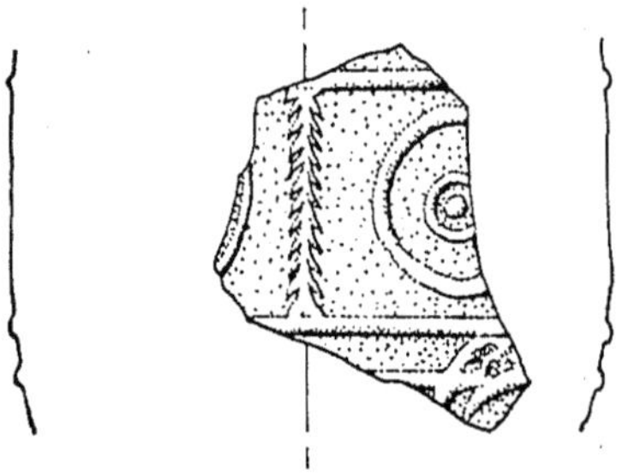

208

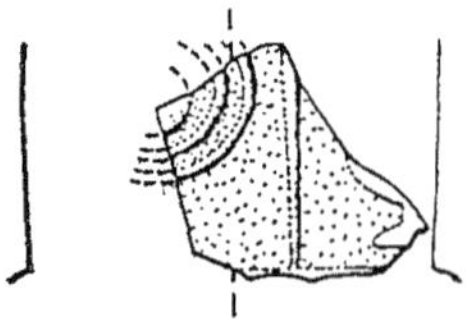

210

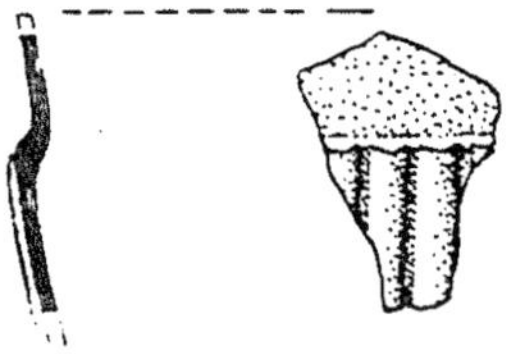

211

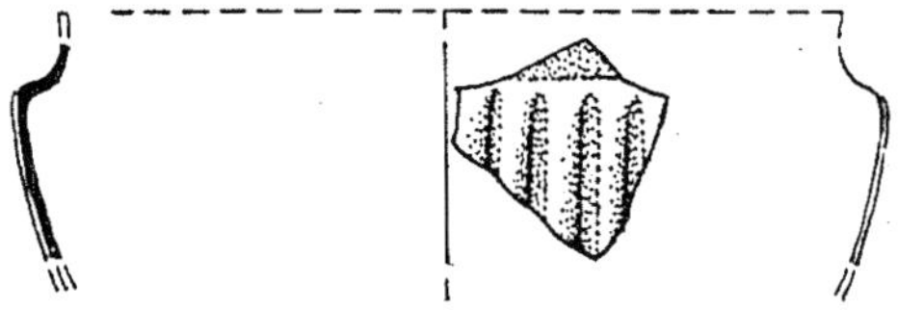

212

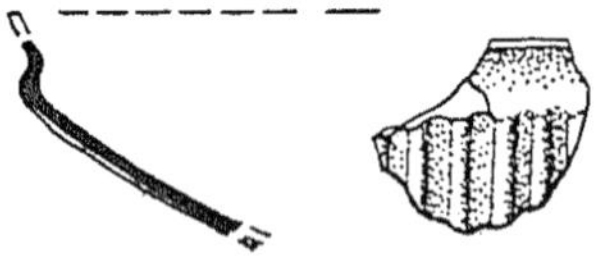

213

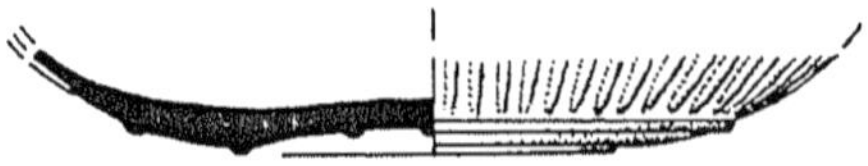

214

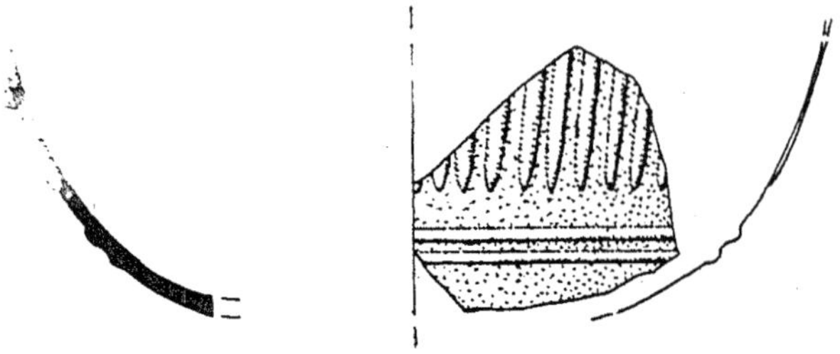

215

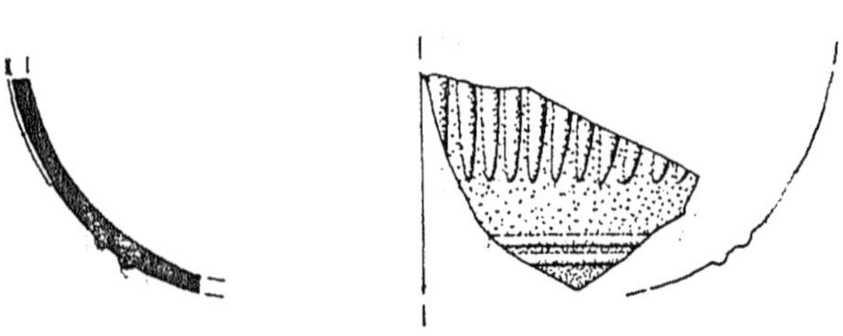

217

Plate 12. Mold-blown decorated beakers and bowls.

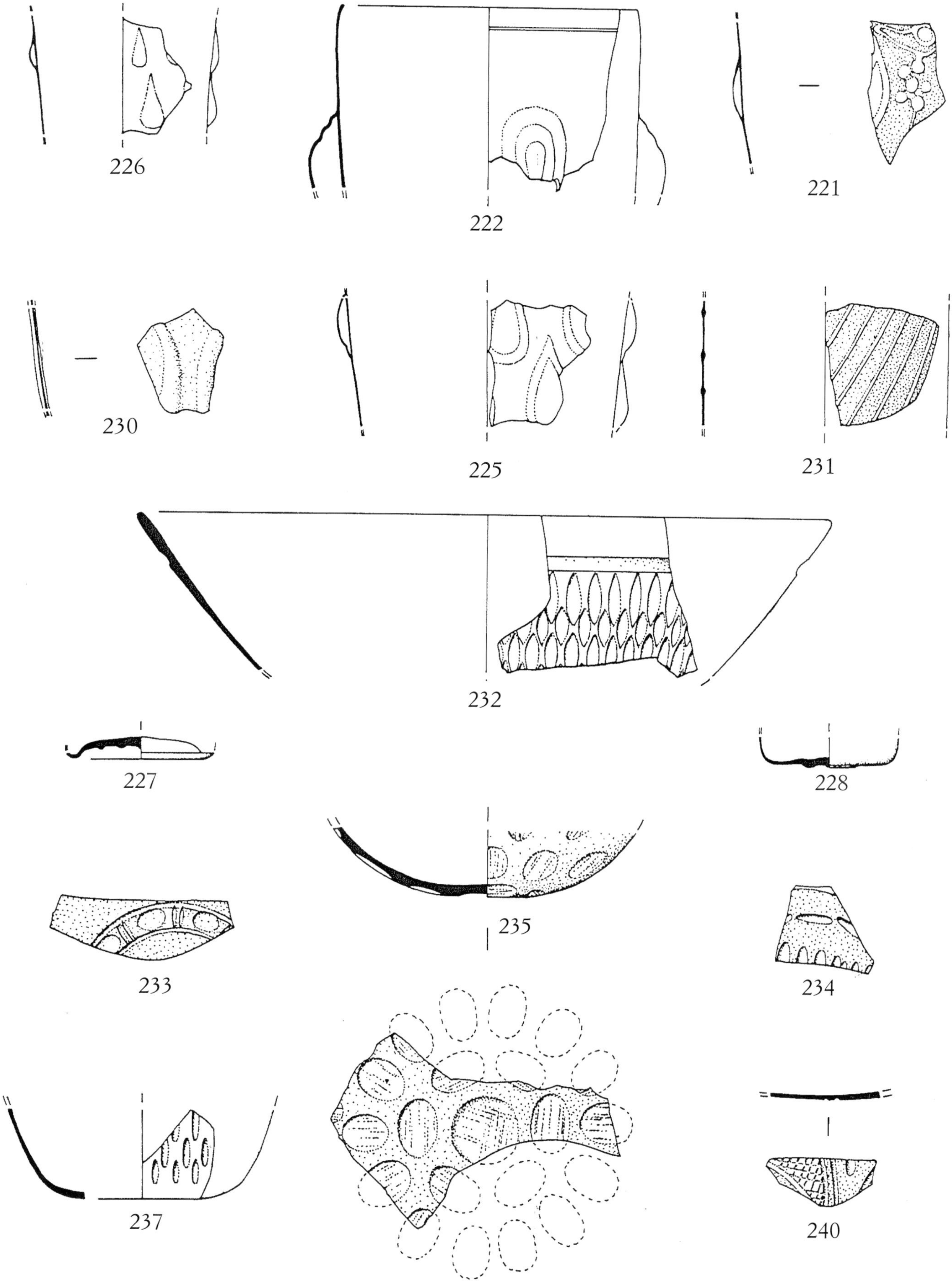

Plate 13. Mold-blown and optic-blown decorated beakers. Cast and blown facet-cut vessels.

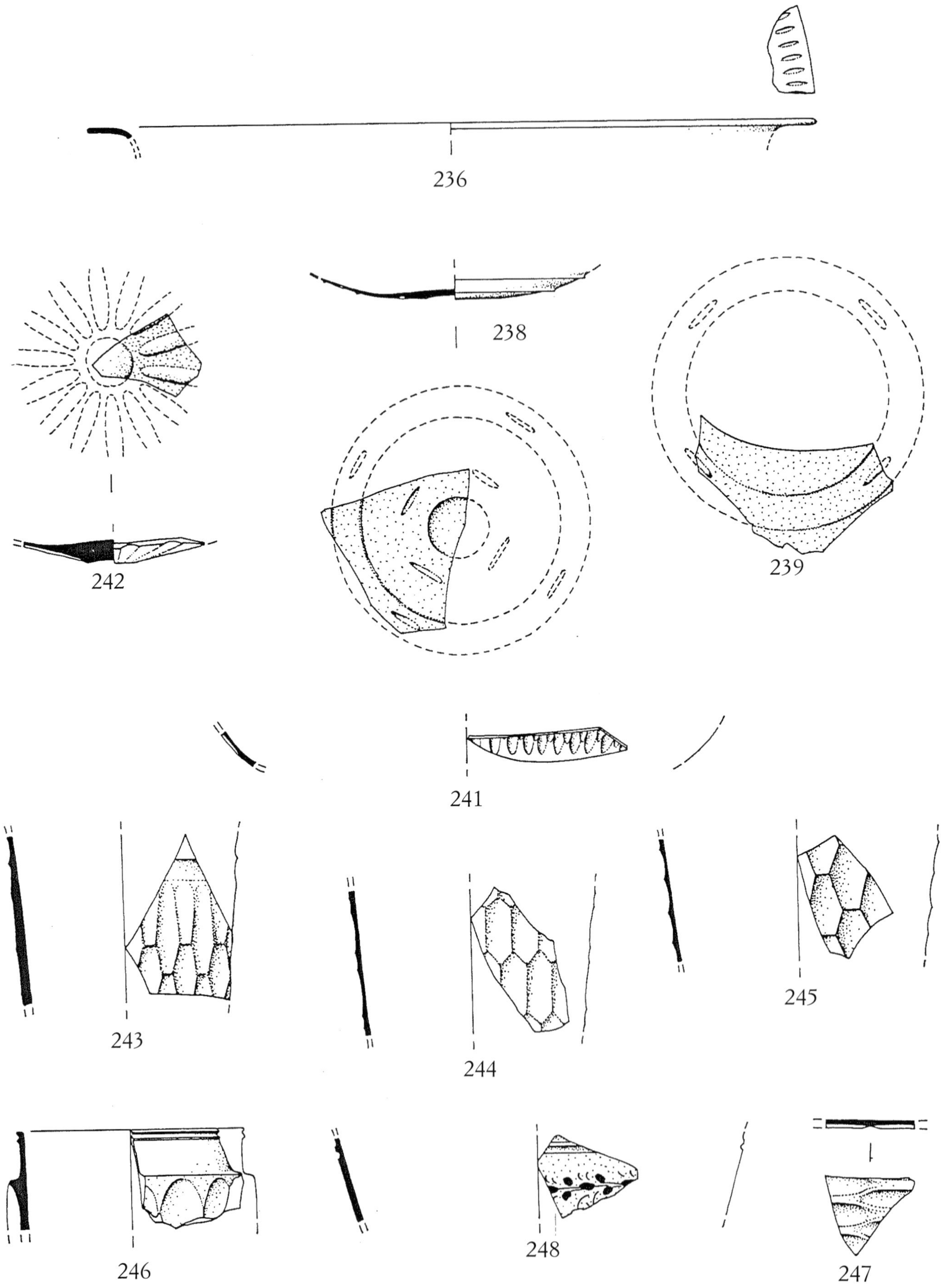

Plate 14. Cast and blown facet-cut and engraved vessels.

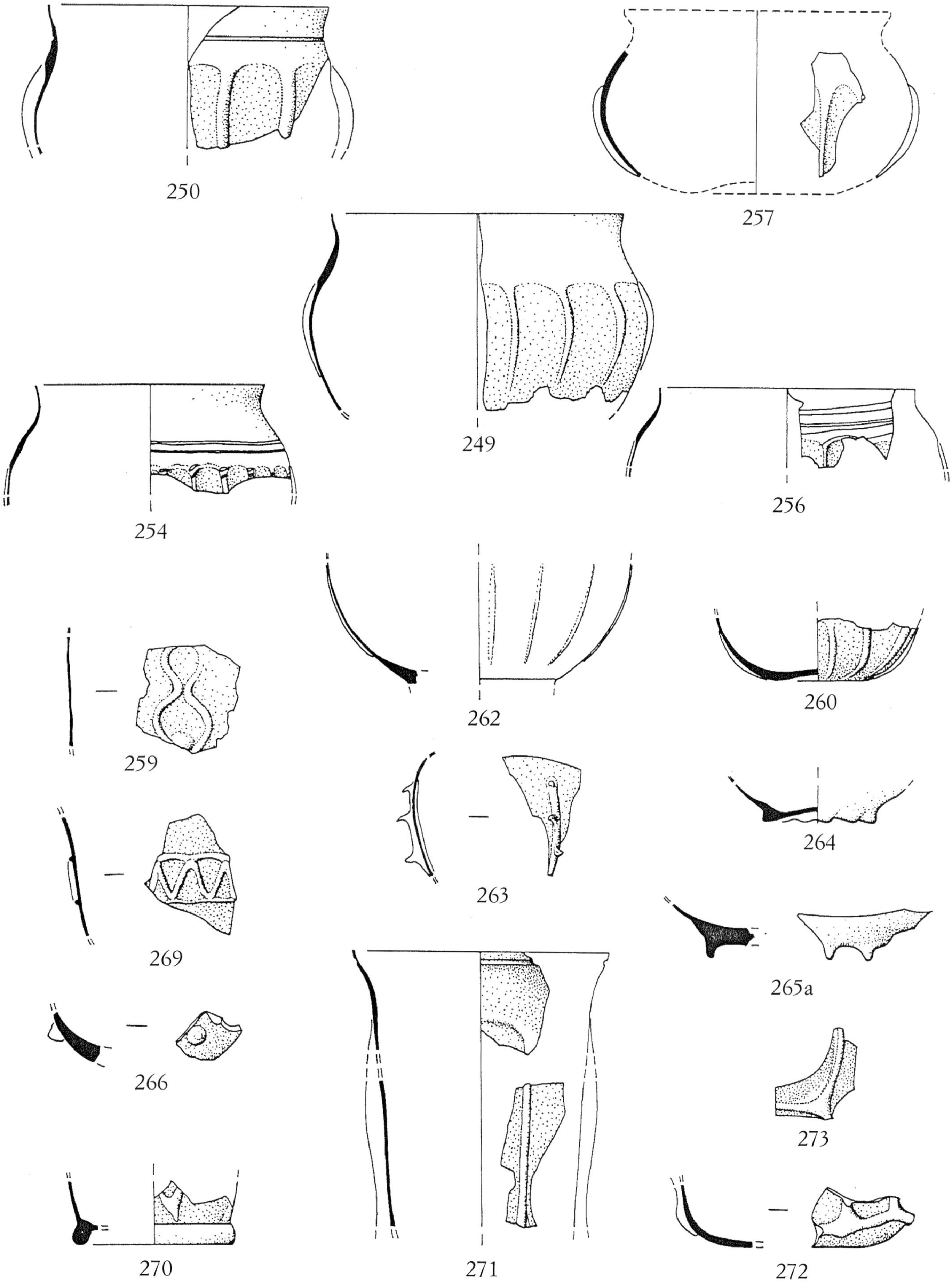

Plate 15. Blown vessels with pinched, trailed, and applied decoration.

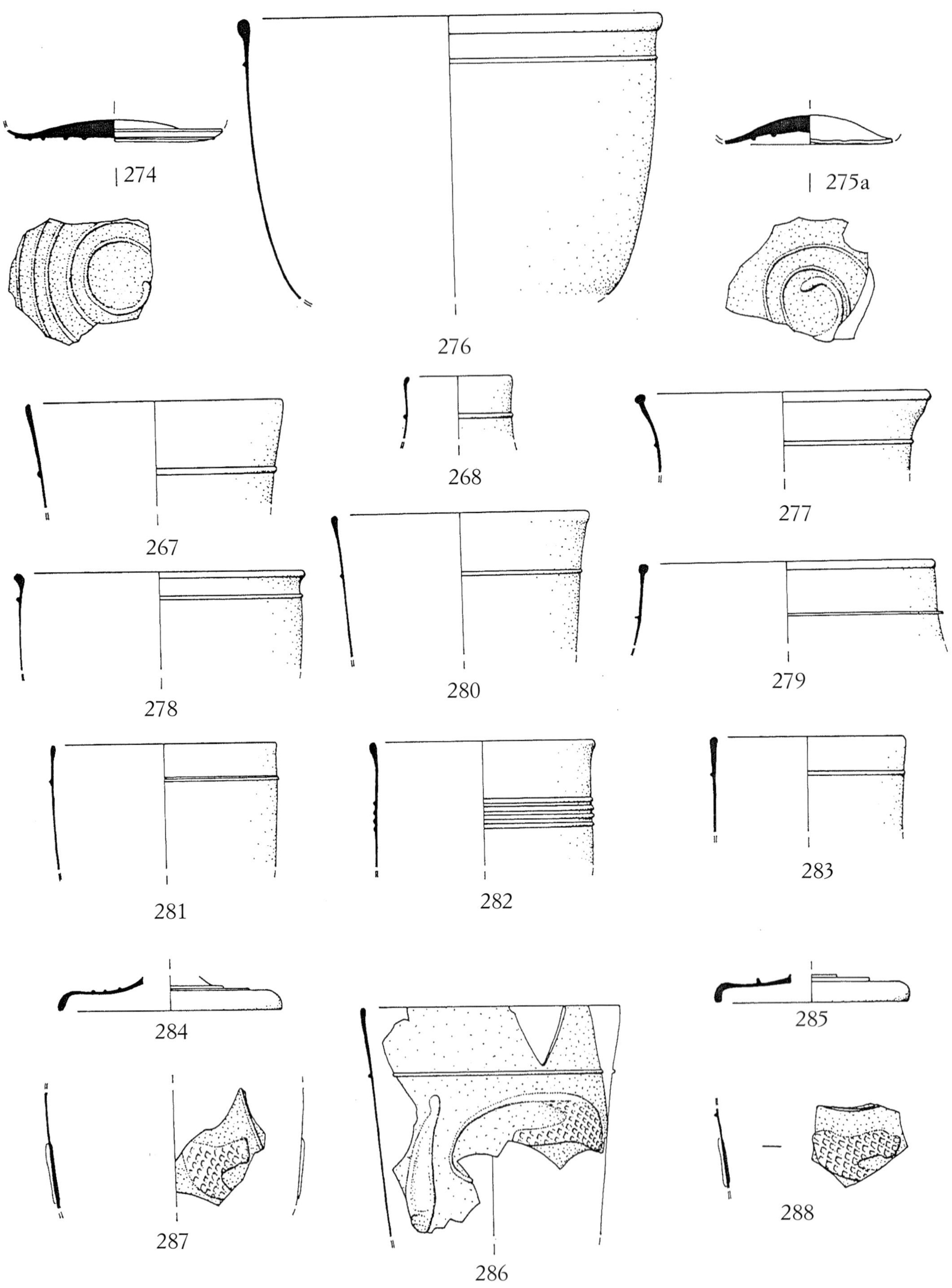

Plate 16. Blown vessels with trailed and tooled decoration.

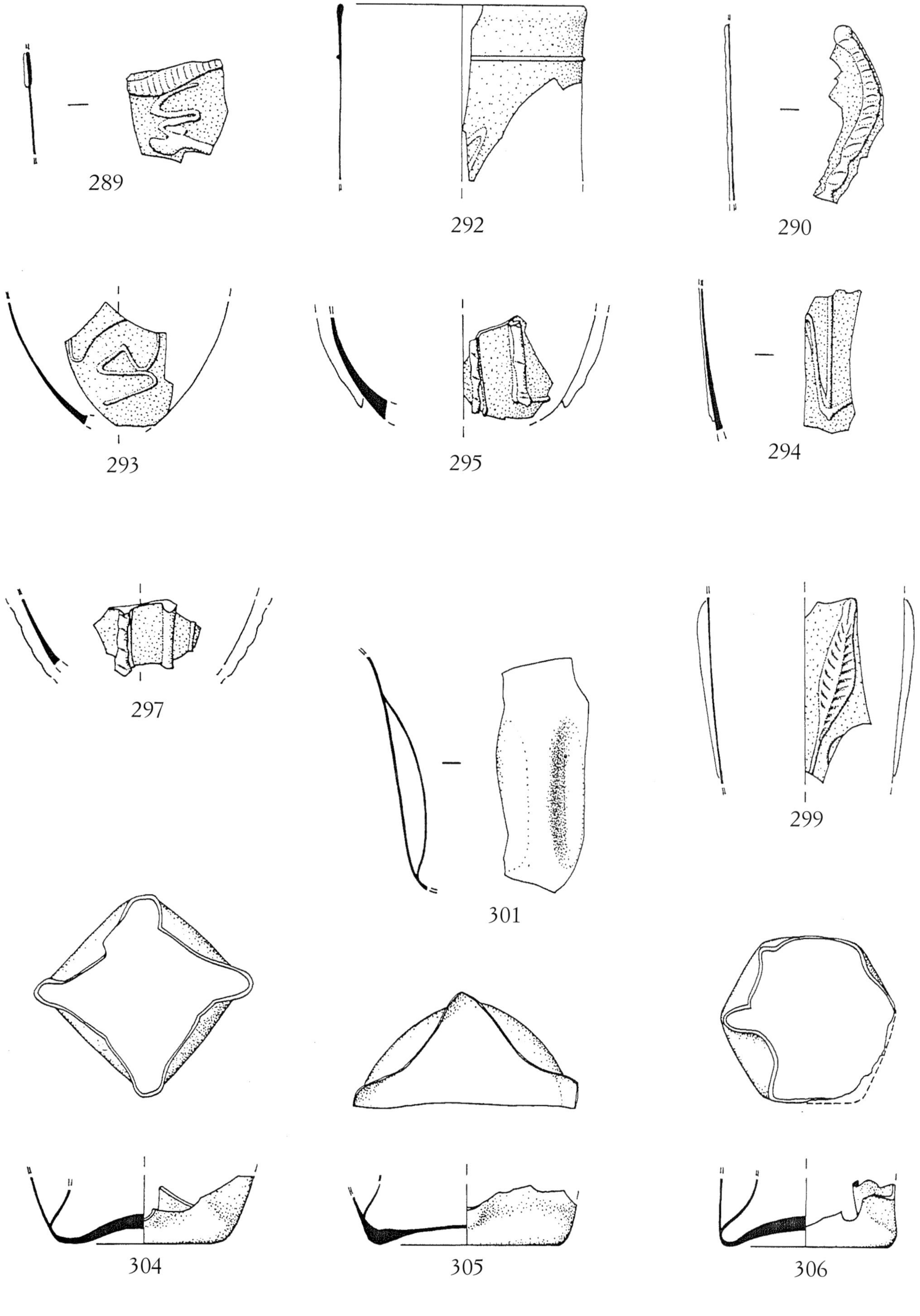

Plate 17. Blown vessels with trailed, tooled, and indented decoration.

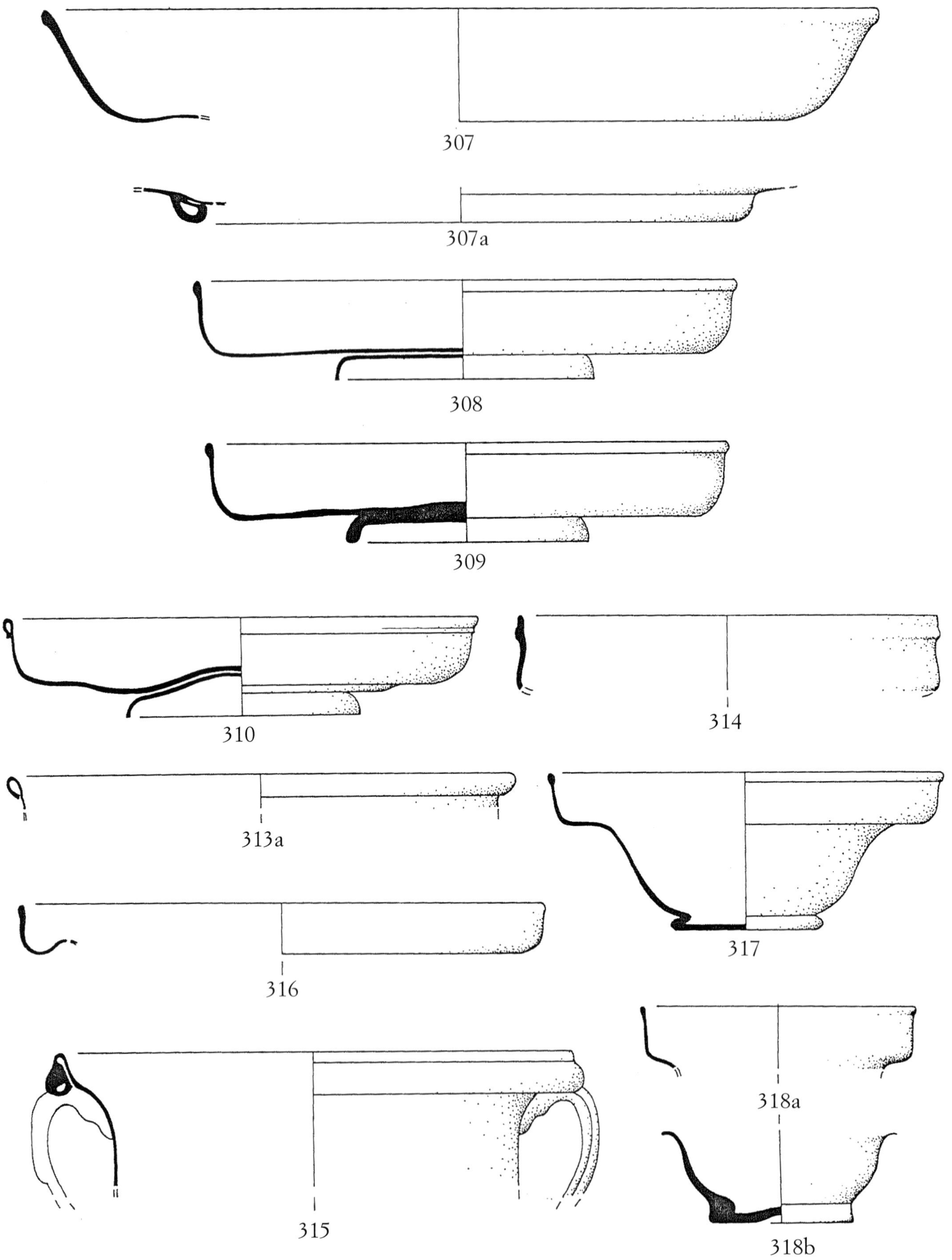

Plate 18. Blown colored fine wares: plates, dishes, and bowls.

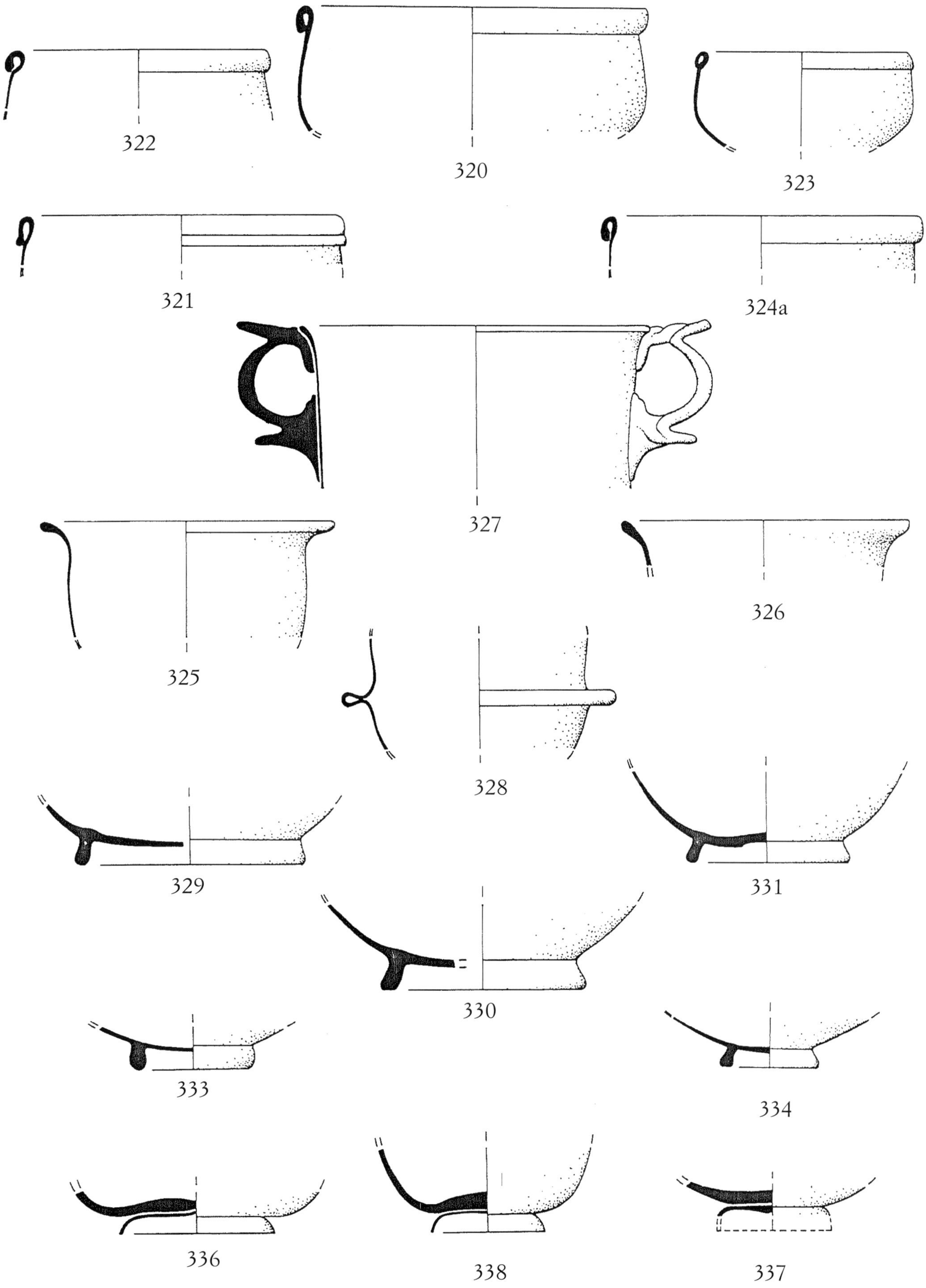

Plate 19. Blown colored fine wares: bowls.

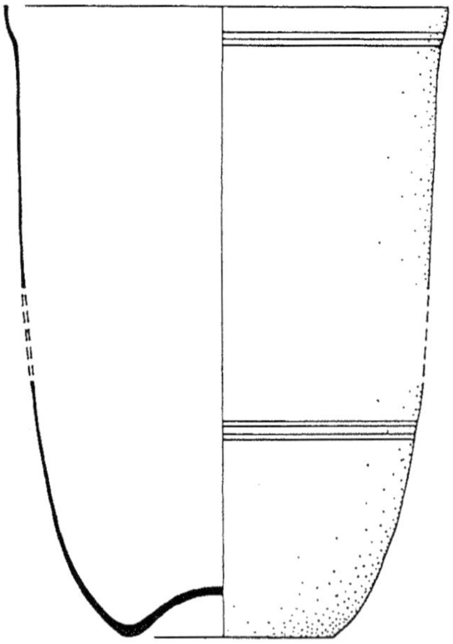

342

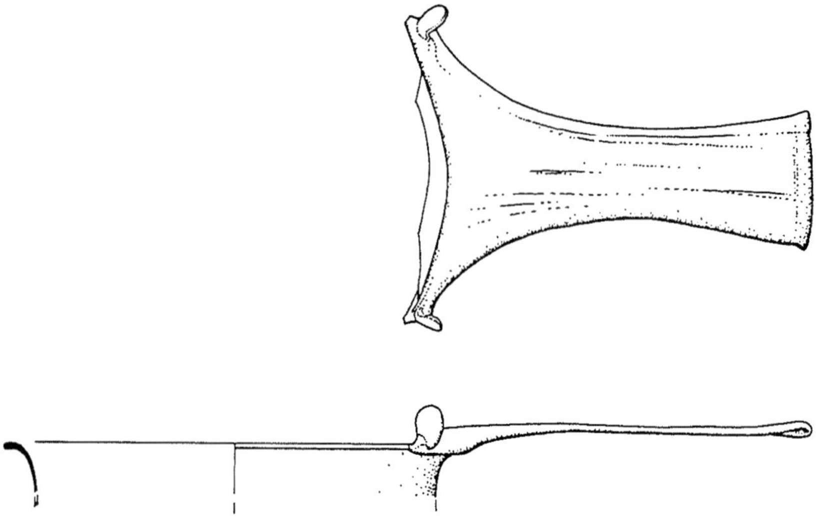

341

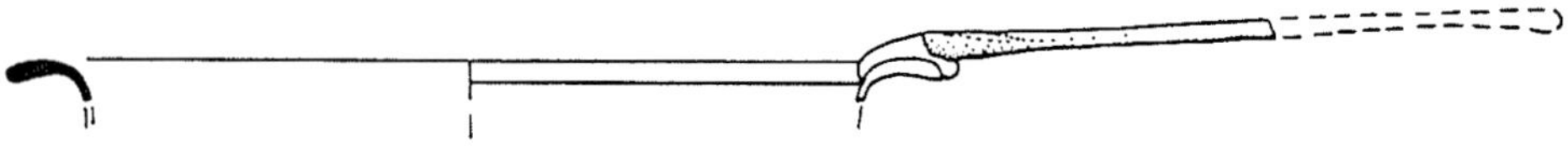

340

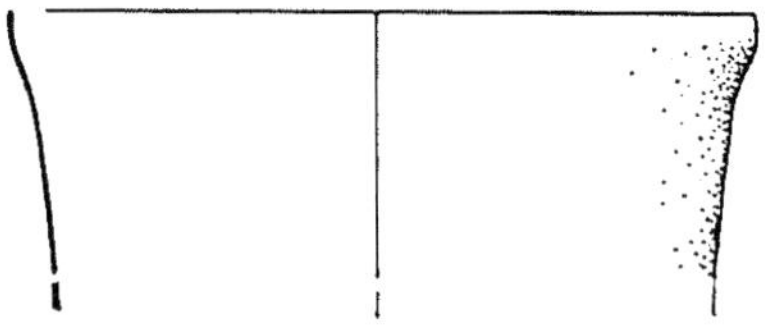

343

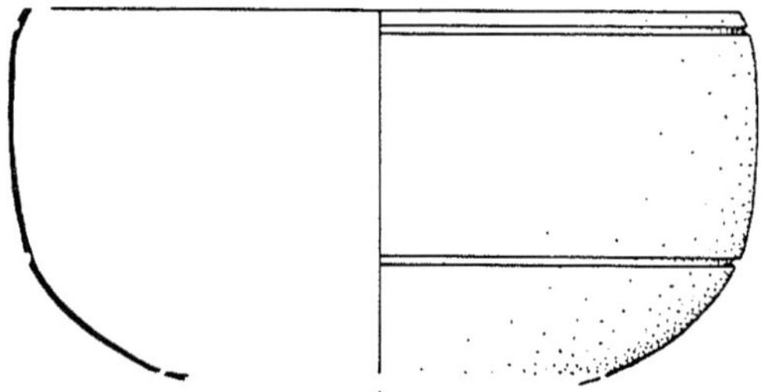

344

347

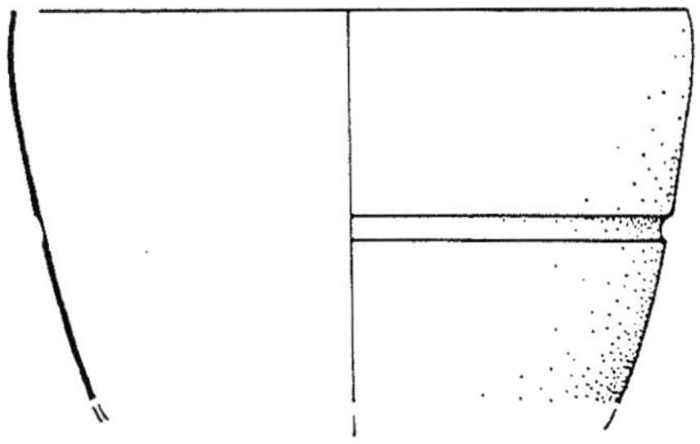

345

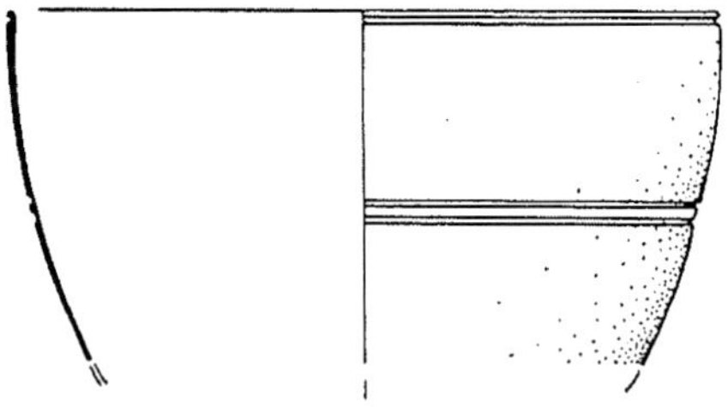

346

348

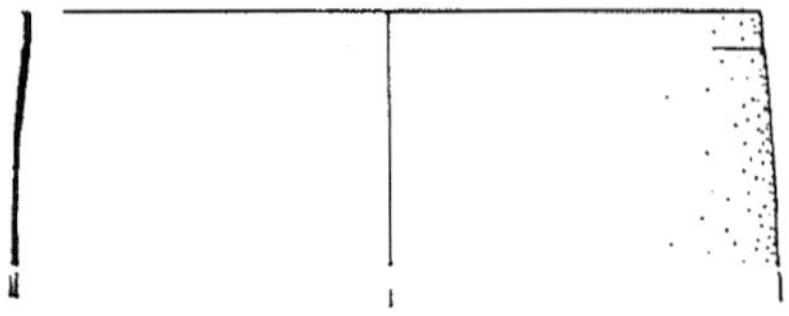

349

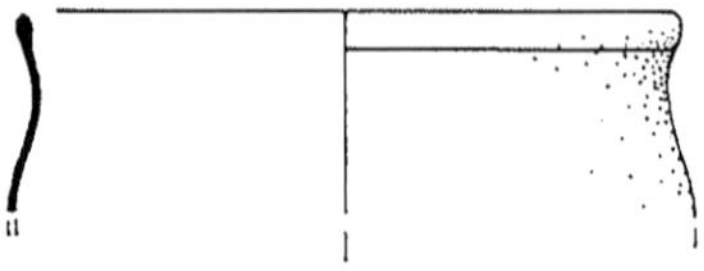

351

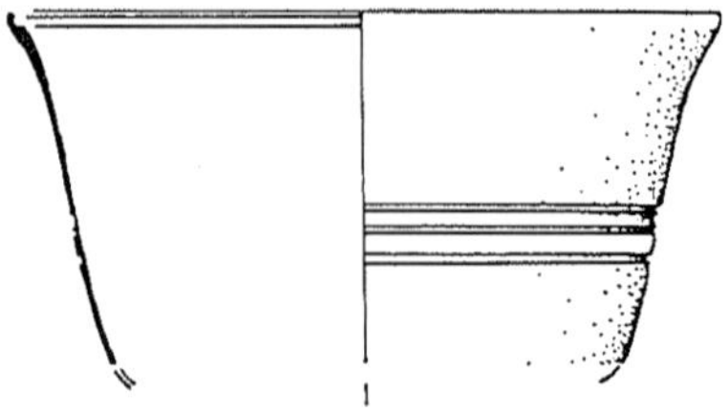

350

Plate 20. Blown colored fine wares: bowls, beakers, and saucepans.

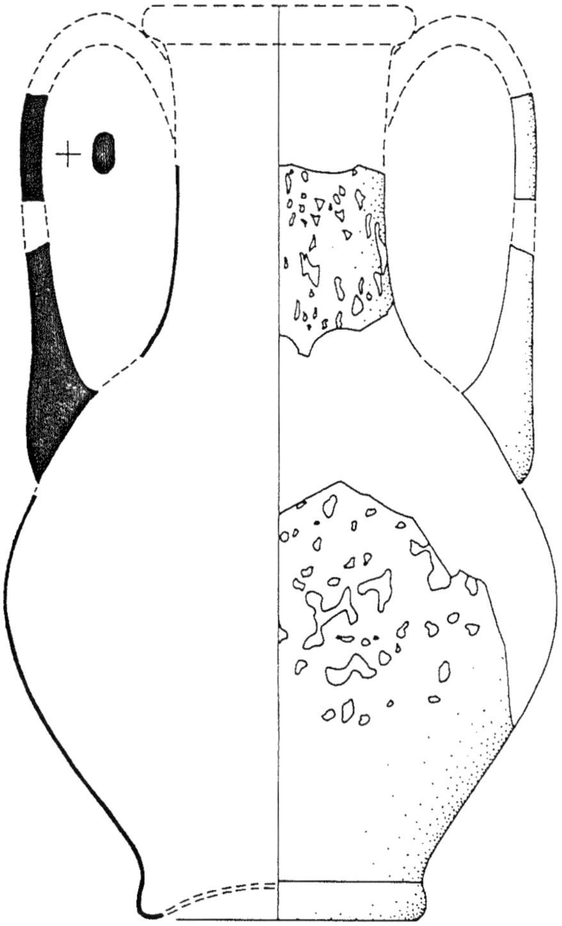

354

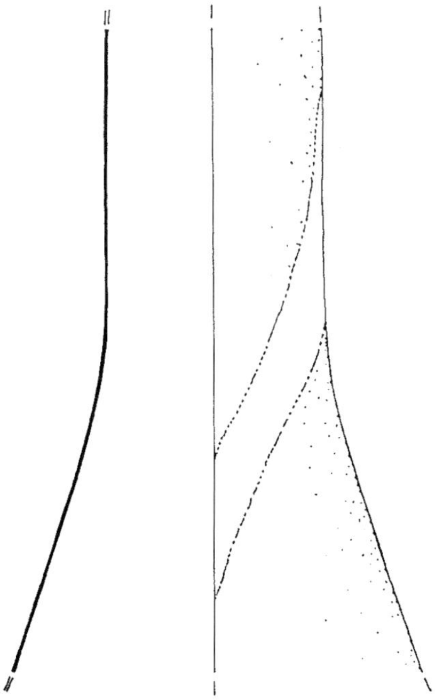

355

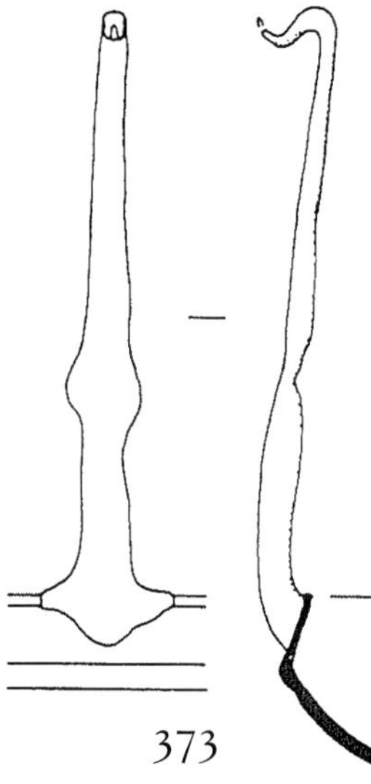
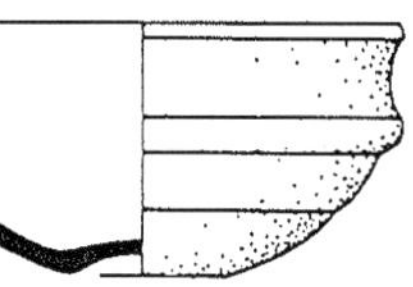

373

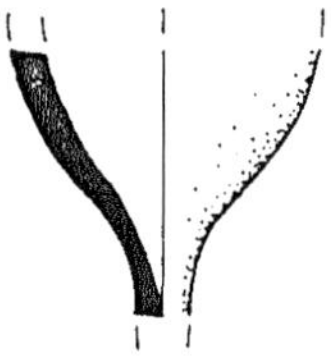

360

361

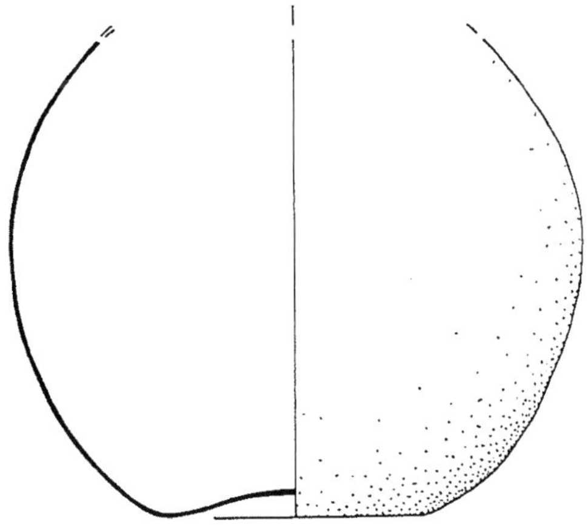

353

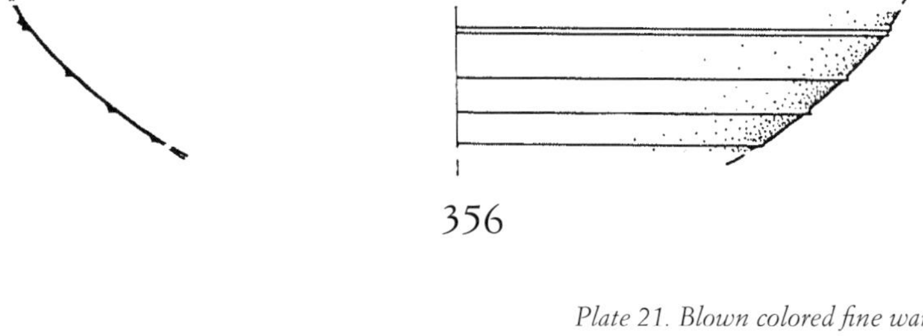

356

359

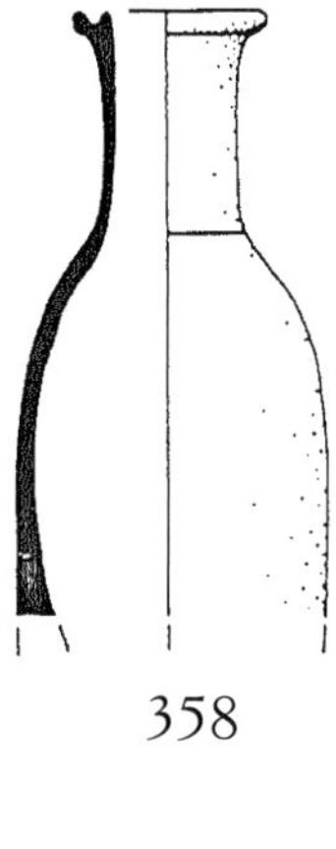

358

Plate 21. Blown colored fine wares: various shapes.

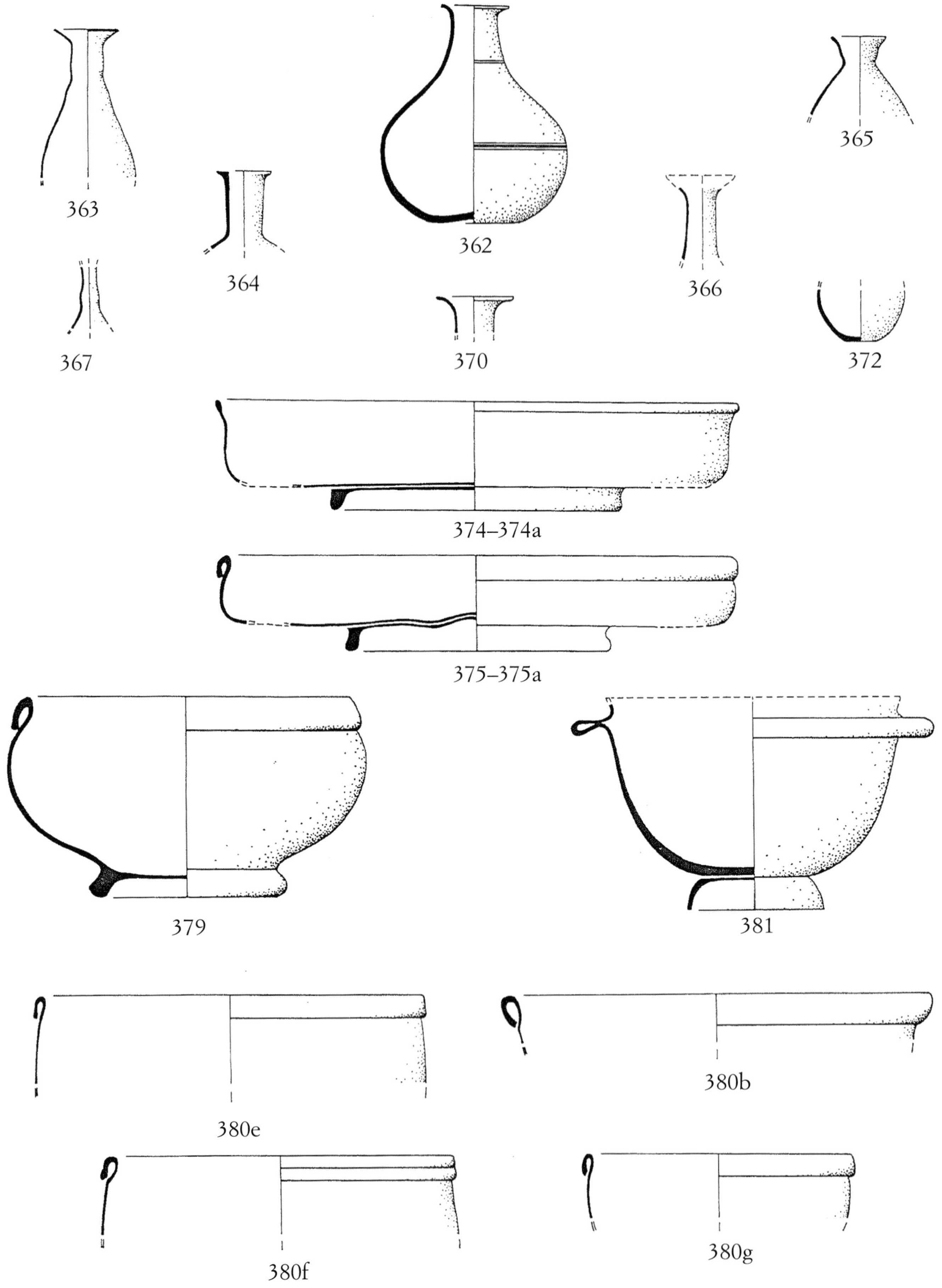

Plate 22. Blown colored fine wares: small thin-walled bottles. Blown, naturally colored dishes and bowls.

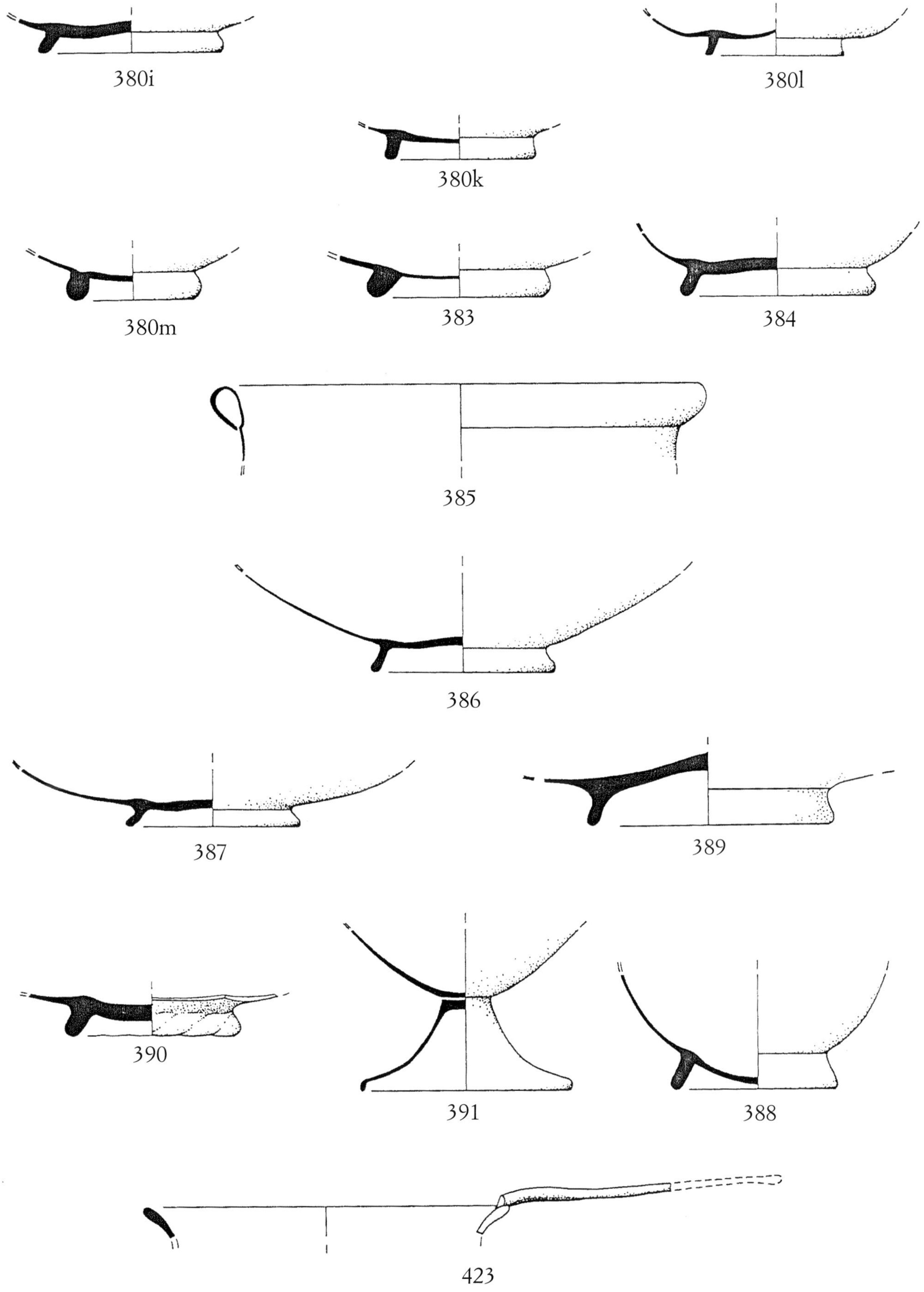

Plate 23. Blown, naturally colored bowls and saucepans.

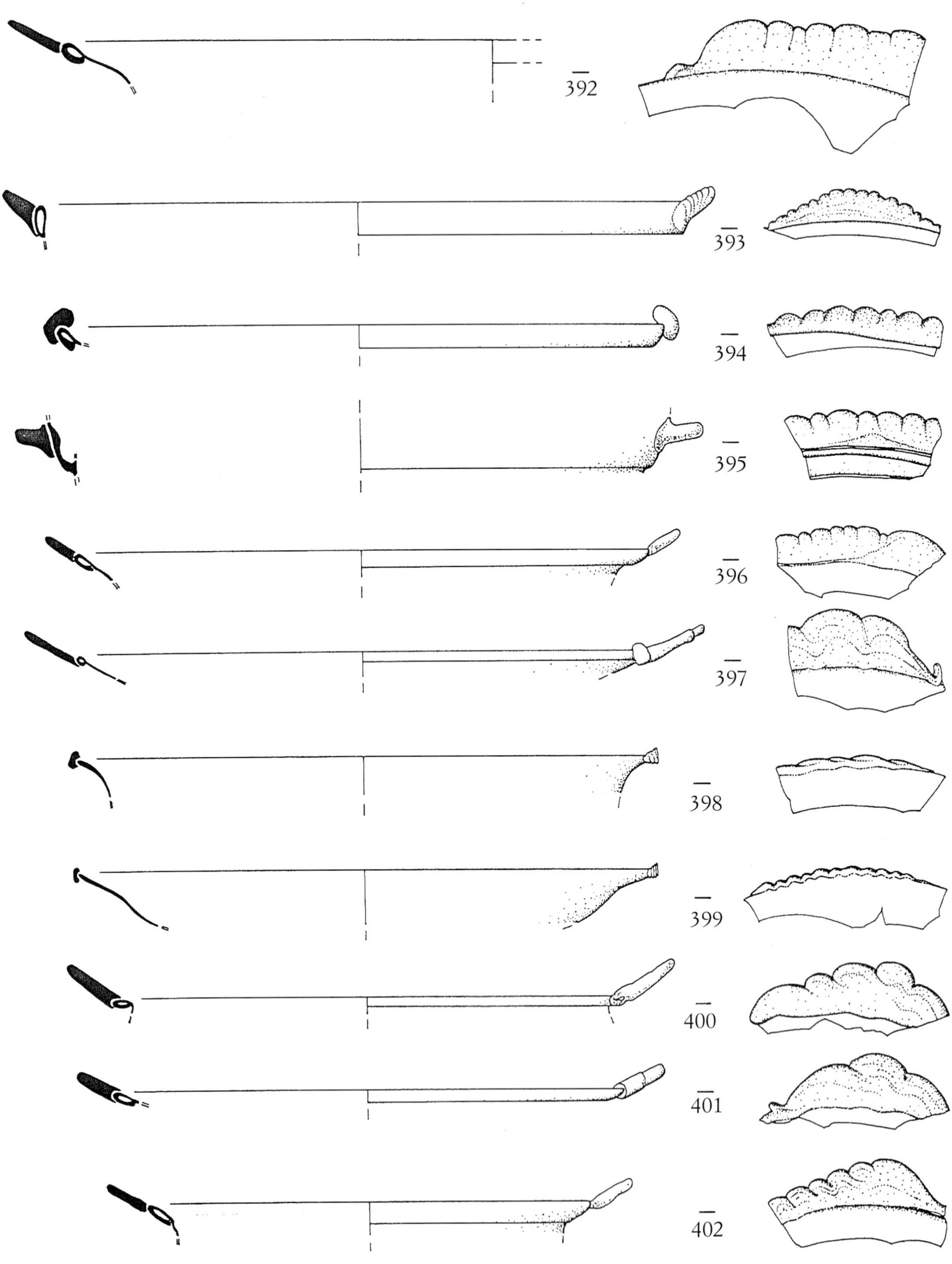

Plate 24. Blown, naturally colored plates and bowls with applied fillets.

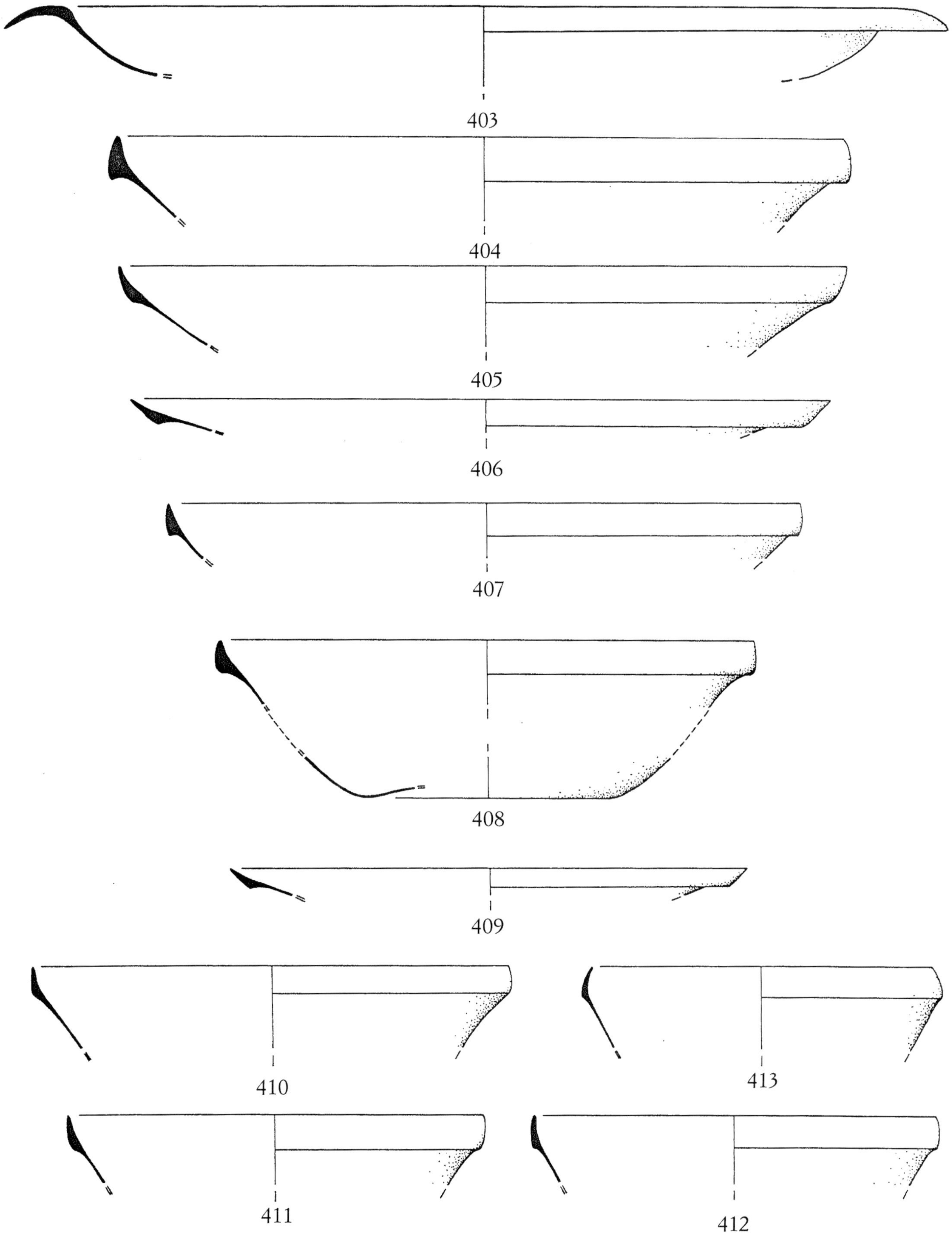

Plate 25. Blown, naturally colored and colorless bowls with thickened rims, triangular in cross section.

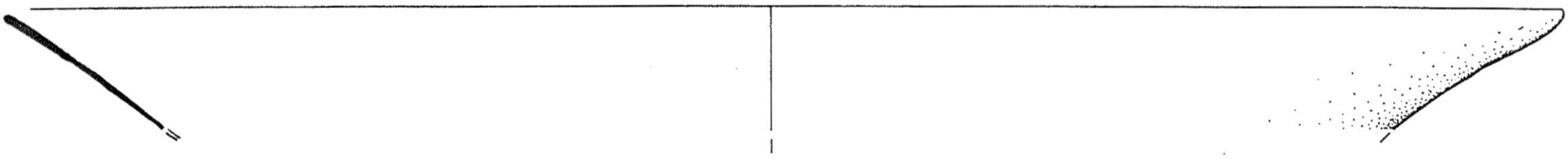

414

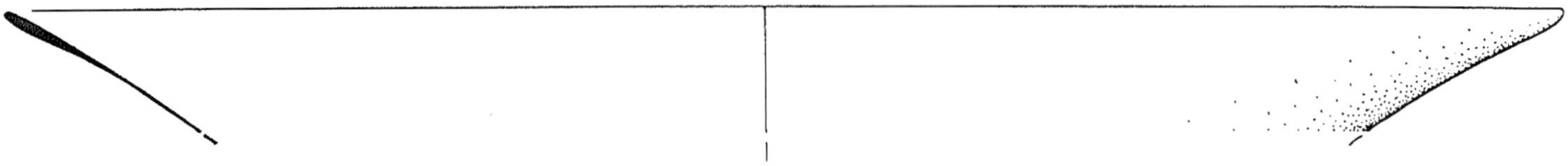

415

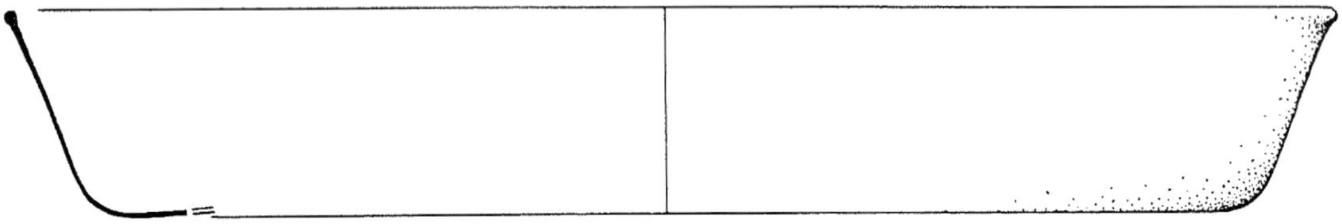

416

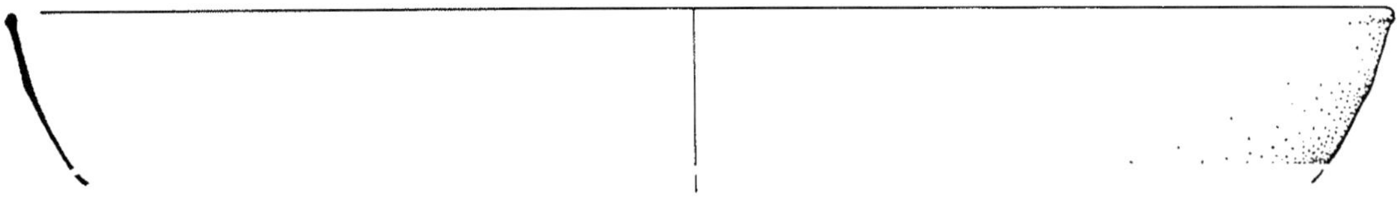

417

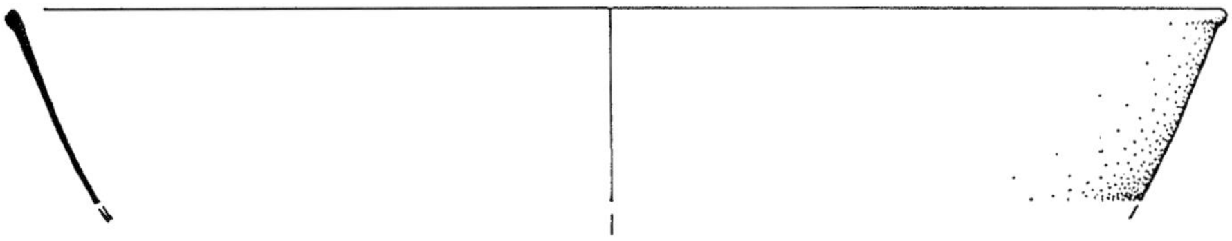

418

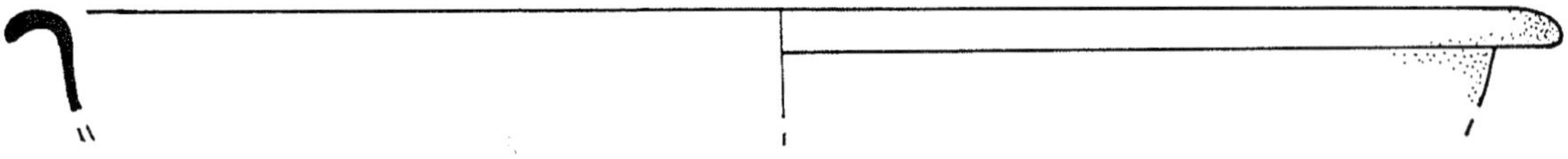

419

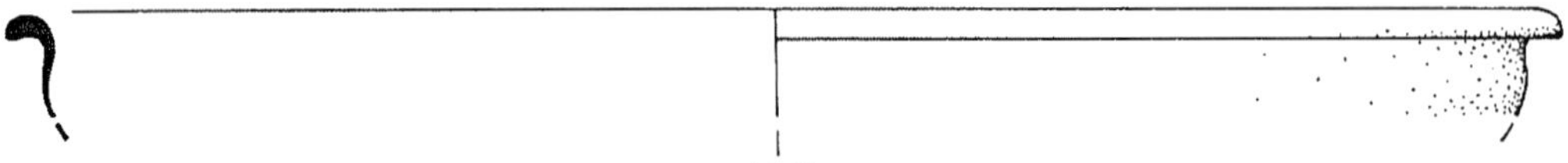

419a

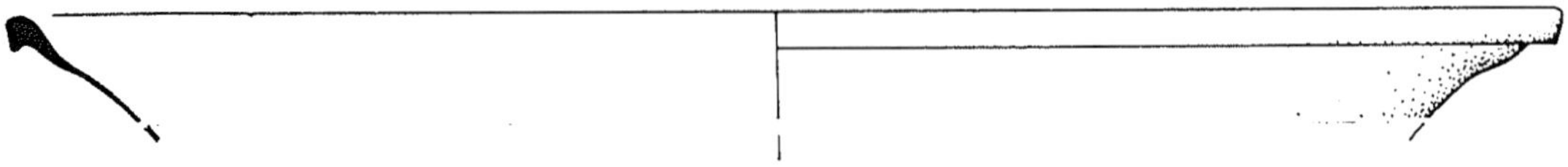

420

Plate 26. Blown colorless plates and bowls.

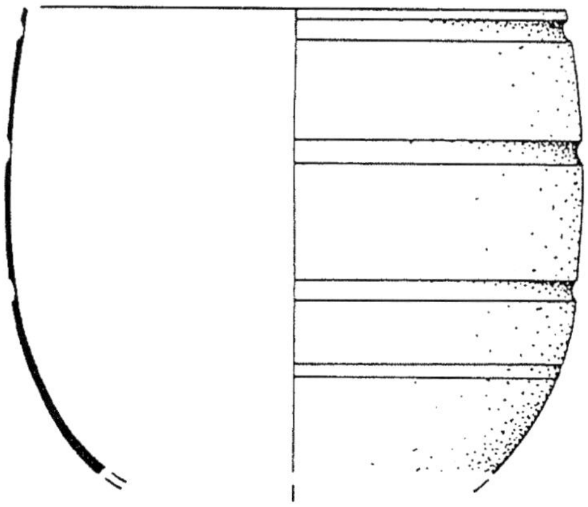

424

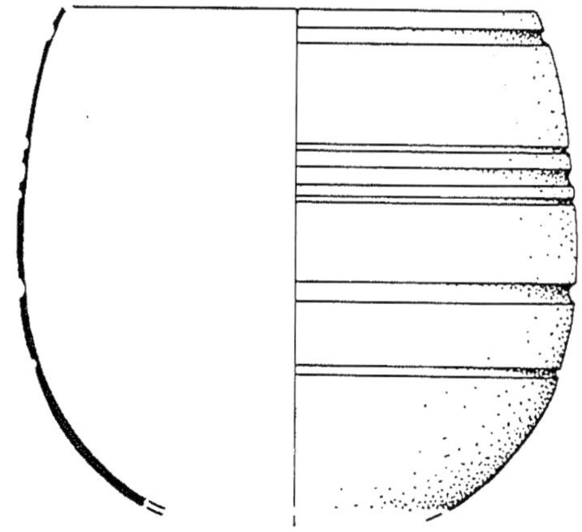

425

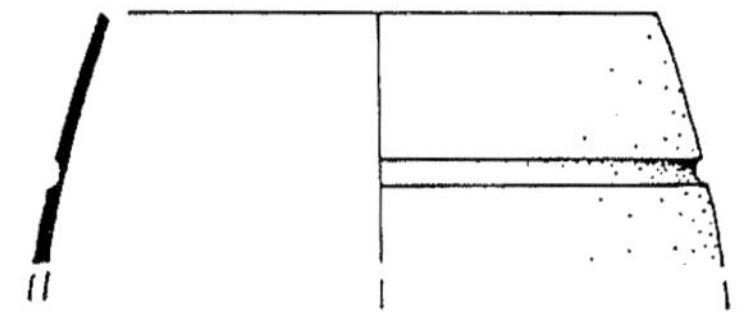

428

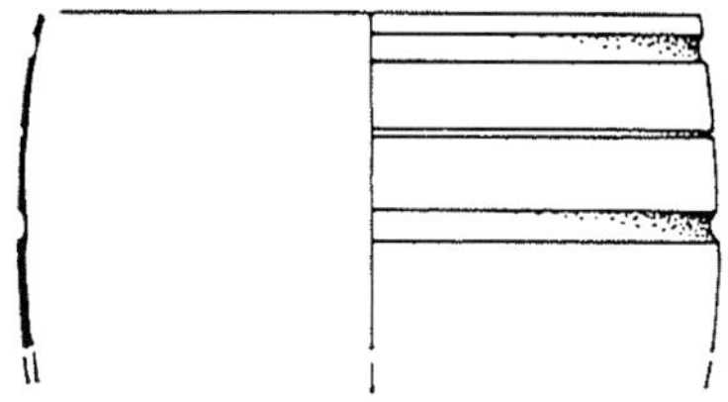

427

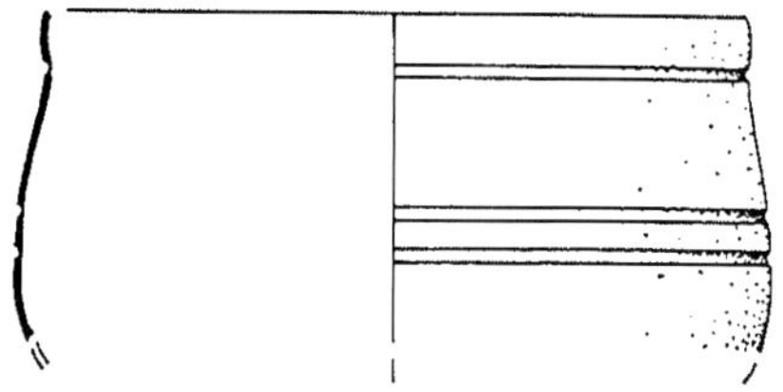

429

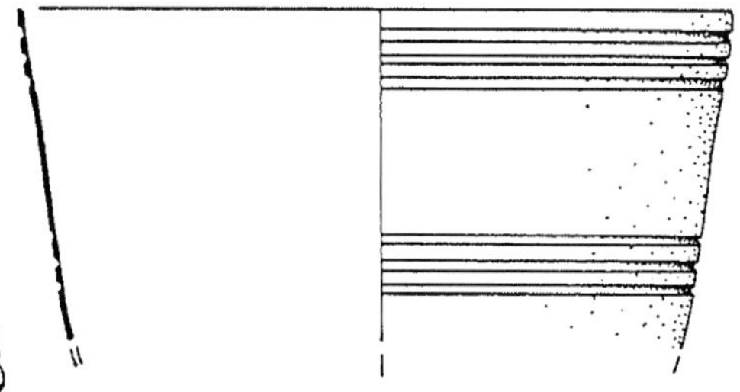

432

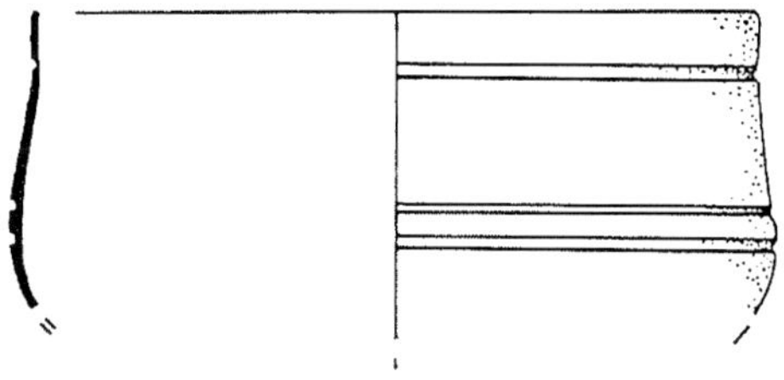

430

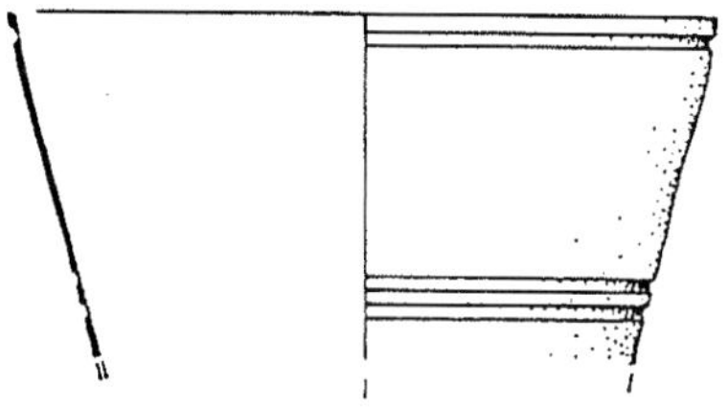

431

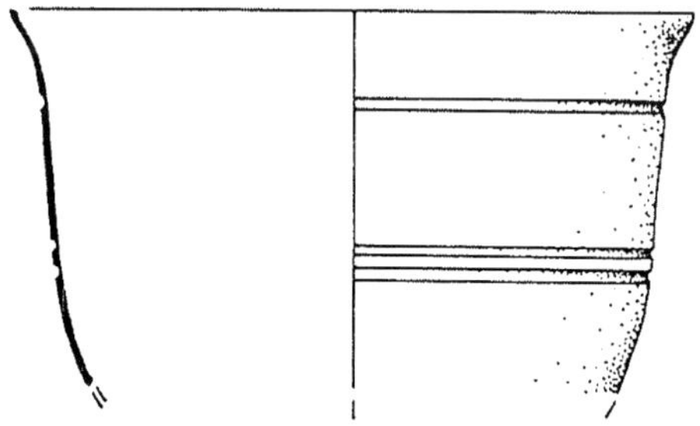

434

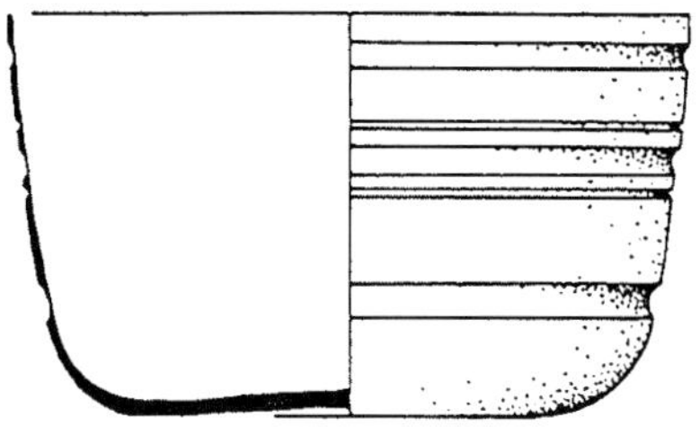

433

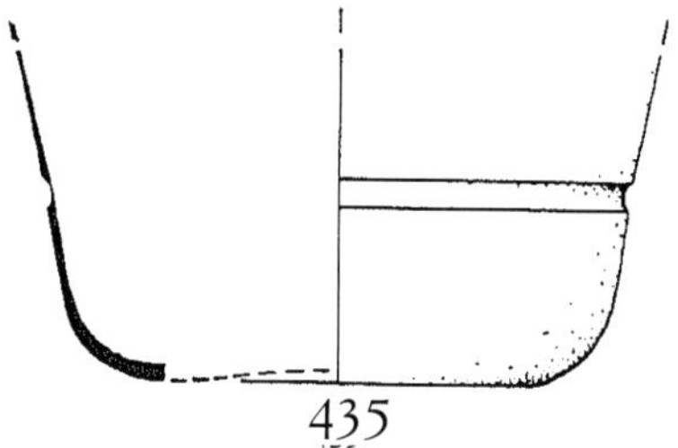

435

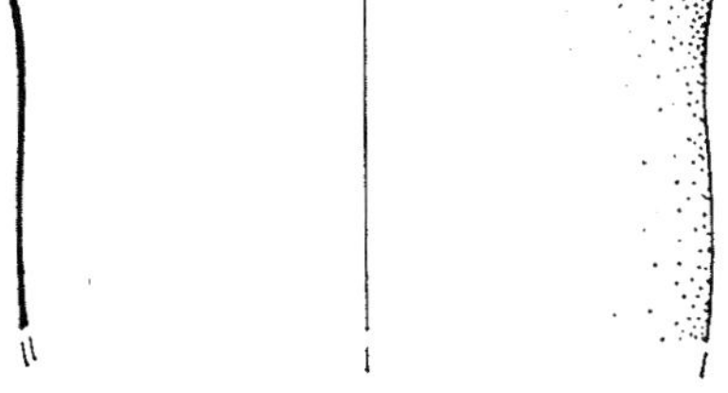

436

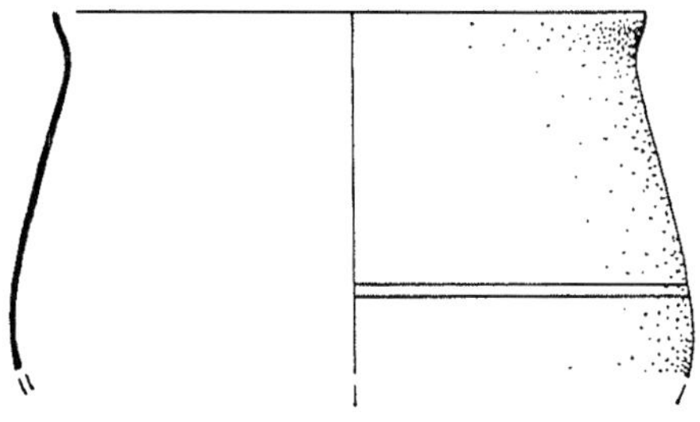

437

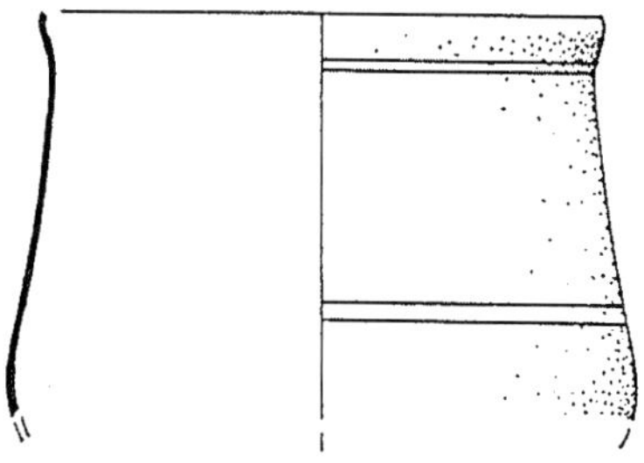

438

Plate 27. Blown, naturally colored beakers.

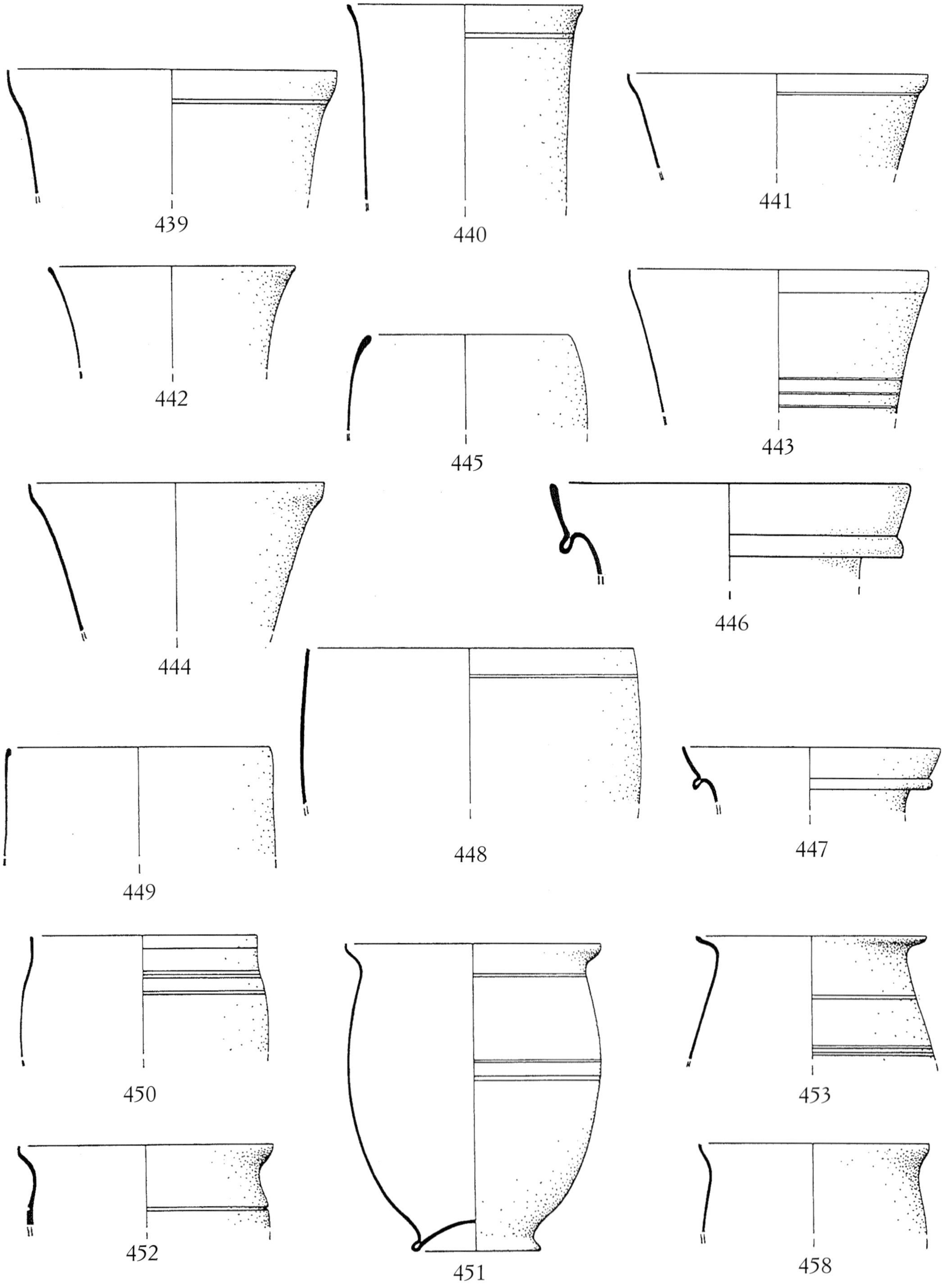

Plate 28. Blown, naturally colored and colorless beakers.

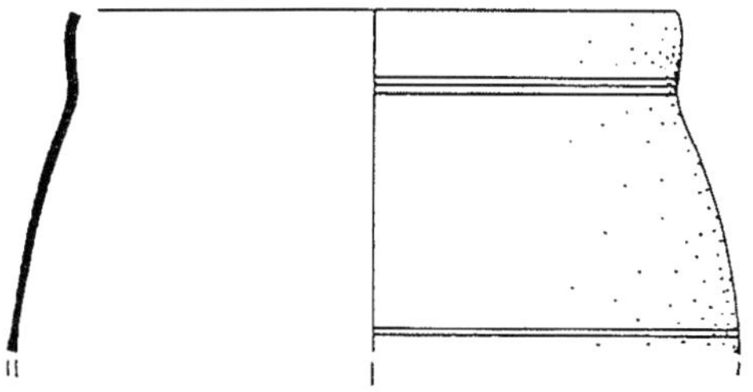

454

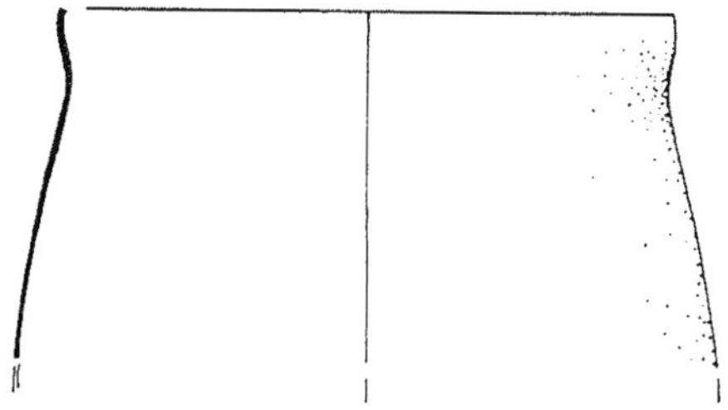

456

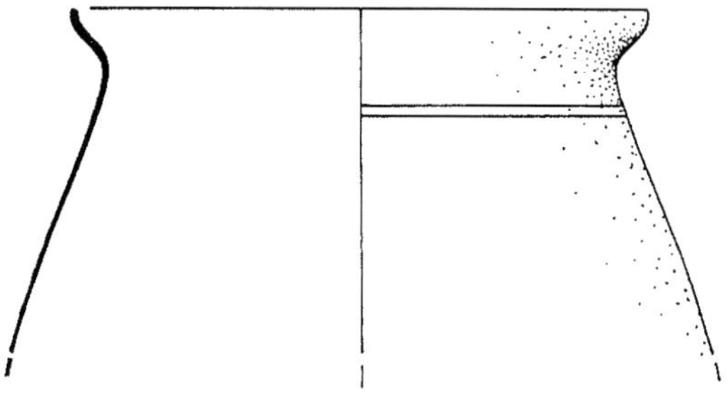

455

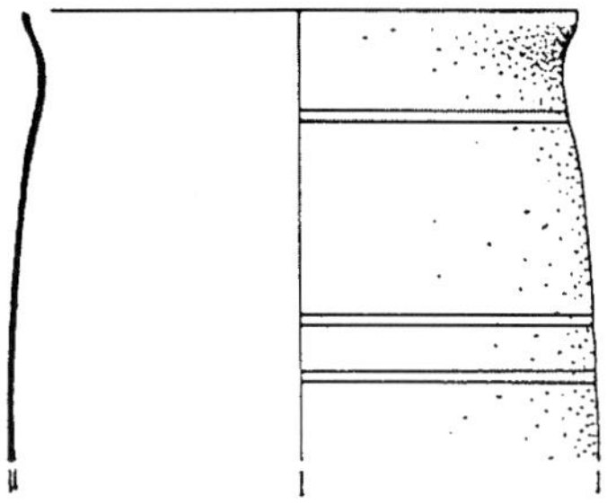

457

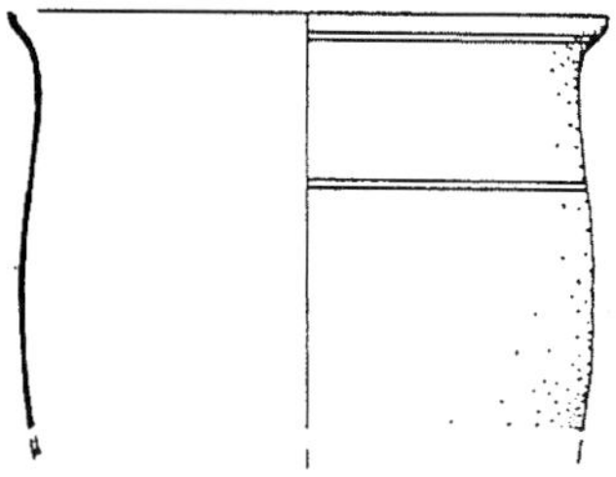

459

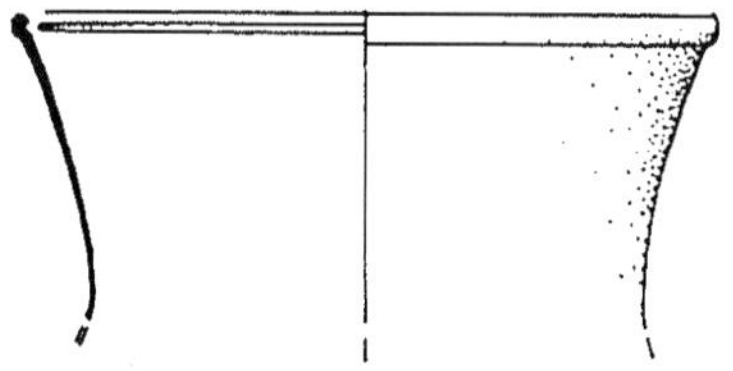

460

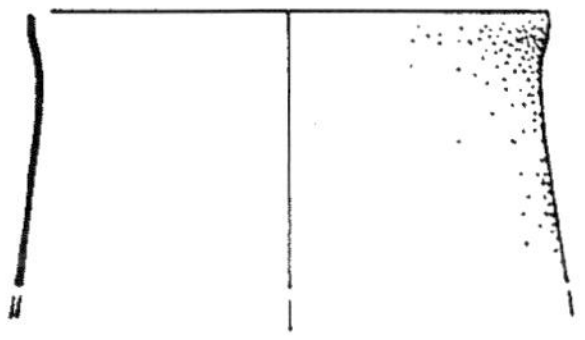

461

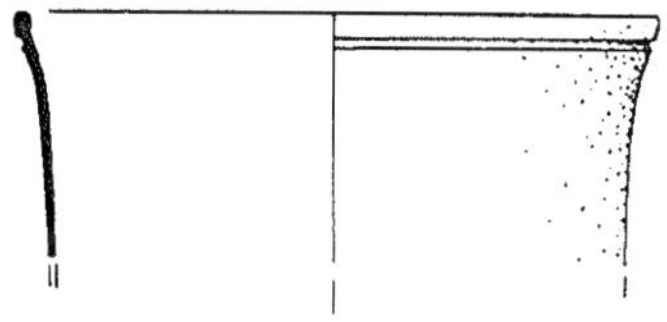

462

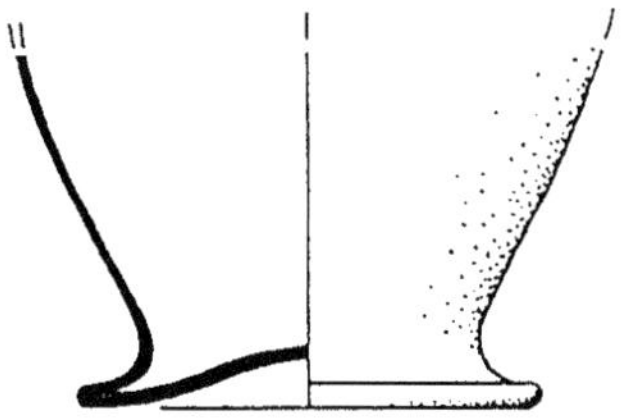

463

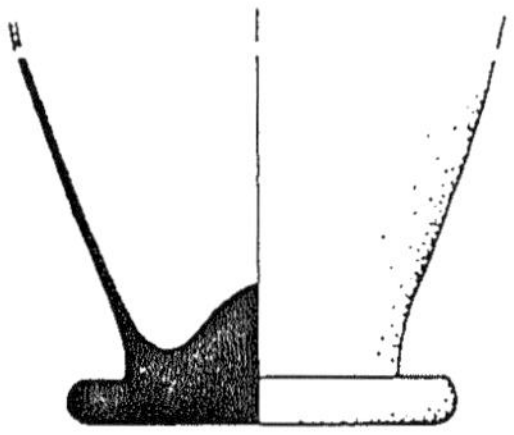

464

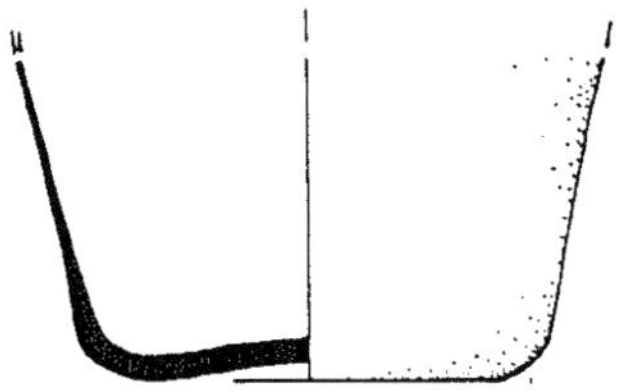

466

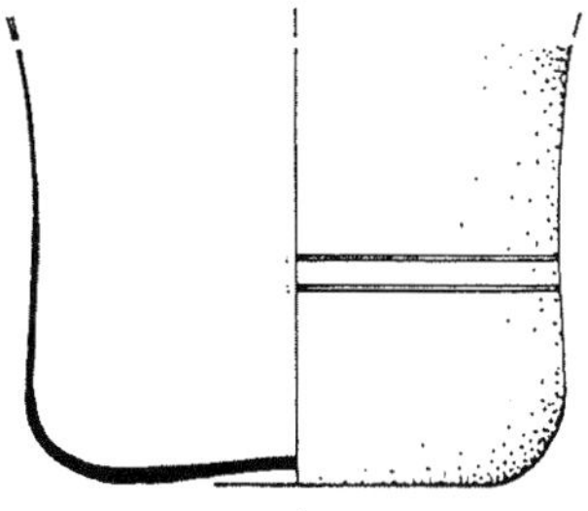

465

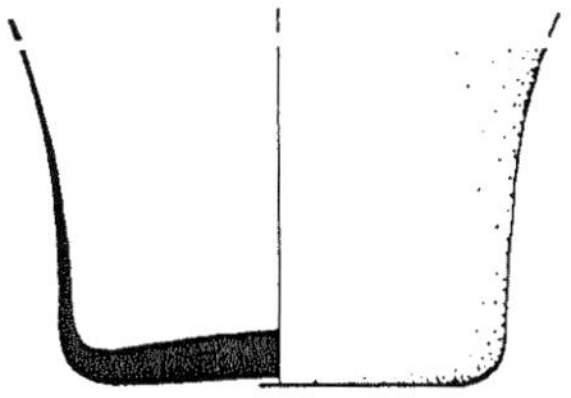

467

Plate 29. Blown, naturally colored and colorless beakers.

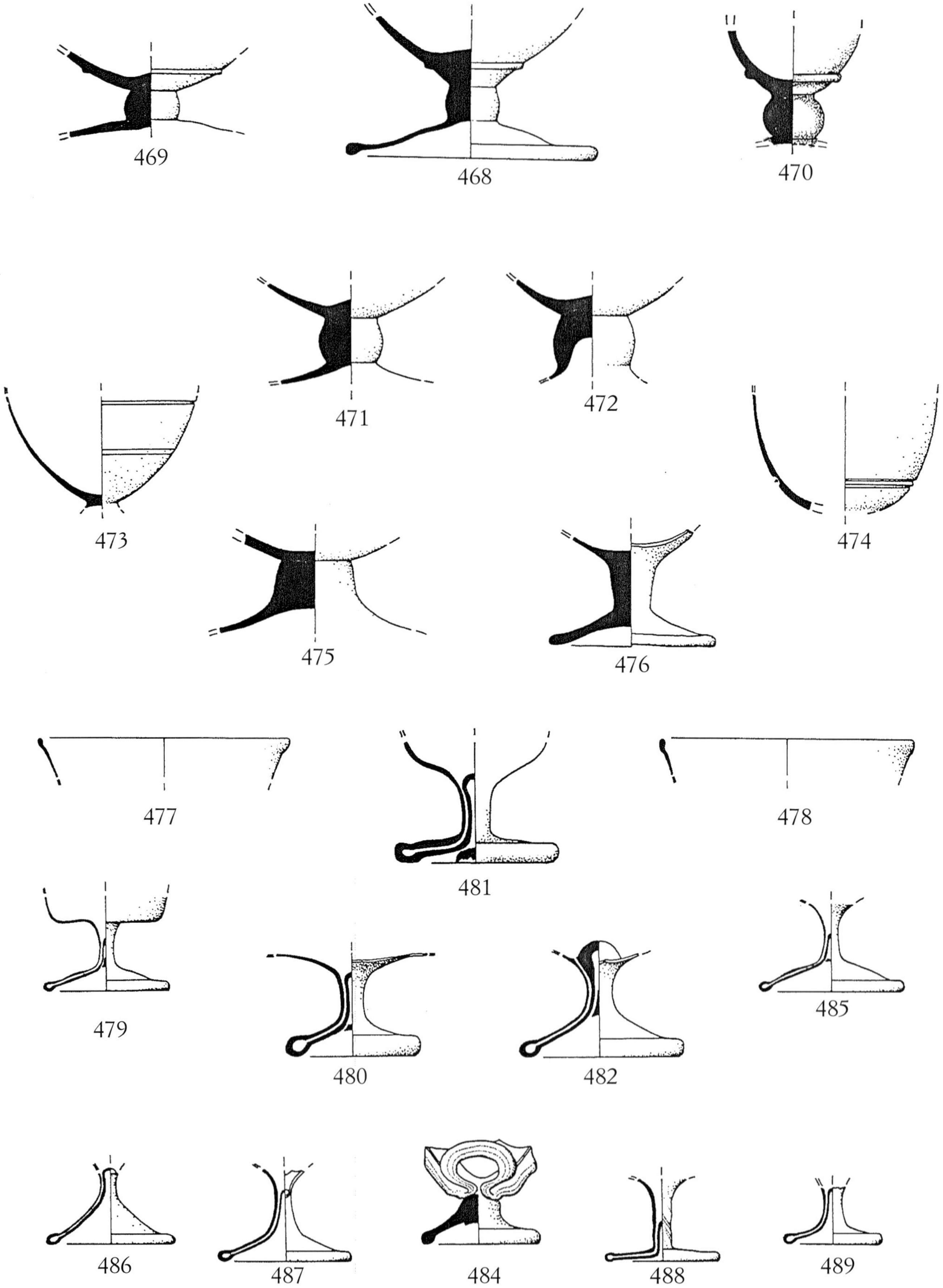

Plate 30. Blown, naturally colored and colorless stemmed bowls and cups. Blown late antique stemmed cups.

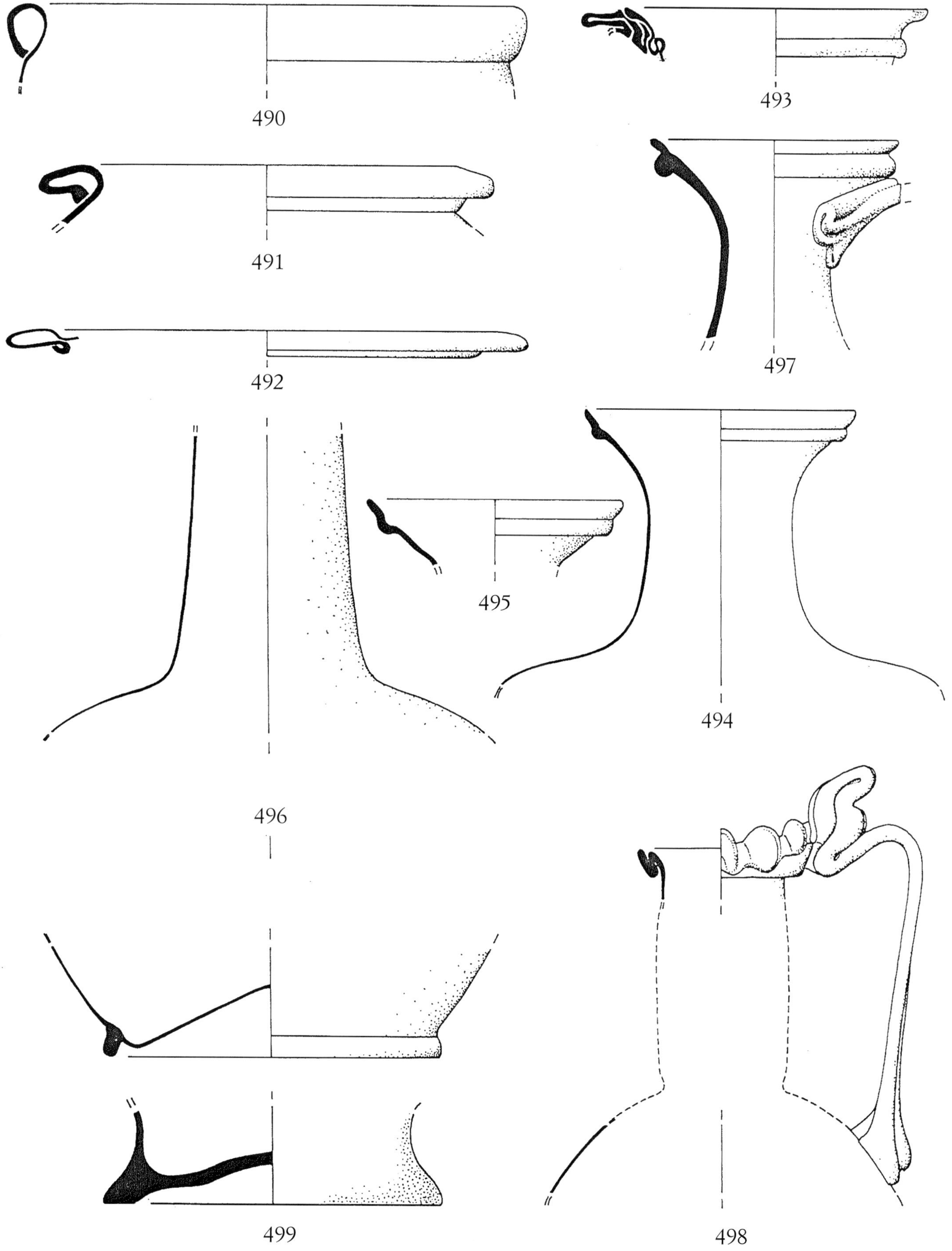

Plate 31. Blown, naturally colored and colorless wide-mouthed jars, jugs, and flasks.

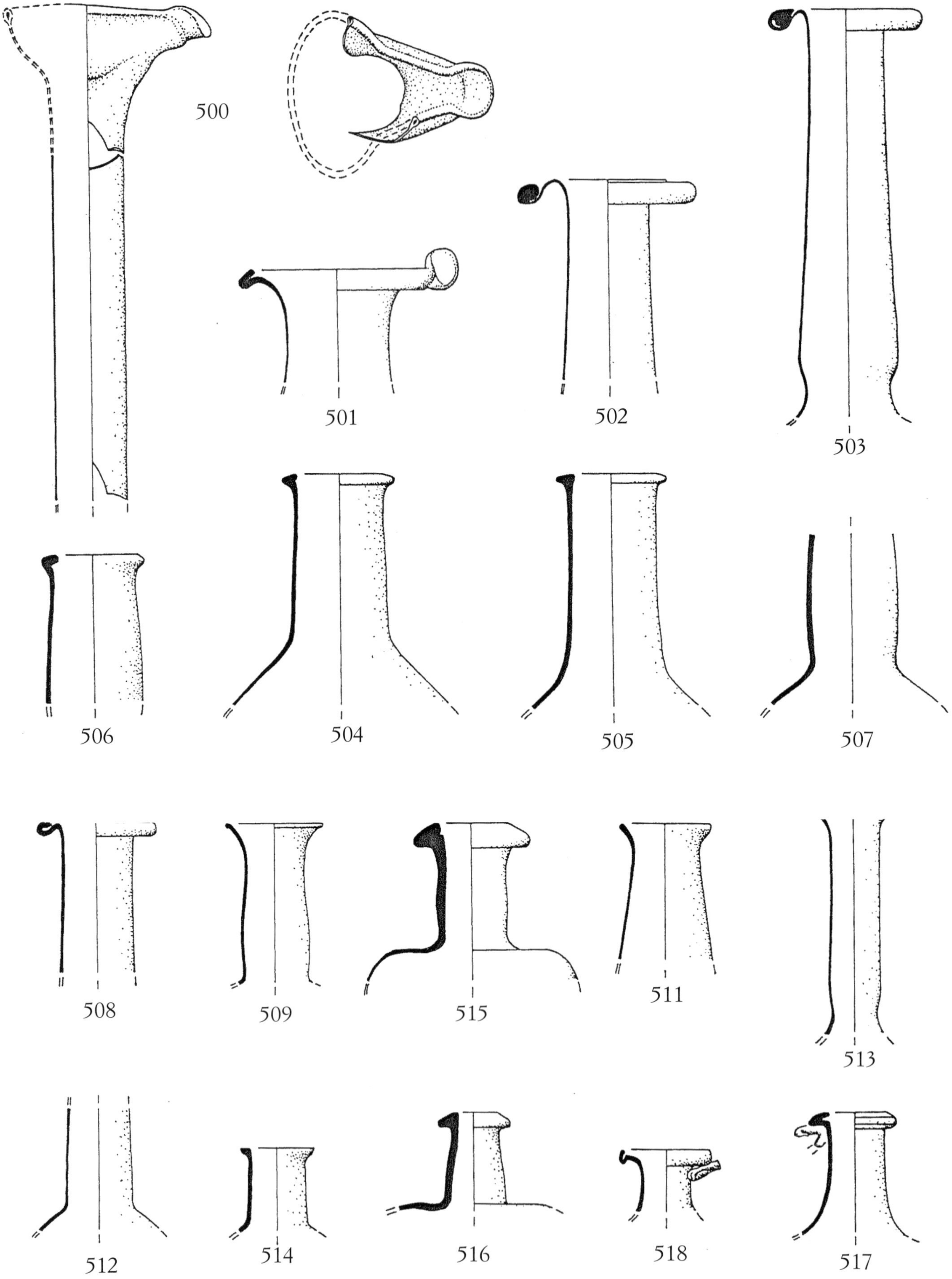

Plate 32. Blown, naturally colored and colorless flasks.

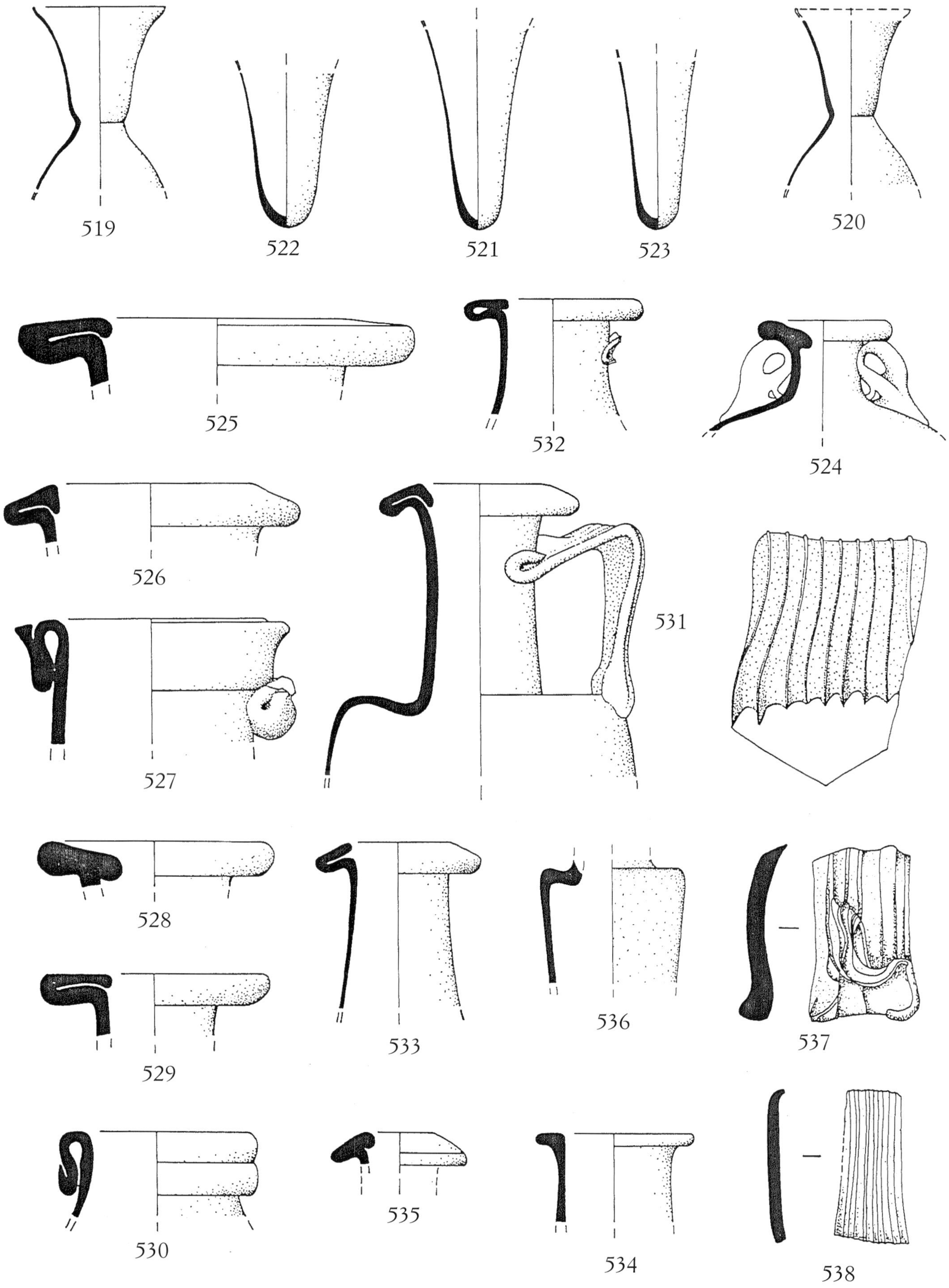

Plate 33. Blown and mold-blown, naturally colored and colorless flasks, jugs, and prismatic bottles.

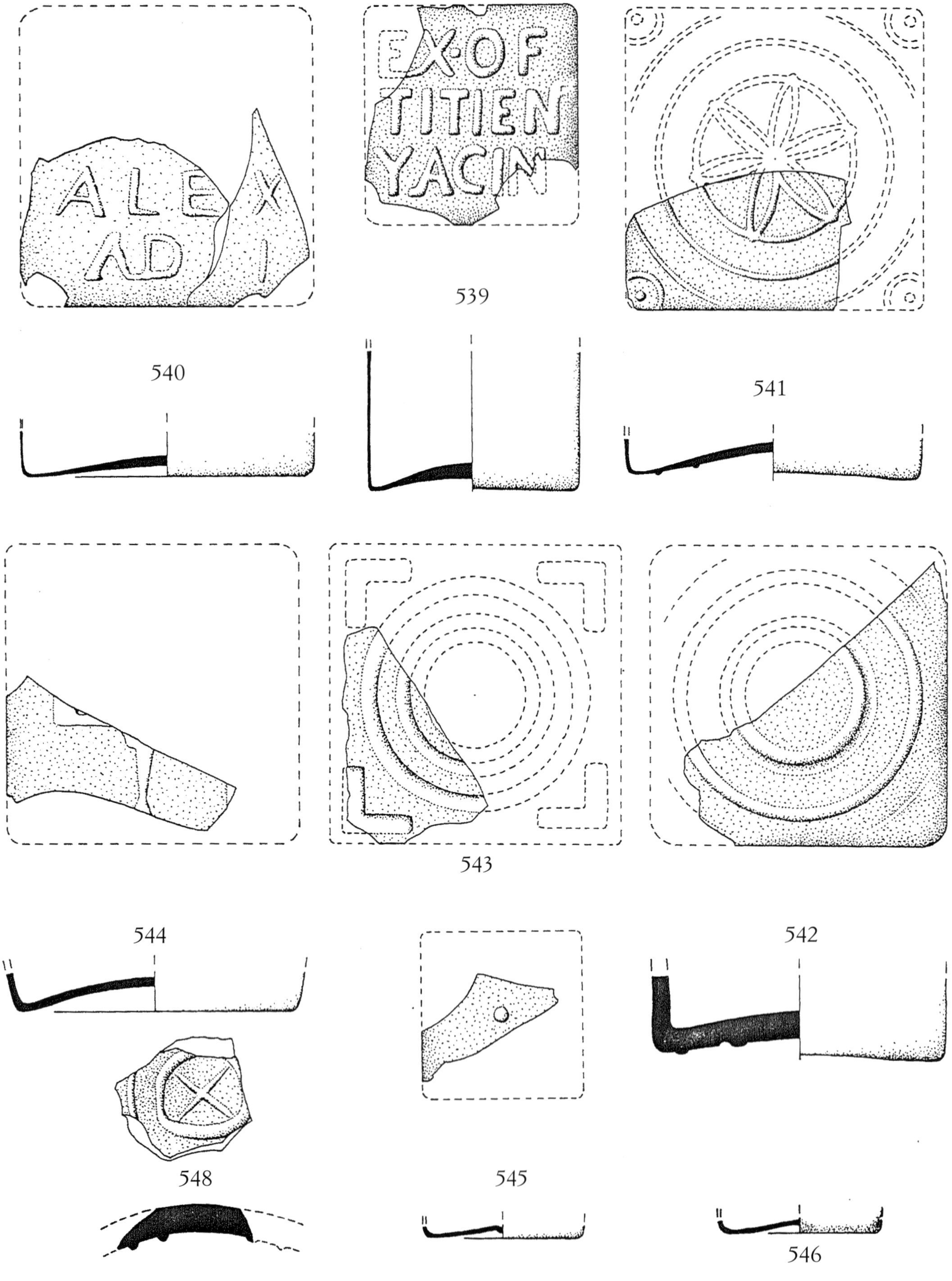

Plate 34. Mold-blown prismatic bottle bottoms.

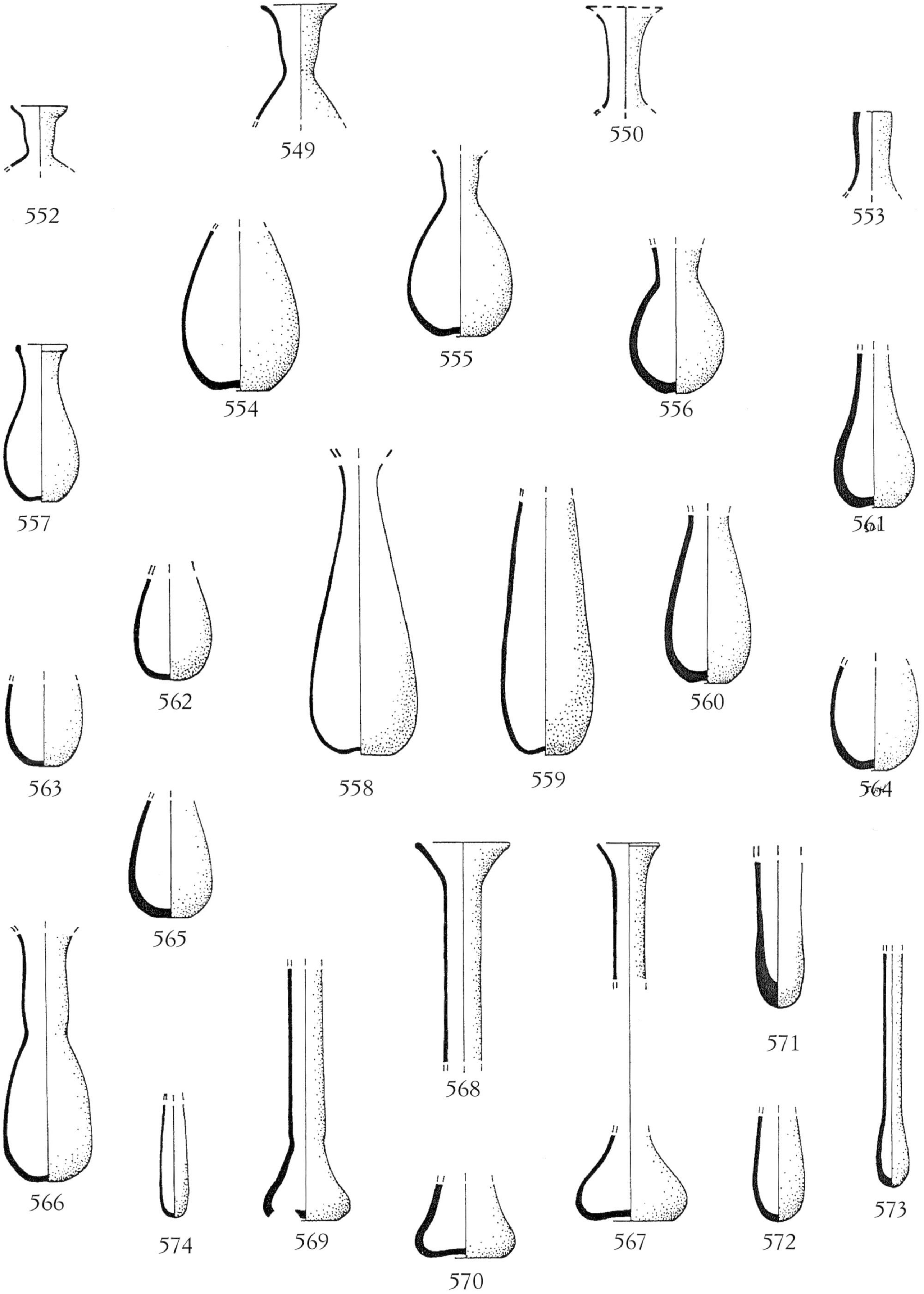

Plste 35. Blown, naturally colored and colorless small thin-walled bottles.

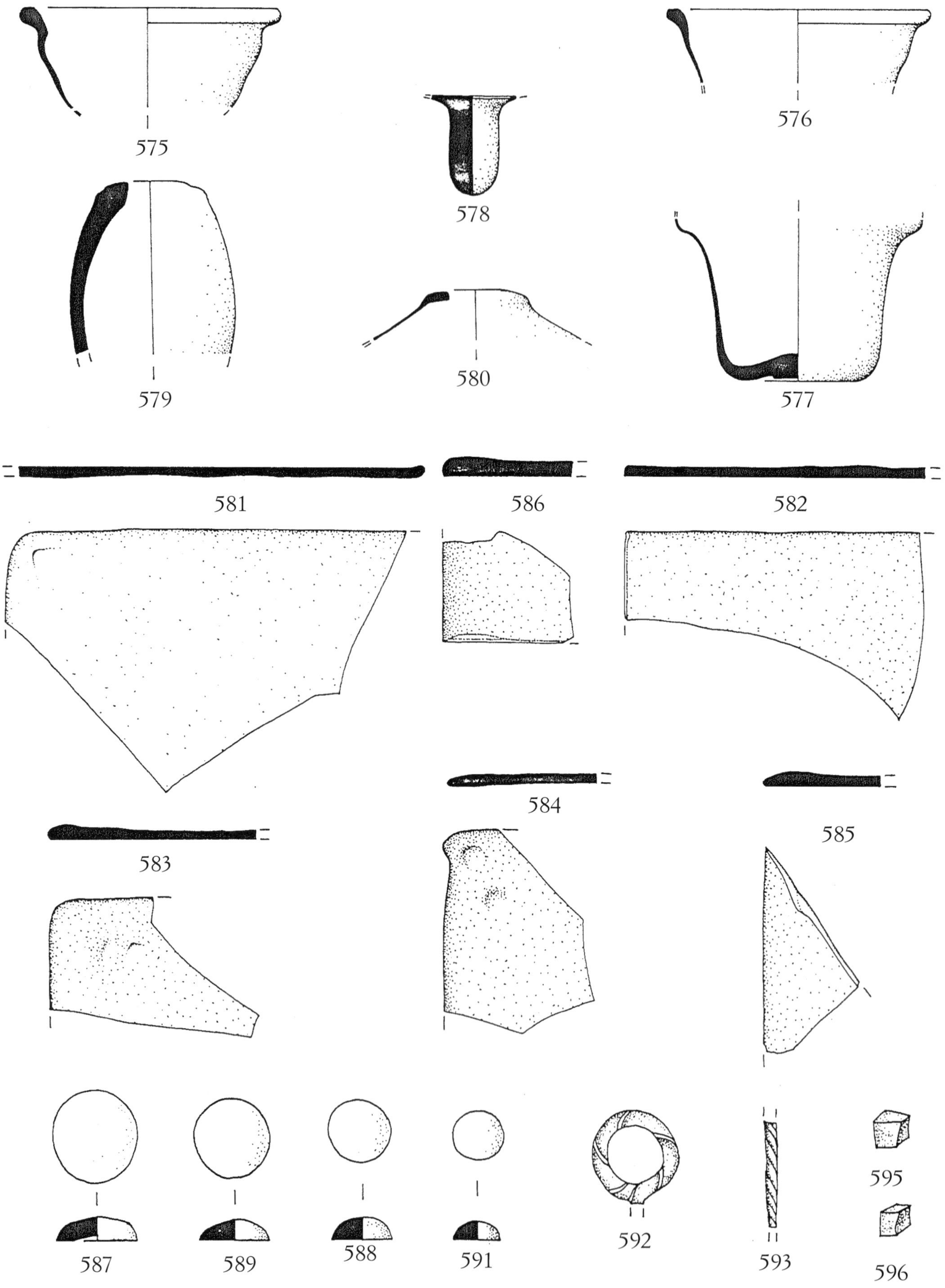

Plte 36. Blown vessels: variants. Cast window glass. Cast and tooled gaming pieces, rods, and tesserae.

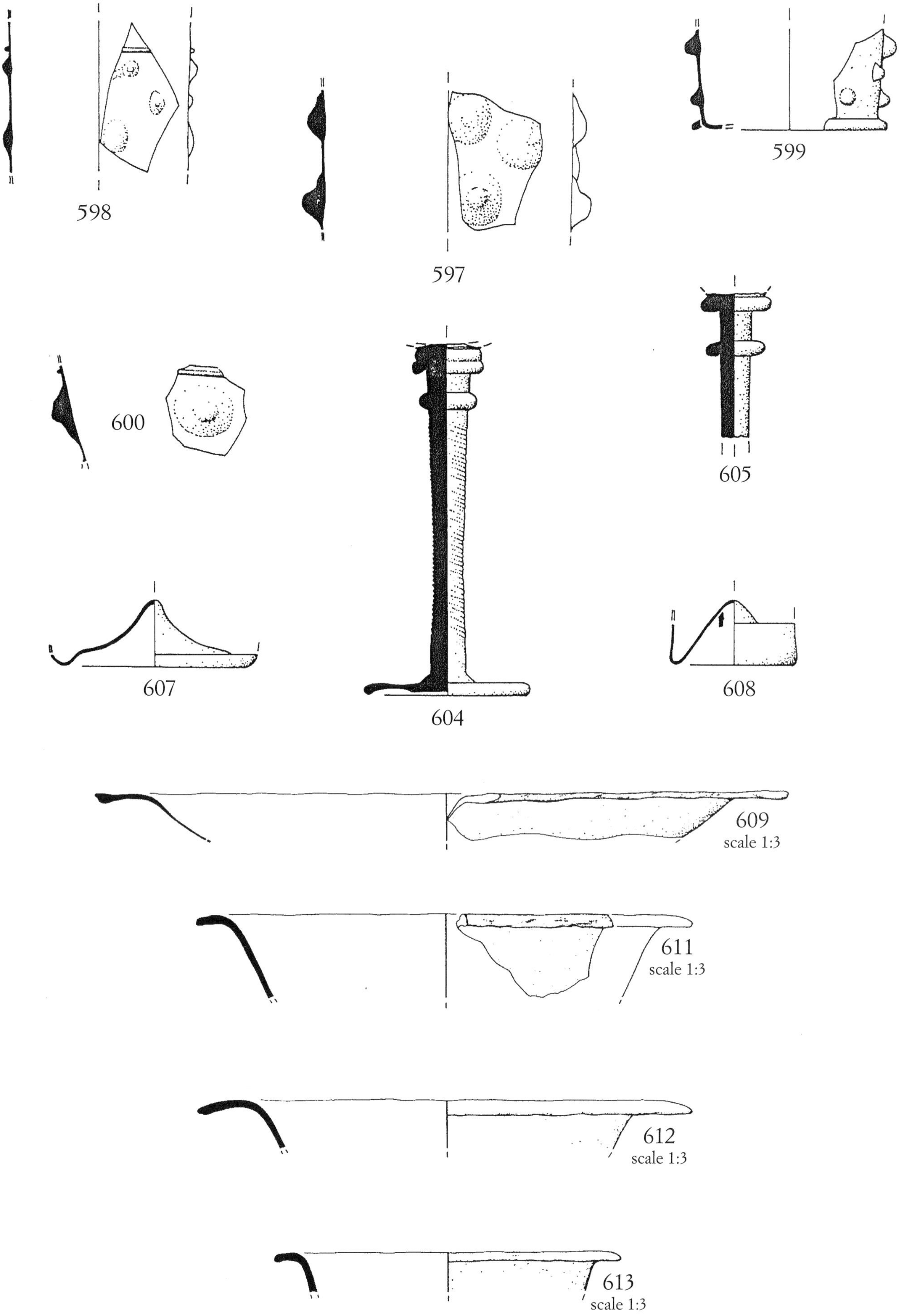

Plate 37. Medieval blown vessels. Cast sagged oversize colorless bowls.